World Cattle II

Cattle of Africa and Asia

UNIVERSITY OF OKLAHOMA PRESS : NORMAN

World Cattle II

Cattle of Africa and Asia

by John E. Rouse

International Standard Book Number: 0-8061-0864-9 in use for Vol. I
Library of Congress Catalog Card Number: 69-10620

Manufactured in the U.S.A. First edition, 1969; second printing, 1972.

Contents

Volume II

PART FOUR: Africa

Africa

Land area (sq. mi.):	11,733,000
Population (1966):	312,000,000
Density (per sq. mi.):	27
Agricultural (70%) (1966):	214,000,000
Cattle population (1964):	123,000,000
Buffaloes (1964):	2,000,000

THE CONTINENT OF Africa is basically a major plateau, 11,733,000 square miles in area. It is dominated by many elevated regions that have a marked influence on local climates. It lies between 37 degrees north latitude and 35 degrees south, which in the Western Hemisphere encompasses the area from Virginia to central Argentina. Joining Asia on the northeast at the Sinai Peninsula and separated from Europe only by the Strait of Gibraltar, Africa is bounded on the north by the Mediterranean Sea and on the east by the Red Sea and the Indian Ocean. The Indian Ocean and the Atlantic Ocean are off the southern coast; the Atlantic also lies to the west.

The Malagasy Republic, occupying the island of Madagascar, is included with Africa in the discussion of cattle since most of her bovine population came from that continent; the island of Mauritius, an outpost farther to the east and unrelated to any continent, also is included.

The African terrain varies widely. South of the narrow coastal plains along the Mediterranean Sea which terminate in rough and mountainous country, the Sahara and Libyan deserts stretch across the width of the continent. Below the desert a wide belt of savanna and rain forest reaches from the Atlantic coast to the high East African plateau. This region of moderate-to-good rainfall extends to the desert lands of South-West Africa, below which is another area of better rainfall that reaches to the South Atlantic and the Indian Ocean coasts.

African deserts are not entirely devoid of plant life. For miles the surface may be practically barren, but these stretches are always broken, sometimes by rocky ridges and sometimes by wadies or draws favorable for water accumulation. In these areas much plant growth is found when moisture conditions are favorable. In the northern deserts there are oases, some of considerable extent, watered by springs or wells, where cultivation under irrigation is practiced. In the marshy spots there the water is brackish when it accumulates, and the plant growth is so inedible that even goats will not consume it. Such areas comprise most of the land surface of Algeria, Libya, and Egypt, each of which is two to four times as large as Nevada. Nomads in their endless travels to feed their stock diligently search out the other desert plant growth.

There are vast areas of medium-to-high plateau savanna with poor-to-excellent rainfall; some of these regions are open, some wooded, with the ever present acacia predominating. In a broad band across the continent from 14 degrees north to 16 degrees south, with the elevation rising from west to east, the rainfall varies from 20 to 60 inches. The rather narrow coastal plain on the Atlantic side varies from desert conditions in the south to 80 inches of rainfall or more as the hump is reached; still farther north, where the Sahara meets the Atlantic, there is no rainfall. The coastal plain on the east is well watered for the most part until it reaches the desert in the north. Along the Mediterranean Sea the coastal plain enjoys good rainfall in the west and becomes arid in the east. The high plateau of East Africa on both sides of the equator enjoys some of the most equable climate on the continent because of the elevation of 3,000 to 6,000 feet.

Until quite recently only the Mediterranean countries in the north had contact with the West. The Arab countries of southwestern Asia were the first slave traders, invading the eastern coast south of the Sahara for this purpose during the first millennium. Western influence was not felt to any great extent in this area until the middle of the nineteenth century, when British and, later, German colonization began, although farther to the south the Portuguese reached Mozambique in 1498. The first contact with Europeans on the west coast was made when the Portuguese established ports of call there in the middle of the fifteenth century. The Dutch followed 150 years later during their period of maritime supremacy. Their initial objective was the acquisition of slaves along the west coast, but the colonizing of South Africa soon followed. At the end of the seventeenth century the English and French started to move in; later came the Belgians in the Congo and the Germans in what was then the Cameroons and in the southwest. Colonization then spread over the entire continent.

In the period between the two world wars, the seeds of independence were germinating in varying degree over all of Africa. They first burst into fruition among the countries along the Mediterranean coast, all of which except Algeria were accorded national independence within ten years after World War II. In the following decade Algeria and most of the other colonial entities in Africa—the major exceptions being Rhodesia and the Portuguese provinces of Angola and Mozambique—joined the ranks of the new nations. Ethiopia, independent for many centuries except for a brief period of Italian occupation between the wars, Liberia, self-governed since her formation in 1847, and South Africa are the only African states with a deeply rooted independence. Rhodesia, formerly Southern Rhodesia until she joined the Federation of Rhodesia and Nyasaland, which soon disbanded, is in a category by herself—independent on her own unilateral declaration although not officially recognized as such by the rest of the world.

Such rapid changes in the political scene have had their effect on the bovine as well as on the human populations. Beautiful herds of dairy cattle have been slaughtered in some countries in the owners' efforts to salvage something before they moved out. Disease-control measures have sometimes been abandoned or have become inoperative for lack of proper administration. In some countries herds have declined from lack of development or even from nonmaintenance of watering facilities, the greatest need of many African cattle, which in some areas go as long as five days between waterings. The incentive for the European grower to develop or even maintain his current operation is often lost as he views the uncertainties of the future in the politically unstable atmosphere. This situation is most unfortunate in areas short of beef and dairy products, not only for local consumption but for a ready export market that could benefit the steadily increasing shortage of foreign exchange in many African countries. Most African cattle, however, more or less remote from these influences continue the never-ending search for grass and water as they have for countless centuries in the past.

Most of the inhabited areas are characterized by either the nomadic herding of cattle, sheep, and goats or by a primitive, shifting agriculture. In the latter case a plot of land is cleared and cultivated for a subsistence crop for a few years, until the mineral content of the soil is near exhaustion. Then a fresh plot is cleared, and the process is repeated. Most of the cattle population is in the care of nomadic or seminomadic tribes. The continual movement to water, the dominating factor in this sort of husbandry, depletes the vegetation along the old, established trails and around the watering places. Because of either the distance from water or the infestation of the tsetse fly, vast areas remain ungrazed or are still utilized principally by game. The social and in some areas even religious significance of cattle ownership dominates the African's way of handling his animals.

Ranch-type cattle raising by Europeans is a major factor only in southern Africa—Rhodesia, South Africa, South-West Africa, and to a lesser extent Angola and Mozambique—and also in Kenya. Even in these countries native cattle usually outnumber those grown by Europeans. Cattle are run by Europeans to some extent in many other countries of Africa, but the over-all volume of such operations is relatively small.

Most importations of European cattle breeds into Africa have occurred since the turn of the century, sometimes successfully, more often not. There have also been minor importations of Zebu cattle from Pakistan, India, and Brazil. All such foreign cattle are here referred to as "exotic"; the indigenous types are termed "native." It is not unusual for a European grower to raise only native cattle, and there are some outstanding developments of indigenous breeds by artificial selection for desired characteristics. Instances of Africans raising exotic cattle are rare, however.

Cultural aspects involved in the handling and the use of cattle by the Africans south of the Sahara vary to some extent from tribe to tribe but in their general concepts are quite universal. Among the tribes of the cattle-owning peoples, the possession and the care of cattle are the main objectives of a man's life. Although this objective is a "status symbol" or a "sign of wealth," such expressions are really inadequate. Throughout the ages cattle have occupied a somewhat analogous position in the most advanced civilizations of their time. *Pecunia*, the Latin word meaning "money," is derived from *pecus*, meaning "cattle." Even the English word "chattel," meaning any movable property, was long synonymous with cattle or any other four-legged property. But cattle are more than simply property to the African owner. His life's pattern is frequently determined by the care of his animals. They in turn are partly, and sometimes wholly, the means of his and his family's and retainers' subsistence. The flesh of the animals is rarely eaten except on ceremonial occasions when they are killed for a feast, or as a means of salvage when one dies a natural death. The milk and, in a few tribes, the blood they yield are the main contributions to the owner's livelihood. The African's primary ambitions are, first, the size of his herd and then to maintain it. While number is important to him, he also has an eye for quality. Whenever there is occasion to dispose of any of his herd, he invariably disposes of the poorest animals.

The nearly universal custom of lobola, the transfer of a number of cattle to the father of the woman a man takes in marriage, is not adequately defined by calling this transfer the "bride price." In many tribes the practice involves not an outright transfer of title but rather a trust arrangement to ensure faithfulness on the part of the bride. If she is unfaithful, the cattle are to be returned. Among the Bantu peoples two head of cattle are usually slaughtered for the wedding feast, and a third is given to the mother to "wipe away her tears." Evidence of good faith on the part of the husband is another implication involved: if he does not care for his new wife properly, he immediately forfeits the title to the remaining cattle. Because work in the African family except for the care of the cattle devolves upon the women, a man should receive some measure of compensation if he loses the efforts of a young and able daughter when she marries. Certainly there are deeper considerations involved than those of the dowry system of many European countries, which is the reverse of lobola.

As the money economy of the European encroaches on the African cattle owner, the latter's animals become a more tangible form of wealth, but he will only part with them for an immediate need. He does not exchange them for money unless the money is the means of supplying that need. He still prefers his wealth on the hoof, and herein lies the greatest handicap to commercial cattle production in Africa. The offtake from African-owned herds is always small, in remote areas practically nil. Most animals simply die from natural causes.

North of the Sahara, cattle are less intimately involved in the cultural lives of their owners. They are grown for milk, for draft, and to a limited extent for meat, which is usually a matter of salvaging animals that are no longer productive. In the Maghrib, which comprises the countries along the Mediterranean coast from Morocco through Libya and is isolated from the rest of Africa by desert, the camel is a source of milk, draft, and meat. In Egypt, which is also isolated from the rest of the continent by desert, the water buffalo nearly equals cattle in number, and the camel is also a factor in agriculture. In southern Africa, where European growers run substantial numbers of cattle, the animals are handled much as they are in the Western world. In these same areas, however, the native Africans also have many cattle, often exceeding in total number the animals of the Europeans. Although the African in the southern part of the continent is beginning to market his cattle to a limited extent, his husbandry practices are much the same as those in other parts of Africa.

ORIGINS OF AFRICAN CATTLE

Prehistoric evidence—diggings and fossils—definitely place the origins of present-day cattle in Asia, Europe, and North Africa. Nothing has been found to indicate that cattle existed in any other part of the world until transplanted by man within the past few centuries. The countless prehistoric and later migrations of the peoples of the Eastern Hemisphere, who took their cattle with them, set the patterns which determined the indigenous cattle breeds living today. Except for the introduction of cattle into North and South America, Australia, and New Zealand within the past 450 years, the movement of cattle by modern man from one locality to another has had little effect on cattle types or breeds. Exotic breeds in some numbers have been introduced into various parts of Africa during the past century. Where these animals have persisted, either they have done so in a comparatively pure form and under unusual care in regard to nutrition and to disease control or they have been absorbed without any marked effect on the native cattle. The following outline of the origins of the cattle now in Africa is largely based on the studies of J. H. R. Bisschop, of South Africa, and on his summation of the work of others.[1]

Egyptian Longhorn (*Hamitic Longhorn*).—By 4000 B.C., before historic times, long-horned, humpless cattle had been domesticated in the Nile Valley and had been husbanded there for many centuries. Information found on tomb drawings has led to the belief that these animals disappeared during the second millennium before Christ. Their progenitors, which had also lived in the fertile Nile Valley, were wild and may have evolved in Asia and reached Africa through the natural movements of wild animals in search of food. After domestication the Egyptian Longhorn was carried with the migrations of its owners as far south as Ethiopia and also westward along the Mediterranean coast to the Atlantic. From here, continued migrations moved these cattle in a pure state, for there were no other types with which they could mix. Such movements took the animals down the Atlantic coast as far south as present-day Nigeria and probably across the Strait of Gibraltar to the Iberian Peninsula. Apparently they were not as hardy as the Shorthorn humpless cattle which followed them and to which the Egyptian Long-

[1] J. H. R. Bisschop, "Parent Stock and Derived Types of African Cattle," *South African Journal of Science*, Vol. XXXIII (March, 1937), 852–70.

horn eventually succumbed in most areas. Egyptian authorities claim that, as evidenced by wall carvings, there was originally a polled breed of cattle in Egypt. Such a type could have been differentiated out of the Egyptian Longhorn by selective breeding. Polled individuals in rather large numbers are found in many African cattle today, and it would not be unreasonable to assume that this tendency existed in the Egyptian Longhorn.

Brachyceros ("Short Antenna" or Shorthorn).—These cattle probably entered Egypt with migrations across the Isthmus of Suez at some time before the dawn of recorded history. Most authorities agree that these animals had their origin in the domestication of the wild cattle of Europe. After they entered Africa, human migrations carried the Brachyceros along the same routes taken by the Egyptian Longhorn, both up the Nile Valley and along the Mediterranean coast, then up and down the Atlantic coast, eventually reaching Nigeria to the south and possibly even England to the north. Being hardier than the Longhorn, they eventually replaced it.

Brachyceros cattle were not humped and were smaller than the Egyp-

Egyptian Longhorn cow as shown on a wall in the temple of Queen Hatshepsut, who died in 1482 B.C.

Brachyceros bull depicted in Queen Hatshepsut's temple.

tian Longhorn. Present-day breeds such as the West African Shorthorn can be traced to these animals; Brown Atlas cattle as seen today along the Mediterranean coast from Morocco to Libya probably also had their origin in the Shorthorn cattle.

For a time both the Egyptian Longhorn and the Brachyceros existed simultaneously in Egypt. Pictures on the walls of the excavated temple of Queen Hatshepsut, dated 1482 B.C., show a Longhorn cow and a Brachyceros bull. Whether there was any widespread crossing of these two types is not known. Such interbreeding certainly could have occurred and had its effect on the representatives of both breeds that were taken from Egypt on the migrations along the northern coast of Africa.

Longhorn Zebu.—The first humped type of cattle to reach Africa, the Longhorn Zebu arrived about 1,000 years after the Shorthorn cattle, entering Africa from Aden and crossing to Somalia probably about 2000 B.C. It is thought to have originated in what is now western Pakistan or in India and, after centuries of slow and often interrupted migrations of its owners across the Middle East, reached the southern tip of Arabia. In the region now comprising Ethiopia and Somalia, the

Longhorn Zebu met the Egyptian Longhorn, and interbreeding produced the African cattle now known as Sanga, which spread to Kenya, southern Sudan, Uganda, and then over much of eastern Africa. This mixture is the base stock of many indigenous breeds of African cattle as seen today.

According to Bisschop,[2] following this invasion of Africa by the Longhorn Zebu, the migration route of the Hottentots carried the animals, without admixture of other types, across Africa and then down to the Cape of Good Hope. Here they were found in the hands of the Hottentots when the Portuguese first reached Africa, six years before Columbus discovered the New World. These cattle, it is claimed, were the progenitors of today's Africander.

What is known of tribal movements indicates that, after leaving East Africa, the tribes with the Longhorn would have encountered few, if any, other cattle en route to the Cape. It is therefore possible that the Longhorn could have remained in a relatively pure state during the many centuries it took for the trek across the continent. The weak point in this theory of the origin of the Africander is the lack of information about why it bears so little resemblance to any of the present Zebu breeds of India. The Africander horn is distinctively shaped—long, horizontal, only slightly curved, and markedly oval in cross section—whereas horns of the Zebu are generally round in cross section, although some Zebu breeds have a horn with a slightly oval cross section. Zebu horns usually extend upward; the horns of the Africander extend outward from the head. Artificial selection, which can modify horn shapes to a remarkable degree, was practiced by many African tribes. The radically different Africander horn could be accounted for as logically by assuming the crossing of nonhumped and Zebu cattle (Sanga) and subsequent artificial selection by the native owners for horn shape, as by assuming selection on the straight Zebu for a preferred type of horn. The Sanga includes several breeds with very unusual horns.

What appears to be a more plausible theory about the Africander's origin is that it derives from the Sanga.[3] Here Sanga is defined as a neck-humped animal (as opposed to the chest-humped), the usual result of a mixture of humpless cattle and Zebus. The hump of the Zebu is set on the shoulders, behind the line of the front legs; the Africander's hump is on the neck, in front of the front legs, which is a typical Sanga characteristic. Also, it seems improbable that, over the many centuries in-

[2] *Ibid.*

[3] I. L. Mason and J. P. Maule, *The Indigenous Livestock of Eastern and Southern Africa.*

volved, one tribe or people in Africa could have kept their cattle free from mixture with the cattle of other tribes when cattle raiding and tribal warfare were continuous throughout Africa. This situation would apply during the period when the Longhorn humpless and the Longhorn Zebu were both present in the region that is now Ethiopia and before the Hottentots started their migration across Africa to the west.

Sanga.—When the Longhorn Zebu arrived in Africa, it reached areas in Ethiopia and Somalia that already carried a population of Egyptian Longhorn cattle. Indiscriminate interbreeding of these two types resulted in the fourth basic type of African cattle, the Sanga. Usually showing most of the typical Zebu characteristics, the present descendants of the Sanga vary greatly in size, conformation, and type of horns; they have a pendulous dewlap and sheath, upturned horns, and invariably carry a hump, although it varies widely in size. Migrations of their owners carried the Sanga across Central Africa below the Sahara nearly to the Atlantic coast and also southward from East Africa to Southern Africa.

Shorthorn Zebu.—This last type of Asiatic cattle to enter Africa in early times apparently arrived about the seventh century of the Christian Era. It also had its origin in India and in many characteristics is similar to Indian Zebu breeds as seen today. Migrations carried it down the east coast of Africa and also across Central Africa to the Atlantic coast. One school of thought holds that the Shorthorn Zebu resulted from the Shorthorn cattle meeting and mixing with the Longhorn Zebu in their treks through the Middle East or Arabia, but the preponderance of evidence would indicate that the Shorthorn Zebu arrived in Africa from India in a relatively pure state. The Sanga type was well established by the time the Shorthorn Zebu reached Africa, and, since the two types met in their migrations in either East or Southern Africa, the mixing which occurred was also responsible for some of the indigenous breeds now in Africa.

METHOD OF PRESENTATION

Because of broad similarities in breeds and types, in cultural aspects, and in methods of handling, the discussion of African cattle has been presented under the following geographical sections:

1. East Africa: Ethiopia, Kenya, Malagasy Republic, Mauritius, Sudan, Tanzania, Uganda, and Zambia.
2. North Africa: Algeria, Libya, Morocco, Tunisia, and the United Arab Republic.

3. Southern Africa: Angola, Mozambique, Rhodesia, South Africa, and South-West Africa.

4. West Africa: Cameroon, Chad, Republic of the Congo, Ghana, Liberia, Nigeria, and Senegal.

Ethiopia

Land area (sq. mi.):	471,800
Population (1966):	22,590,000
Density (per sq. mi.):	48
Agricultural (90%) (1962):	20,350,000
Per capita income (1966):	$59
Cattle population (1964):	25,300,000
Year visited:	1965

THE EMPIRE OF ETHIOPIA lies from 4 to 17 degrees north of the equator at the beginning of the high plateau region of East Africa. It is a beautiful, mountainous land, in area equal to Texas, Oklahoma, and New Mexico combined, and has a population density greater than that of Texas. The borders are with Sudan in the west, Kenya in the south, the Somali Republic in the south and east, and Afars and Issas, formerly French Somaliland, in the east. There is a coastline on the Red Sea in the northeast.

Substantial evidence exists to help support the claim that the country can trace a continuity of Christianity back farther than any other nation now in existence, even though the type of Christianity involved has in the past countenanced some unusual aspects. Hamitic and Semitic peoples predominate in the population. Ethiopia is by far the oldest independent country of Africa, having successfully held off the nineteenth century contention of the colonizing powers of Europe for African territory. During the five years of an abortive occupancy by the Italians that ended in 1941, many good roads and irrigation works were established in the country.

Primitive agriculture and nomadic husbandry of livestock are the means of livelihood for most of the population. Coffee is the largest cash crop. The cattle population in relation to the human population is one of the highest in Africa, but little economic use is made of this factor. Large areas are well adapted to cattle raising but usually are inefficiently utilized. Much of the country enjoys a pleasant climate because of the elevation of 6,000 to 10,000 feet on the plateau area, which has a fair rainfall pattern. There are extensive grasslands, although the ever-present acacia tends to be dominant wherever overgrazing occurs. Cattle provide for the subsistence and principal occupation of the nomadic tribes and are also an important element in the life of the settled rural population.

CATTLE BREEDS

Cattle in Ethiopia are commonly known by the name of the districts in which they are found or by the name of the tribe that runs them. They are indigenous breeds and types resulting from tribal selection which goes back for centuries. A limited number of European milk breeds found principally around Addis Ababa were first introduced during the Italian occupation. More recently there has been some experiment-station work with Western beef breeds.

Borana.—This is the same breed of Shorthorn Zebu ancestry as the Boran in Kenya. As seen in Ethiopia, it is predominantly light grey in color, although reds and some blacks are not uncommon. A grey bull is usually much darker on the shoulders and rump than on other areas of the body. The hump is well defined, larger in the male but quite promi-

Borana bull.

Borana cow.

nent in the female. The horns are short and stubby, and there is a tendency to be polled. Compact and of typical Zebu conformation, the animal varies widely in weight according to nutrition. On good feed a mature bull weighs well over 1,000 pounds and a cow 900 pounds; but most Borana cattle in native hands do not reach these weights because of low feed availability, particularly during the dry season.

The Borana probably originated along the northern frontier of Kenya, which borders on Ethiopia and is the homeland of the Borana tribe. There is a similarity in conformation, hump, size and shape of horns, the polled tendency, and coloring between the Borana of Ethiopia and the Kenana of Sudan.

Arusi.—Cattle locally called Arusi are in the Goba area of Arusi Province, southeast of Addis Ababa. They are distinctive but hardly uniform enough to be classified as a breed. They are smaller and less compact than the Borana, and their thick horns, often two feet in length, curve upward and are widely separated. Characteristics of the animals include a pronounced slope to the rump, a good-sized hump, and a pendulous dewlap and sheath. A mature bull as marketed rarely weighs as much as 900 pounds. The color is predominantly grey, although red to black animals are seen.

The Arusi probably traces to the Sanga or to a Shorthorn Zebu–Sanga cross.

Arusi bull.

Shawa.—Named for the province where they are run, these cattle are found in the grazing area southwest of Addis Ababa.

A small, humped animal, the Shawa varies in color from grey to black. The mature Shawa cow weighs only 500 pounds. The Shawa appears to be a nondescript and degenerate type, and although held in large numbers by the tribesmen, is of minor commercial importance.

Ogaden.—Harar Province, in eastern Ethiopia along the Somali border, is the home of the Ogaden, which is similar to the Shawa although somewhat larger. The animal is run by nomadic tribes.

The predominant color is light grey, but there are dappled-grey individuals and some that are nearly black. Their horns are rather short, often stubby. A mature cow weighs as much as 750 pounds.

Shawa cattle on trail to water.

Ogaden herd of a nomadic tribe. Photographed east of Harar, near the Somali border.

Exotic Breeds.—Of the European breeds introduced near Addis Ababa, the Friesian predominates, with the Ayrshire a poor second. With careful attention being given to prophylactic measures, these breeds are successfully maintained in commercial dairy herds and are also crossed with native cattle to improve milk production. Milk is in short supply in Ethiopia as in most other African countries now feeling the effects of Western civilization as they begin to have a money economy. The price of milk sold to retail stores is 16 cents a quart. The total number of exotic dairy cattle in Ethiopia is estimated at 2,000 to 3,000 head.

MANAGEMENT PRACTICES

The cattle of the country are continuously grazed under the constant attention of herdsmen. The animals do well in the rainy season but rapidly weaken in condition as the grass dries up. Except for those with dairy herds near Addis Ababa, there are few European cattle owners in Ethiopia.

In the south, as far as the Kenya border, the cattle owners are entirely migratory, moving seasonally with the grass growth. Household effects and housing—collapsible shelters formed of U-shaped hoops that are covered with stretched skins and woven grass mats when set up—are transported by camels. Camps are made at intervals of a few days to a few weeks, depending on the grass. The diet of these people consists principally of milk. The cattle in the south are predominantly of the Borana breed and usually are well cared for. They are watered infrequently, however, sometimes during extremely dry spells going as long as five days without water. The wants of the owners are few and are easily supplied by exchanging a few head of stock as required. Their herds, held as a matter of prestige, are largely maintained intact or are increased, and they are drawn on in sizable numbers only when a new wife is wanted. In the case of a prominent headman, ownership may reach 1,000 head or more.

Jijiga, a town in eastern Ethiopia near the Somalian border, is the center of one of the major native cattle areas. Flat plains stretch from the hilly country to the west and, with the 20-inch normal rainfall, grass growth is good although poorly utilized because of inadequate watering and lack of grazing control. The Jijiga wells are one of the widely separated watering places. They are dug wells, six to eight feet in diameter, with the water surface fifty to eighty feet below ground level. One man at the bottom fills a goatskin or a hand-carved wooden bucket and tosses it to the next man, standing on a ledge two feet above his head, who in turn throws it to the third man, and so on until it reaches the top man, who empties the water into a trough. This work is done to a continuous, joyful chant. Crude mechanical devices employing a pulley and rope to raise a hand-carved bucket are coming into use.

One well of the eight in operation waters a maximum of 800 head of cattle a day or five times that many sheep and goats. All day long, herds of 25 to 100 head or more belonging to one owner are trailed in and back to grass after watering—a practice repeated every three days. Herds brought in for watering are usually in reasonably good condition, and, although having been without water for three days, they stand

Plant growth at the end of the dry season under controlled grazing. A herd of native cattle is pictured in the background. Photographed at Abernosa Experiment Farm.

Across the fence from Abernosa, where native herds graze at the will of their owners.

quietly for hours awaiting their turn to drink, with only several young boys controlling even the larger herds. Such docility, often seen in African cattle, is a result of having herdsmen in constant attendance.

This periodic stock watering is the occasion for social diversions. Stock owners, who have been out of touch with humanity while on the savanna, gather in groups and refreshments in primitive form are available from small boys.

In the Rift Valley and the mountainous areas where streams are more numerous and rainfall is heavier, water is not as much a problem as it is in the savanna country of the south and west. Even here, however, there is continuous overgrazing in the areas adjacent to water. The limiting factors in stock raising in Ethiopia are stockwater and lack of disease control. Provision of good watering facilities and then the control of grazing, which this would permit, could double the cattle population. A convincing illustration of this is seen across the fence of the government experiment farm at Abernosa in south-central Ethiopia. The rotated pastures under fence on the farm provide ample growth for the dry season; the common grazing lands outside the fence are grubbed to the ground.

Bulls are run with the cows the year round and are left entire except for those wanted for draft in populated areas. These bulls are generally castrated at two to three years of age, often by pounding the testicles between two rocks, although use of the knife is becoming more common. Calving rates are low, less than 40 per cent, and death loss is high among young calves.

Except for the milk consumed by their owners and some blood used by the tribes in the extreme south, the large majority of the herds goes unutilized. Natives consume a small amount of cattle flesh, usually from animals which have died of natural causes. Cattle are slaughtered, however, for weddings or funeral feasts, or, in the more populated areas, for special holiday celebrations. Although the Ethiopian probably does not display the affection for his cattle that the Negroid African does, there is no authentication for stories claiming that the Ethiopian slices meat off the rumps of live animals to eat raw on festive occasions.

European milk breeds were introduced by the Italians who established dairies near Addis Ababa, which are still being maintained. Because of the moderating effect of the elevation on the climate, these exotic breeds do quite well, provided careful attention is given to prophylactic measures against African cattle diseases. A few of the better dairies with Friesian herds average close to 10,000 pounds of milk a lactation on herds of 100 head or more. Such an enterprise, however, is operated by an Italian who has remained in the country rather than by a native Ethiopian.

MARKETING

Cattle markets are located in the larger towns. In the small towns cattle, sheep, and goats are sold in the same market; Addis Ababa has separate cattle and sheep markets. Sales are usually to traders who resell for slaughter directly to the abattoirs or to farmers to go back to the country. On a big day in Addis Ababa 1,000 head, all of which have been trailed to the market, changes hands; and each transaction is a bargaining procedure. Many of the animals sold for slaughter are in poor condition because the nomadic owner is reluctant to part with his better stock. A 700-pound bull in fair condition being sold for slaughter brings approximately $60.00. Cattle from areas relatively near Addis Ababa often grade low-good. It is illegal to slaughter females capable of reproduction.

Cattle section of the livestock market in Harar, in eastern Ethiopia.

ABATTOIRS

Much of the meat of the country comes from local slaughtering in which one or two animals are dressed out by the village butcher. Most slaughtering facilities are municipally owned, but the Addis Ababa abattoir is a privately-owned share company. An average of 300 head of cattle, principally bulls, are killed there daily along with some sheep and a few hogs. The facilities are old, but processing is handled in a reasonably sanitary manner. Carcasses are sent to the butchershops while still warm. Hides are stretched and tanned, and bone meal, blood meal, and glue are also produced. This yield of by-products is unusual in Africa.

In co-operation with the Ethiopian government, an English firm has set up the National Meat Corporation of Ethiopia, as modern an abattoir as can be found in all of Africa, at Shashamane in the Rift Valley 200 miles south of Addis Ababa. The plant is located in one of the better cattle-producing areas with good rainfall and stockwater. Cattle are purchased locally for about 5 cents a pound liveweight, not much more than half the amount the same animal would bring in Addis Ababa. Difficult and costly transportation is the reason for this price differential. Despite the fact that this plant is surrounded by good cattle country with a large cattle population, it is difficult to obtain the desired number of animals for slaughter, simply because the native owners are reluctant to part with their stock. In 1965 the kill was only about 160 head a day and, because of the shortage of cattle, the plant was operating only four days a week. Three hundred head a day on one shift is the rated capacity. Everything in connection with this operation is conducted in accordance with the high standards typical of English abattoirs the world over. This starts with a three-week quarantine of all animals to be slaughtered. The marketing arrangements are unusual and illustrate the extent to which tax angles can affect the cattle industry throughout the world. The tenderloins are sold to the upper-class trade in Addis Ababa; the rest of the carcass goes into sixty-six–pound packages of frozen, boned beef. These are shipped 400 miles by refrigerated truck to Djibouti; the seaport on the Gulf of Aden, in what is now the French territory of Afars and Issas (formerly French Somaliland). The meat then goes by refrigerated ship to Gibraltar, where it is processed for extract and corned beef and reshipped to England, thereby avoiding import tariffs since Gibraltar is a crown colony and exempt from such duties.

CATTLE DISEASES

Most of the diseases and many of the parasites that plague African cattle are present in Ethiopia, but perhaps the losses are not as severe as they are in some of the surrounding countries. Except for the dairy herds the cattle population consists almost entirely of indigenous animals which have developed a high degree of tolerance to endemic diseases. Foot-and-mouth disease occurs, but animals usually recover after a setback; and, since cattle are held largely for prestige, some retardation in their growth is not a serious penalty.

Outbreaks of anthrax and rinderpest are more serious and receive more attention from the Government Veterinary Services. Control measures, however, are poorly administered and are far from effective. There are very few veterinarians in the country, so "vaccinators" who have been given a minimal training in giving inoculations do much of the work. Vaccines are diluted when the supply runs short and often compensation is accepted for giving preference to a herd.

In some areas rinderpest is treated with a primitive vaccine consisting of urine from an infected animal being diluted with water and then introduced into a laceration made on the nose of the animal to be treated. The procedure leaves a terrible scar but is said to be quite an effective prophylactic measure.

Because of moderate temperatures at the high elevation of most of the country the tsetse fly is not a problem. In the southwest there are small areas of infestation, however. Under present conditions and barring epidemics, loss from disease is not too important since most of the cattle population does not serve any commercial purpose.

GOVERNMENT AND CATTLE

Except for disease control the government's major activity concerning cattle is the programs managed by the Oklahoma State University team employed by the United States AID, under the sponsorship of the Ministry of Agriculture. These programs involve feeding tests on Borana cattle, keeping a demonstration dairy herd, and making efforts to upgrade native cattle in crossing experiments.

Although the economy today could not support much of a cattle-feeding industry, Ethiopia, with its fair rainfall pattern and the capability of producing more grain than needed to supply the requirements

Finished Borana steer, 1,010 pounds. Photographed at Debra Zeit Experiment Station.

of a population of only moderate density, could eventually have a sizable development along this line. The experiment station at Debra Zeit feeds Borana steers, purchased from the native growers at six months of age, castrated, and then finished out to weights of 950 pounds at eighteen months of age. Gains of as much as three pounds a day are obtained on a ninety-day feed test.

A small artificial insemination center, the only one in the country, has been established at the Agricultural College farm of Haile Selassie University, in Alemaya. Using frozen semen from the United States, the center has obtained a conception rate of 50 per cent on the first insemination. Here a model demonstration dairy also has been set up on a scale that could be applicable to a small owner.

A program to determine the best exotic beef breeds for upgrading local cattle has been undertaken, using frozen semen of Santa Gertrudis, Charolais, Holstein-Friesian, Angus, and Hereford bulls, all imported from the United States. Results in 1965 indicated that the Santa Gertrudis breed gave the most promising results. With the good beef potential that the native Borana breed has shown, however, such crossing would not appear to be as promising a line of endeavor as selection within that native breed for a good beef animal, as has been done in Kenya.

OUTLOOK FOR CATTLE

Pastoral practices in Ethiopia will undoubtedly continue for some time along the same lines they have followed for centuries in the past. Change in the cultural life of the people will have to precede any marked change in stock raising. If the economy of the country continues to improve as it has in recent years, even though the rate of advancement is slow, cattle raising is certain to increase in importance. This probably will first be seen in increased dairying near the larger towns, where the change to a money economy takes place more rapidly.

Until the nomadic cattle owners have a greater desire and need for money, the large potential that exists for raising beef cattle will not be utilized. There are some indications that this monetary quest is beginning to occur. As the government has become more insistent on tax collections, cattle have been taken in payment. To a limited extent, the cattle owner has begun to sell the number of cattle necessary to obtain the funds to pay his tax rather than to turn his prized animals over to the tax collector.

There is one uncertainty, however, that involves any long-range improvement in cattle raising as well as all other activities in the country.

Emperor Haile Selassie I has accomplished much for his country since he returned to power in 1941. Although recently he has attempted to transfer power to a parliamentary system, he was seventy-six years old in 1968 and no competent successor was in the offing. There is a growing feeling of unrest among the younger element of the population for a change to a more democratic form of government. What will happen when Selassie no longer rules only the future can determine, but progress in cattle raising will be decided by the outcome.

Kenya

Land area (sq. mi.):	225,000
Population (1966):	9,645,000
Density (per sq. mi.):	43
Agricultural (75%) (1964):	7,230,000
Per capita income:	$82
Cattle population (1965):	7,235,000
Offtake (3.5%) (1963):	250,000
Year visited:	1965

THE REPUBLIC OF KENYA rises from a narrow coastal plain in the southeast that extends for 250 miles along the Indian Ocean to the East African plateau. The land area is equal to that of Missouri, Oklahoma, and Kansas combined; and the population density is slightly smaller than the average of these three states. The northern borders are with Ethiopia and Sudan; the Somali Republic lies on the east, Tanzania on the south, and Lake Victoria and Uganda on the west. Lying directly across the equator, with about 4 degrees of latitude on either side and with most of the country being at elevations of 5,000 to 8,000 feet, Kenya enjoys one of the most pleasant climates in Africa. Rainfall varies from 20 to 40 inches annually over some of the best grazing land on that continent.

Ninety-seven per cent of the population is made up of the numerous African tribes, including Bantu, Nilotic, and Hamitic peoples. Asians, predominantly Indian but with a minority of Arabs, account for 2 per cent of the population. Europeans, mostly British, numbered 56,000 in 1962 but have decreased since then.

Late in the nineteenth century British colonists began to move into the area that became the colony of Kenya. During the following seventy years a highly productive agricultural economy was developed in which cattle raising was a large factor. There was a deterioration in agriculture generally after independence was attained in 1963. The large productive farms of the Europeans were purchased by the government under the Africa Settlement Plan and broken up into small plots for native farmers. Initially this procedure was not very successful. The step from scratching seed in the ground with a stick to using mechanized husbandry methods was more than they could have been expected to take at once. European settlement officers were then installed to advise and administer the areas under the Settlement Plan, and results were materially improved. The settlement program had not affected many of the stock ranchers in 1965; the large ranches still remained intact, and cattle and sheep raising by the European owners continued much as they had in the past. The government apparently realized the loss in revenue that would ensue if the management of these high-producing enterprises was eliminated. The rancher, with an acute knowledge of what had happened to the farmer, thought he saw the handwriting on the wall and, with little confidence in the future, began to formulate plans to leave; by 1967, however, this attitude appeared to be lessening. Considerable foreign capital—Italian, British, and American—has recently been put into ranch operations.

CATTLE BREEDS

The indigenous cattle of Kenya fall into two groups—the well-recognized Boran breed and a native, nondescript type. In addition to these cattle there are exotic dairy breeds which were introduced by the British settlers.

Boran.—When the British began active colonization in Kenya, they

Improved red Boran bull.

Improved red Boran cow.

first brought in Hereford, Red Poll, and Ayrshire cattle with the idea of upgrading the native stock. After a number of years of disappointing results in working with these northern breeds in the tropical climate, the cattlemen among the colonists learned that the indigenous Boran cattle, the outstanding beef breed of East Africa, would outproduce both the imported breeds and their crosses on native cattle. The disease tolerance and resistance to heat stress of the indigenous breeds of East Africa are attributes which the European breeds cannot be expected to acquire in a few generations. Base stocks for herds of Boran cattle were obtained from the tribesmen in the surrounding country. Selection was started within the breed for beef characteristics, and a superior meat animal has resulted. The program took on such proportions that a society for the Boran breed was established in 1951.

The native Boran of Kenya is the same animal as the Borana of Ethiopia. It is chest-humped and either solid red or solid whitish-grey in color. As run by the tribesmen in northern Kenya, the females do not exceed 900 pounds when mature at six years of age and the bulls rarely reach 1,100 pounds. The Boran is a naturally slow maturing, well-muscled type and characteristically has small, thick horns and a tendency to be polled. In any large Boran herd a few animals will be found that are naturally without horns. The hump is rather large, prominent

Improved grey-white Boran bull.

Improved grey-white Boran cow.

even in the female, and the dewlap and sheath are large and pendulous.

Forty years of selection for early maturing and good fleshing characteristics have resulted in the improved Boran of the European grower. Quality of the beef is comparable with that of European breeds. Mature animals weigh at least 200 pounds more than native cattle. Steers at three and one-half years of age are marketed off the grass at 1,100 pounds, as compared with 900 pounds for four- to five-year-old animals purchased from native herds and put on the same pasture.

Herds are sometimes separated by color, that is either red or white, although ordinarily no color distinction is made. Dehorning is quite generally practiced.

Nearly finished Boran steers, slightly over three years old, 1,025 pounds.

Native Cattle.—In the more populated areas most cattle of the African are nondescript. In western Kenya there are some small cattle, rather uniform in size, with cows weighing 600 to 700 pounds. The color can be either solid black or solid white or can be various mixed colors. The horns are of medium size. The hump is small.

Exotic Breeds.—Excellent dairy herds of the European milk breeds were established by colonists in populated areas. The Friesian predominated, but there were also good representations of Guernsey, Ayrshire, and Jersey breeds. The population of dairy cattle exceeded

Native cow in western Kenya.

400,000 at one time but has fallen considerably since Kenya gained its independence. Large numbers of the animals were sold for slaughter as the European farmers attempted to realize something from their holdings before they moved off. Some of these dairy herds are still being maintained pure since the breeding program has been continued by the Central Artificial Insemination Station, at Kabete.

MANAGEMENT PRACTICES

Tribesmen have always controlled the majority of the cattle of Kenya. Commercially these animals are of much less importance than the improved herds of the European ranchers. The usual African practices in running cattle result in poor development of the animals in the native herds. In some areas during the dry season, stock is watered only every three days. Insufficient feed occasioned by severe overgrazing in the vicinity of the watering places is common; this is also true in the limited pasture areas around the villages. These conditions, coupled with the dominant ambition to own as many cattle as can be accumulated on the common grazing lands, lead to undernourishment.

Foremost among the native African cattle raisers are the Masai. As colonization proceeded, they were moved to established tribal areas in the drier, hilly country in the south, with its 20-inch annual rainfall.

Depending on his affluence, a Masai owns from 10 or 15 head to several hundred. His cattle today are better cared for than those of other tribes. They are taken out to graze in the morning and returned in the evening to the owner's kraal, a circular barrier formed of thornbush, for milking or the drawing of blood, as well as for safekeeping from predatory animals or from theft. The Masai depends on his cattle for food to a greater extent than most other African cattle owners. He is not a meat eater, although cattle are slaughtered for food on ceremonial occasions and the flesh of an animal that has died of natural causes is not allowed to go to waste. An animal killed by a lion, however, is never touched. Although cereals are now being consumed to an increasing extent, the milk and blood the animals produce are a major part of the Masai diet. Girls or women do the milking, using a receptacle, usually a gourd, that has been washed with cow's urine to aid the coagulation of the milk; men take the blood by inserting a concave spear point into the animals jugular vein to provide a channel through which the blood can flow into a container. Grown men, instead of the small boys used by other tribes, are the herders and are always in attendance.

As found by the early white travelers, the Masai cattle were rather small, compact, usually black animals with a hump less prominent than that of many other African cattle in this region. Contact with the colo-

Masai herd containing a number of Boran cattle.

nists on the borders of the tribal areas and the forced migrations have resulted in considerable admixture with other breeds. Cattle acquired by theft also make their contribution to the Masai herds. Raids to acquire livestock have been a way of life with these people for ages and are of constant occurrence. Just as a Masai youth once was not considered a man until he had killed his lion with a spear (still a practice in remote areas), now he must prove himself by going with other young warriors on raids to obtain cattle.

As a result of the dilution of the Masai's cattle, there is now nothing distinctive in their animals. Castration of bulls not desired as sires is generally practiced, but only in the remote areas is the old method of pounding the testicles between two rocks employed.

The better European rancher follows modern management practices. He is generally located in the "White Highlands," so called because it is the area in which the white man settled on coming to Kenya. Here the wide, grassy slopes at elevations of more than 5,000 feet lead to Mount Kenya. Pastures are fenced and systematically rotated. Stockwater is provided by wells, often with pipelines running to the different pastures. No hay is put up since the dry growth between the rains is sufficient to carry the cattle through in good shape until the grass becomes green again.

Although there is a twelve-month growing season, normally with 30 inches of rain, the White Highlands area is not well adapted to cultivated crops because of the precipitation pattern. It is, however, an excellent cattle country. This region has been the center of the outstanding development of the indigenous Boran breed by selection within the breed for a beef animal. Three per cent bulls are usually run with the cows the year round. With no seasonal variation in temperature, there is no occasion for a breeding season. In the best operations calf crops of 88 per cent are obtained, and death losses are only about 2 per cent. Calves are usually weaned when approximately eight months of age. When they are three and one-half to four years old, steers are marketed off the grass at weights averaging 1,000 to 1,100 pounds. Some growers keep progeny records on herds of several thousand cows; the cows, all number-branded, are identified with their calves at weaning and the marginal cows culled. Prophylactic measures against the endemic African cattle diseases are rigidly followed.

A central artificial insemination station is maintained at Kabete, on the outskirts of Nairobi. Established in the colonial days, it is being continued by the Kenya Ministry of Agriculture but is still under European direction. Approximately 50,000 cows annually are artificially bred by the several substations the center supplies. The rectal technique of insemination is used. Friesian, Ayrshire, Jersey, Hereford, and Sahiwal bulls, excellent examples of their breeds, are used at the center and are well maintained. An interesting development is the use of coconut milk as a diluent, which permits storage of semen at room temperature which is specified as between 73 and 77 degrees F. Egg yolk citrate and antibiotics are used in the diluent. Semen prepared and stored under these conditions with an initial motility rating of 70 to 80 per cent will drop in most cases to only 50 per cent at the end of five days. For warm climates in which cooling facilities are frequently not available, the advantage of this method of preparing semen is obvious.

MARKETING

Purchasing livestock for slaughter and operating the abattoirs is a government monopoly administered by the Kenya Meat Commission, an agency established in colonial days and continued by the present government. The commission buys all the cattle sold for commercial slaughter, and the large European ranchers contract for delivery months in advance at the fixed price for dressed carcass as set by the government. The commission also has buyers throughout the country who do their best to get the African owners to sell some of their cattle. These cash-on-the-spot sales usually involve only a few head or even a single animal from one owner who has a pressing need for money. Such sales are all on a per-head basis, appraised by an eye estimate of how the animal will grade and its weight.

The system of grading and pricing introduced by the British in many of their African colonies is still employed. Most African countries do not recognize by grading and price differential the quality of either a carcass or of a live animal for slaughter. The established practice in Kenya is to pay the producer for his cattle on the basis of the cold-dressed weight (CDW) of the carcass. Fixed prices are established a year in advance by the Kenya Meat Commission and remain in effect until changed. The price paid to the producer is the same as that at which the meat commission sells to the retailer, the processing cost of the abattoir theoretically being recovered from the offals and hides.

All of the former British colonies in Africa retain the influence of the uniform system which was developed in colonial days for marketing and processing cattle. Sanitation and inspection were usually exemplary and have usually been reasonably well maintained by the new African governments.

Grades and prices as of January 1, 1965, were as follows:

Grade	*Price per pound CDW*
Baby beef: Not more than two incisor teeth and otherwise meeting first-grade requirements.	$ .23
First: Not more than four and one-half years old, quarters well fleshed, uniform fat covering not exceeding three-fourths inch. Maximum weight for each carcass 650 lbs., minimum for steers 400 lbs., some marbling in rib eye required.	.20
G.A.Q. (Good Average Quality): No age requirement, maximum carcass weight 700 lbs., CDW minimum for steers 359 lbs., more fat covering permitted, less marbling required.	.18
F.A.Q. (Fair Average Quality): Moderate fleshing of quarters and less fat cover required, carcass weight 280 to 800 lbs.	.16
Third: Inferior in fleshing to F.A.Q. grade.	.145
Fourth: Poor conformation and inferior fleshing, bruised or damaged carcasses of the better grades.	.13
Manufacturing A: Carcasses of first four grades but containing seven to twenty cysticercus cysts (measles).	.09
Manufacturing B: Carcasses which would grade third or fourth but contain seven to twenty cysticercus cysts.	.07

Manufacturing Grades A and B are permissible only at the Athi River and Mombasa abattoirs, which are equipped to manufacture extract and canned meat. If the inspector suspects the presence of cysts, both quarters are thoroughly slashed to discover the number present, making the meat unfit for purposes other than manufacturing.

In comparison with USDA standards, first grade would be equivalent to low-choice, G.A.Q. or F.A.Q. to good.

This price schedule returned the rancher $115 (equivalent to 10 cents a pound liveweight) for a 1,110-pound steer in 1965—a price the European grower was quite happy with.

ABATTOIRS

Kenya has two major abattoirs. In 1965 the largest, at Athi River, fourteen miles from Nairobi, was killing an average of 500 head of cattle and an equal number of sheep and goats daily. The equipment is thoroughly modern and sanitation is exemplary. Although the plant is directly under the control of the African government, management and supervision of the abattoir are still in the hands of the European personnel who were in charge before independence. A similar plant, killing an average of 300 head of cattle daily, is in operation at Mombasa, Kenya's principal port on the Indian Ocean. The output here is mainly for export.

Beef is in short supply in Kenya. There is a sizable export market for frozen meat to the Moslem Red Sea countries. Foot-and-mouth restrictions prevent the export of uncooked meat to most European countries, but sizable quantities of corned beef and beef extract are exported to England. More cattle of any grade obtainable could readily be absorbed by the existing market. The paradox is that there are cattle in the country in numbers more than sufficient to supply the demand if the African herd owners would only sell them.

CATTLE DISEASES

Most of the endemic cattle diseases of East Africa persist in Kenya, including the trypanosomiasis of the tsetse fly. Many of the disease problems stem from the wild game. The numerous antelope species are often carriers and hosts, particularly of East Coast fever, anthrax, and rabies. Foot-and-mouth disease is apparently not much of a problem except with newly introduced exotic breeds. If outbreaks occur in acclimated herds, they do not cause serious loss. Ticks and internal parasites are a constant problem; some areas dip or spray entire herds every week or ten days. In the Masai herds there are losses from pleuropneumonia, anthrax, and black quarter (blackleg), of which the wildebeest is the dominant carrier. *Cysticercus bovis*, locally termed measles, is often found in the cattle of the whole country. The larva of the tapeworm finds its way to the muscle structure of cattle and if consumed by man matures in the digestive tract.

Apivag, a venereal disease of bulls, has recently been identified at the research station in Kabete. The testicles harden and shrink and the animal becomes sterile for life. Indigenous bulls are largely immune to the disease, but it is very virulent in exotic cattle.

Because of the systematic control measures they employ, the well-managed European ranches ordinarily have small disease loss. Calves are vaccinated for blackleg at four to six months, then every twelve months until they are three years old. All cattle are vaccinated against rinderpest and are sprayed every two weeks for external parasites. Heifer calves are vaccinated for brucellosis at weaning, usually when they are about eight months of age.

The English provided the dairy herds with even more thorough prophylactic treatment, but this has been materially relaxed under African ownership.

GOVERNMENT AND CATTLE

The Government Research and Experimental Station at Kabete has made marked progress in the preparation of vaccines and the control of many African cattle diseases. Vaccine and serums produced there are recognized as superior products and are used throughout East Africa. Rinderpest, one of the most virulent of African cattle diseases, can be controlled by one inoculation of a vaccine developed at Kabete.

Mention has been made of the government control of marketing. This was still being handled in an orderly manner in 1965. It is a fertile field, however, for political intrigue and graft, as has been seen in other African countries, and the future could hold something quite different. Some irregularities were said to be cropping up in the cash buying of native cattle.

OUTLOOK FOR CATTLE

As it is in much of the rest of Africa, the future of cattle in Kenya is tied to the cultural attitude of the tribesmen and to the political policies of the future. With a cattle population approaching 8,000,000 head, the annual offtake including tribal and illegal slaughter is estimated at 250,000, slightly more than 3 per cent. The major part of this offtake is from the European-owned herds which are substantially outnumbered by the native-owned cattle. If the tribesmen could only be induced to part with a few per cent annually, the economic production could be doubled and the owner could still keep his numbers up. Better management in grazing practices would permit a further sizable increase. Neither of these changes can take place until the nomadic and semi-nomadic cattle owner is ready to accept a money economy, and he shows little inclination to do this. The highly productive European herds can only be continued if the present ownership and management are allowed to remain. Some growers have already left because of the uncertainties of the future, although their operations for the most part have fallen into other competent hands. But even under present conditions there is some retrogression because of the rather widespread wait-and-see attitude. A political whim, perhaps only a threat to purchase the ranches, as was done with the large farms, could change this feeling overnight to a get-out-with-what-you-can movement.

The future of cattle in Kenya will be determined by the turn such imponderables take.

Malagasy Republic

Land area (sq. mi.):	227,000
Population (1966):	6,335,000
Density (per sq. mi.):	28
Agricultural (85%) (1964):	5,400,000
Per capita income (1958):	$74
Cattle population (1964):	7,500,000
Offtake (3.9%) (1963):	291,000
Year visited:	1966

THE MALAGASY REPUBLIC occupies the island of Madagascar, lying in the Indian Ocean 250 miles off Mozambique, 13 to 26 degrees south of the equator. It is 1,000 miles long and from 200 to 300 miles wide and covers an area seven-eighths the size of Texas. From a low, wide tropical coastal plain on the west, with a rainfall pattern varying from 15 to 35 inches, the land rises gradually to a central plateau region 2,500 to 6,500 feet above sea level, which enjoys the equable climate of high altitudes in the tropics. This plateau becomes mountainous in the east then drops to a rain-forest belt along a narrow coastal plain on the Indian Ocean. The eastern part of the plateau area is the most heavily populated; the western and southern plains are the cattle country, inhabited largely by nomadic tribes whose principal occupation is cattle raising. The population density is a little less than two-thirds that of Texas.

Culturally and politically Malagasy is something of a transitional step between Africa and the Far East. In the unrecorded past it was reached by Malay people. Two theories exist about the route they took: one is that they came up the island chain to southeast Asia, crossed southern Asia, and went down the coast of Africa and over to the island; the other is that they followed a more direct route, breaking off from the Malayan islands and reaching Madagascar by crossing a wide stretch of the Indian Ocean. Whatever their route, they were the ancestors of the most important element of the present, heterogeneous population. The Malayans were followed by Arab and African peoples. All of them, and later the French, interbred in varying degree and were the source of today's inhabitants. The Merina tribe (largely of Malayan extraction), which lived primarily in the plateau area, largely controlled the island at the end of the nineteenth century. These tribesmen established a line of royalty and had quite a stable government. During the nineteenth century the British worked their way into becoming the major influence on the island's external affairs without assuming direct control, but in 1895 the French occupied the capital and soon took over completely. They did not, however, solidify their position in the rather thorough manner that was accomplished in their African colonies. They relied on the coastal peoples, primarily African-Arab in blood, as their tool of colonization; and the resentment this policy created among the plateau Merinas still exists. The unrest which precedes the move for independence in Africa began to take active form in 1956 and complete autonomy was obtained in 1960.

Today France, whose support has maintained the economic and financial structure since independence, appears to be taking that protection out of Malagasy. The latest evidence came in 1966 with the withdrawal of sugar export supports, an important earner of foreign exchange. This development does not conform to the usual French-African policy of retaining factual control of former colonies and does not augur well for Malagasy. The cultural and racial differences between the coastal peoples and the more-advanced plateau Merinas will probably not be peacefully resolved.

The agriculture of Malagasy is widely diversified. Rice, the largest crop, is the staple article of diet and the entire crop is consumed locally. Normally the island is self-supporting in rice; but in 1965, a bad year because of cyclone losses and a highly destructive rat infestation, a sizable quantity of rice had to be imported. The largest foreign exchange earner of agricultural crops is coffee, followed by vanilla, then by beef, either frozen for the French army or sent live to Mauritius. A small amount of fresh meat is sent to France by air. Beef and live cattle trade is a poor utilization of the cattle potential. Sugar, sisal, and tobacco make up nearly one-third of the total agricultural yield.

CATTLE BREEDS

There were no indigenous cattle on Madagascar. The cattle now on the island are descendants of those brought from Africa in the past. They are nondescript and incapable of classification into either breed or type, the only common characteristic being that they are humped. Usually the horns are rather large and heavy, curved outward then inward toward the tips. Some animals, however, have rather short horns.

Madagascar bull showing the triangular hump often seen.

Triangular hump on Tanganyika Zebus. Photographed at Dodoma Cattle Market in central Tanzania.

Madagascar cattle bear more resemblance to those of Tanzania than to the animals of any other African country. Although not a distinguishing characteristic, a rather small hump that in profile exhibits an equilateral triangle is common; the same type of hump is often seen on the native cattle of Tanzania. Arabs were firmly established on the Indian Ocean coast of what is now Tanzania when the Portuguese arrived there in 1500 and also were an important element in the Madagascar population at that time, as they had been for a lengthy period before. It is reasonable, therefore, to assume that the first cattle to reach Madagascar were brought from that part of Africa and that the present-day cattle are descended from them. Madagascar cattle have longer and thicker horns than the Tanganyika Zebu, perhaps because of tribal selection. The tribes raising cattle on the Madagascar plains are known to like large horns. In their religious rites, which are associated with a form of ancestor worship, horns from the animals slaughtered for the ceremonial feast are used to decorate the grave, the number and size indicating the importance of the deceased.

In the past the French have made attempts to introduce exotic breeds —the Sahiwal and the Red Sindhi from India, the American Brahman, and some of the European milk breeds—but nothing of note has come from these infusions. There are a few Friesian cows in small dairies near Tananarive.

MANAGEMENT PRACTICES

Malagasy has a high population of cattle in relation to human inhabitants. The government places cattle numbers at 9,500,000 and the human population at 6,300,000, or a ratio of 1.5 to 1. Other sources give 7,500,000 as the total number of cattle. In 1965 a head tax was collected on a reported 6,000,000 animals. This figure is admittedly low—just how low is only a guess—for cattle are maintained largely by nomads who fail to gather their entire herds for the tax count and hide large numbers in the bush.

Although cattle in varying numbers are found throughout the island, the principal growing areas are on the western and the southern coastal plains. These sparsely populated regions, with a rainfall varying from 15 to 35 inches, are poorly watered except for the stream beds.

Cattle are maintained in the hands of their nomadic owners as accumulated wealth. The appropriation of another's cattle is not uncommon—a custom rather than a crime. Some tribes require a young man to steal at least one cow before he is eligible to marry. The lobola practice of Africa does not exist. Cattle are more generally consumed for food, especially for celebrations and feasts, than in Africa. Human burial, consisting of winding the body in a silk cloth and laying it on a shelf in an above-ground tomb, is accompanied by the slaughter of cattle for a general feast for the mourners, after which the horns of the animals killed are placed around the tomb. For years after burial this procedure is repeated on each anniversary of death; the body is removed, rewound, and replaced during the course of which it is informed of family and tribal happenings of importance during the past year. Other occasions, such as weddings and important births, call for feasts and the slaughter of cattle. All in all, the numbers so consumed are appreciable but defy an approximation and are not included in the estimated offtake. These cattle raisers are largely self-sufficient and have no need for money except when taxes must be paid or a few articles of commerce are desired. They are completely satisfied with their way of life and are without incentive to sell their cattle except for the above purposes.

In the area around Tananarive small farmers buy a few head of thin steers or bulls and feed them a light grain ration before selling the animals for slaughter. There are a few small dairies on the outskirts of Tananarive, the total production from which is 5,000 quarts a day. The two largest dairies have herds of Friesian cows, one with 50 head. Most operators milk from 5 to 10 Madagascar cows. A distributor picks up the milk, ties the milk can on his bicycle, and ladles it out to the housewife's container.

Oxen are frequently used for transport. In some areas they are employed for draft in cultivation; in others all field work is done by hand or by driving loose cattle around in a flooded rice paddy to work the soil into a mud for planting.

MARKETING

Most of the commercial cattle are grown on the western side of the island, where they are bought by traders and trailed more than 200 miles to the main cattle sale held every Friday at Tsiroanomandidy, eighty miles by road northwest of Tananarive. Here they are bought by SEVIMA, the Societe d'Exploitation de la Viande a Madagascar, a privately owned company which has the principal abattoir on Madagascar, or by individual butchers, who have them processed in the municipal abattoir at Tananarive, where they are trailed after purchase. The total weight loss in trailing to market and again to Tananarive is 70 to 100 pounds. Even though the cattle are held for a rest period of a week or two at Tsiroanomandidy, because of the concentration, adequate pasture on which they can recuperate is not available.

Slaughter cattle passing through a village on the way to Tananarive.

Sale is by the head, the buyers judging by its appearance the grade to which the animal will dress out. The price of a No. 1 grade animal is about 7 cents a pound, or $55.00 for an 800-pound animal. For the lowest grade, No. 4, the going price in early 1966 was equivalent to 4 cents a pound, or $30.00 for a 700-pound animal.

ABATTOIRS

The SEVIMA abattoir, in Tananarive, kills 30,000 head of cattle annually. It was built fifty years ago but still presents a very acceptable operation. Slaughter is by sledge in a killing box. Carcasses are hoisted for skinning and eviscerating. Splitting is done by power saw. A sizable volume of canned meat, principally corned beef, is processed. The plant is powered by wood-fired boilers since Malagasy has no coal or oil. On arrival the cattle are sorted according to the grade they are expected to dress out to and are held by herdsmen to insure a backlog for continuous operation. The grass on the holding area is grubbed to the ground, and until they are slaughtered, the animals are on little but water.

Carcasses are graded into four classifications. Representative weights and the USDA equivalent of the grades employed are as follows:

Grade No.	*Weight Range, in Pounds*	*USDA Grade*
1	700 to 850	Low-good
2	650 to 700	Standard
3	600 to 700	Low-standard
4	Less than 600	Poor canner

The few cattle, which have been mentioned and which are processed through the municipal abattoir by the French butchers, usually weigh 900 pounds or more and would grade high-good. This beef is termed "Boeuf de Fosse." The No. 1 grade is known as "freezing quality" and comprises only a small percentage of the kill. All animals so graded are shipped by rail to another SEVIMA abattoir, in the east coast port town of Tamatave, where 80 head a day for six months of the year are killed. These carcasses are boned and frozen for export to the French army. The facility operates only six months because during the dry season only a few No. 1 animals are available since the growers are reluctant to sell during this period. The hindquarter of the No. 2 grade is sold to the better local butchershops, that of the No. 3 grade to shops with a more price-conscious clientele. The balance of these two grades goes into canned meat, as does all of the No. 4 grade.

CATTLE DISEASES

Tropical cattle diseases do not seem to be a serious factor in cattle raising in Malagasy. Foot-and-mouth disease, rinderpest, and the tsetse fly are not known. Tuberculosis is present, and there is some brucellosis—how serious is not known. Having a widely scattered cattle population which follows nomadic growing practices and consists of cattle which have acquired a tolerance for such diseases as are encountered is undoubtedly a major factor in minimizing bovine disease problems in the Malagasy Republic.

OUTLOOK FOR CATTLE

Although there are many obstacles to the development of a Western-type cattle industry in Malagasy, an American-financed scheme to adapt a large acreage to modern cattle-raising methods is in the planning stage. The plan includes the construction of an abattoir and a freezing plant at Morondava, a port on the western coast. Such an arrangement might be the first step to an integrated cattle industry which could enjoy a healthy export market. Development along the conventional lines of grower to market to abattoir has serious roadblocks, however. First is the inborn cultural practice of holding cattle for wealth and the lack of incentive to sell them. The current offtake could readily be doubled if animals were sold at three to four years of age instead of the present average age of seven years or over. Lack of transportation facilities—roads suitable for trucking or railways serving the growing areas—not only results in sizable losses in the cattle that are marketed but is a deterrent to the sale of larger numbers. Nomadic management methods, overgrazing in the watered areas, and no pasture control or improvement are also major factors, for they affect both cattle numbers and market weights. Such handicaps involve long-range planning for correction, and this is not visible on today's horizon. Also the political future is involved; risk capital would be required in large amounts and would be difficult to obtain under the political uncertainties that appear in the offing.

Mauritius

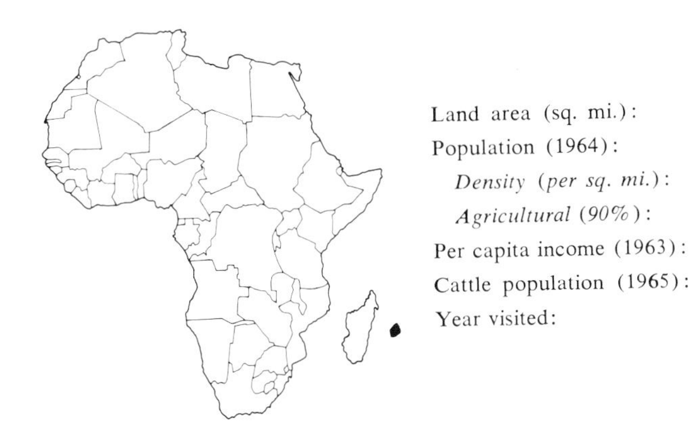

Land area (sq. mi.):	720
Population (1964):	722,000
Density (per sq. mi.):	1,000
Agricultural (90%):	650,000
Per capita income (1963):	$248
Cattle population (1965):	44,000
Year visited:	1966

MAURITIUS occupies the gorgeous little island of that name in the Indian Ocean, 500 miles east of Madagascar at 20 degrees south, 57 degrees east. Ringed by coral reefs, the land rises in all directions from the coastline to a central plateau 1,000 to 2,000 feet in elevation. Spectacular lava protrusions as tall as 2,700 feet and often nearly perpendicular surround the coastal plain. Rainfall increases from less than 50 inches along the coast to 200 inches at a few inland points. It is in the hurricane belt and direct hits occur with serious damage. The vegetation is luxuriant. Formerly largely rain forest, half the land is now under cultivation. Sugar is the life blood of the economy, although there are also sisal for the sugar bags and a few tea and tobacco plantations. Everything that grows flourishes. The staples of diet, largely rice and flour, are imported. One of the highest concentrations of population in the world—multiracial in the extreme, including both Moslem and Hindu Indians; French, British, African, and Chinese descendants; and various mixtures of these—live together quite peacefully on the surface but with some deep-rooted antagonisms underneath. In an area of 720 square miles, half the size of a Louisiana parish, the population density is 1,000 persons a square mile. The self-supporting economy shows a healthy trade balance every year. Although Mauritius was long a British colony, the majority of her European population is French and this is the common language. More than 90 per cent of the population is Indian.

Portuguese explorers first visited the island in the early sixteenth century and found the Arabs already there. The Portuguese attempted no colonization; later the Dutch attempted some settlement but abandoned the island during the first quarter of the eighteenth century, when the French moved in and established extensive sugar plantations. The island then became known as the Ile de France. In something of a brushfire episode during the Napoleonic Wars, the British took possession in 1810, and the little island again became known as Mauritius, the name which the Dutch had given it.

The wave of independence which swept over Africa and then to Malagasy finally reached Mauritius. An isolated island with an economy that is hanging on one thread, the highly mechanized sugar industry which is dependent on British support, Mauritius was as ill prepared to fly its own flag as any community could be. Nevertheless, independence within the British Commonwealth of Nations was acquired early in 1968.

CATTLE BREEDS

Except for a few species of bats and birds, there was no indigenous wild life on the island before man's habitation. In the eighteenth century during the development of their sugar plantations, the French imported cattle to supply their households with milk and beef. These animals are

Mauritius White bull.

said to have come from areas adjacent to Normandy, where the Norsemen could have introduced their cattle, some types of which are often white and have a strong polled tendency. Another theory advanced is that the progenitors of the Mauritius cattle reached Normandy from Britain and were the same as those of the British White Cattle seen today, which also produce polled individuals. None of the present breeds in Normandy, however, show any resemblance to the Mauritius cattle.

Since at the time there were no other bovines on the island, these White Cattle were maintained in a pure state by the French plantation owners, and today their descendants furnish the island milk.

Mauritius White.—Although a distinctive breed, these cattle do not have a locally recognized name; they are called both White Cattle and Creole Cattle. As a means of identification they are herein referred to as Mauritius White Cattle. These fine-boned cattle have the conformation of good dual-purpose, milk-meat animals, although such selection as they have been subjected to has been for milk production. The color is usually a solid white, but black and white, occasionally red, red and white, and light tan are seen. Although most Mauritius Whites are polled, some have small horns, not more than six inches long, lying

Mauritius White cow.

loose and close to the skull. Black ears or white ears ringed with black are often seen, and there is also a tendency toward a protruding eyeball. As usually kept by her Indian owner, a cow rarely weighs even 800 pounds but readily reaches 900 to 1,000 pounds on better feed at the government experiment station; steers on a good fattening ration at the experiment station reach 1,100 pounds at two years of age and show a good beef conformation except for a tendency to have a rather narrow rump.

Mauritius White steer, two years old, 1,100 pounds.

Zebu.—Mauritius has a few mixed Zebu cattle raised for slaughter and run on pasture. Like the Mauritius White Cattle, these animals are also referred to locally as Creole Cattle. They originated with animals brought from Madagascar for draft in the days before the sugar plantations were mechanized; there have also been fairly recent importations. A good type of Zebu draft animal is used for cart work and is often seen on the road. The color is usually white.

MANAGEMENT PRACTICES

When slavery was abolished in 1830, laborers from India were brought in for the plantation work. After completing a period of indentured

Mixed Zebu cattle on pasture.

Zebu draft ox.

labor, the Indian was compensated with money. He then established his village and started keeping a cow to obtain milk for his own use and to augment his meager income. Because the only cattle available to him were the Mauritius Whites, there was no possibility for admixture with any other type of cattle until later, when the Zebu was brought from Malagasy as a transport animal. Even then, however, there was no crossing of the two breeds because that would have resulted in a decrease in milk productivity.

In a small shed next to his dwelling, the Indian laborer kept a cow and her calf. Once a year a lead rope was put on the cow and she was walked to one of the village bulls; the rest of her life was spent in the shed, which had no windows and was kept as nearly as possible in complete darkness to discourage flies and insects. The purity of the breed had to be maintained under these conditions until the recent introduction of artificial insemination. For the 200 years the Mauritius White Cattle have been so isolated, natural selection under man's superimposed conditions has had its effect, and a hardy animal with an exceptional ability to withstand climatic stress and to utilize coarse feed has resulted. Even today the cane-field laborer lives in the same village, which now has some modern aspects, and maintains his cow and calf as did his forebears.

Bull calves, as raised by the Indian one- or two-cow dairyman, are sold for slaughter at one year of age in a half-starved condition, weighing only about 300 pounds, because after they are a few weeks old they are allowed only enough milk to keep them alive. Heifer calves are fed a little better and are grown out for sale as milk cows. Grass cut from the roadside or from the cane-field borders is brought to the animals daily on the heads of the owner's wife and children.

Mauritius White bull calf, one year old, 300 pounds, belonging to an Indian dairy owner.

Because this tropical growth has a moisture content of nearly 80 per cent, more than 100 pounds must be brought in daily for the cow, plus some for her calf. During the cane-cutting season, which lasts about five months, the cattle eat large quantities of cane tops, an even poorer ration than the grass. Concentrate is fed sparingly, if at all. Manure is saved and sold, bringing $2.00 a ton.

These household dairies average six pounds of milk a head daily, excluding both the limited quantity allowed the calf for the first few months and the owner's meager consumption. Compared with this, the Mauritius White herd at the experiment station in Curepipe averages 9,500 pounds a lactation on good roughage but with no grain supplement. The household dairies produce 120,000 pounds of milk a day from the 17,000 cows in milk at one time. A middleman on a bicycle contracts to buy the milk on a take-or-pay basis and collects it from individuals who own cows.

Over the years the breeding practice of taking the cow to a bull owner in the village probably led to some degree of selection for milk production. A cow owner discussing production with his neighbor would learn that a certain bull's progeny was giving more milk than he had obtained from the offspring of a different sire. Logically the next time his cow was to be bred he would take her to the bull that had worked such a wonder. In any event, this method of raising them has kept the Mauritius White Cattle pure and has developed an exceptionally well-acclimated animal, hardy and capable of producing milk on an extremely deficient diet.

A few Europeans are running Zebu cattle on hilly ground, too rough for cultivation, to produce beef. In the best operations pastures are fenced and rotated every seven days to prevent erosion, which could easily become severe because of the heavy rainfall. Steers at two to two and one-half years of age are now being marketed off the grass at 800 pounds. Although all of the land capable of cultivation on Mauritius is now so used, this is slightly less than half the total surface. Careful use of the hillsides as pasture could make the island self-supporting in beef.

A number of years ago an artificial insemination center was established by the ministry of Agriculture and Natural Resources. Sixty per cent of the milk cows in the country are now so bred. In 1964 a government program for selection within the Mauritius White breed for milk production was abandoned, and Friesian semen flown in from Kenya was used for breeding in an upgrading program with the objective of increasing milk production. The semen is prepared by dilution with coconut milk, the procedure developed in Kenya; stored at room temperature; and used for seven days after collection. A Friesian first cross on the Creole under proper nourishment will undoubtedly show an increase in milk production. Considering all factors such as heat stress, close confinement, higher death loss, and, especially, the poor diet the Mauritius milk cow has to live on, however, it is doubtful whether this practice will effectively increase milk production except in the experiment station, where the ration will be ample.

ABATTOIRS

The only large abattoir on the island is at Port Louis. At this plant 5,000 local cattle and 7,000 head imported from Malagasy are killed annually. The price paid by the local butcher averages approximately 15 cents a pound liveweight; dressing percentage is not much more than 50 per cent. Beef retails for 42 to 50 cents a pound. A village of any size may have its own abattoir, which usually consists of a small building with a cement floor, water, and good drainage. Sanitation is as good as can be expected under these conditions.

CATTLE DISEASES

Mauritius' remarkable freedom from cattle disease naturally follows from her husbandry practices. The milk cows are kept completely isolated except for the annual walk to the bull, and artificial insemination is now eliminating this procedure. The small number of Zebu cattle, except for the few herds being run for beef, are also kept closely confined as work animals.

An importation of Friesian cattle from South Africa in 1925 was followed by an outbreak of contagious abortion which was overcome largely by slaughter. All of the government herd of White Cattle, which had been selected for milk production for twenty years, was sacrificed. The cause was not identified as brucellosis until 1948. An outbreak of piroplasmosis in 1938 killed most of the Zebu cattle, but the Mauritius White milk cows, on account of their complete isolation, were not seriously affected.

The island is said to be entirely free of rinderpest and foot-and-mouth disease. Zebu herds running on pasture are dipped regularly for ticks.

GOVERNMENT AND CATTLE

By 1964 the experiment station of the Ministry of Agriculture and Natural Resources at Curepipe had built up a beautiful herd of 75 Mauritius White cows, the result of nine years of selection for milk production, and had six bulls on progeny test. This breeding program was being abandoned in 1966, and the entire herd was crossed with Friesian bulls by artificial insemination. As a control measure, not even a segment of the cow herd was kept pure. The selection program had been based entirely on milk productivity, with no attention being paid to horns or coloring; yet all but 6 of the 75 cows were solid white and only one had developed any horns, these being quite small. One of the best bulls on the abandoned progeny test, however, was a solid light-cream color. He was an animal of excellent conformation, weighed 1,500 pounds, and had the characteristic black ears.

Maintained pure for more than 200 years because of the isolated conditions under which it has been raised, the Mauritius White breed is remarkable. Its progenitors came from France; yet no breed resembling the Mauritius White is seen in France today. A possible connection with earlier types that existed in Scandinavia or in Britain has never been authenticated. A good response to elementary selection for milk productivity has been demonstrated. It is a tragedy for this old breed to disappear in the wake of a bureaucratic notion to improve milk yields.

OUTLOOK FOR CATTLE

Mauritius could become self-supporting in beef production by developing the hilly areas, which cannot be cultivated for sugar cane, into pastures. Bush clearing, fencing, and sowing improved grasses would be necessary. Legumes could be grown but would require phosphate fertilization. Supplemental to such a program, the bull calves of the Indians' milk cows could readily be grown out to good weights on such pastures after they had been established. Milk production of the village owner's cow could be materially increased if only good roughage could be found for her. Because it would have to be imported, there is no place for grain feeding in the present sugar economy. A small start is being made in the use of the rough country for grazing, but there is no evidence of any move to get better feed to the village milk cow.

Sudan

Land area (sq. mi.):	967,500
Population (1966):	13,900,000
Density (per sq. mi.):	14
Agricultural (90%) (1964):	12,500,000
Per capita income (1964):	$91
Cattle population (1962):	7,000,000
Year visited:	1965

THE REPUBLIC OF THE SUDAN, in northeastern Africa, begins in the Libyan Desert and ends at the rain-forest belt in the south. The northern border is with the United Arab Republic; the eastern with Ethiopia and the Red Sea; the southern with the Republic of the Congo (Kinshasa), Uganda, and Kenya; and the western with Libya, Chad, and the Central African Republic. The country lies in latitudes approximately the same as those that reach from southern Mexico to Venezuela in the Western Hemisphere. The area is about twice that of Texas, Arizona, and Nevada combined.

Since the beginning of history the region that is now Sudan was coveted by the Egyptians and the numerous nations that have occupied Egypt. Periods of occupation and control were broken from time to time by local kingdoms regaining power. At the end of the nineteenth century, Britain and Egypt established control of Sudan; and from then on British influence was paramount until independence was fully established in 1956. The country has exhibited little stability in government since then. The peoples in the northern part of the country are Arabs and Nubians, between whom there has been considerable mixture. The more primitive tribes in the south are Nilotic and Negroid. There are very few Europeans.

Containing three-fourths of the population and dominated by Arab influence, the Moslem north is divided into several political groups, each seeking control. The minority black element in the south has been exposed to Christianity but remains mainly pagan and illiterate. This sector is opposed by all the Moslem parties in the north and sees independence from the rest of the country as its only salvation.

Sudan's economy is basically agricultural and pastoral. The principal drainage is by the Nile system from south to north; the Blue Nile enters the country on the Ethiopian border, the White Nile comes from Uganda, and the two rivers converge at Khartoum. One-fourth of the land surface in the north is classified as desert. Below this is an arid belt, where the rainfall is 10 inches and then gradually increases to 60 inches in the extreme south. Cultivation is mostly confined to bottom lands along the Niles; much of the rest of the country, about 500,000 square miles, is adapted in varying degree to stock raising.

CATTLE BREEDS

Well adapted to their environment, the cattle of Sudan are almost entirely of indigenous breeds. There have been only minor introductions of exotic breeds, mostly confined to government farms and experiment stations; but a few private farms have some imported animals.

Kenana.—This breed is generally considered to have resulted from interbreeding Sanga cattle with the Shorthorn Zebu during tribal migrations before recorded history. The Kenana is rather large for African

Kenana bull.

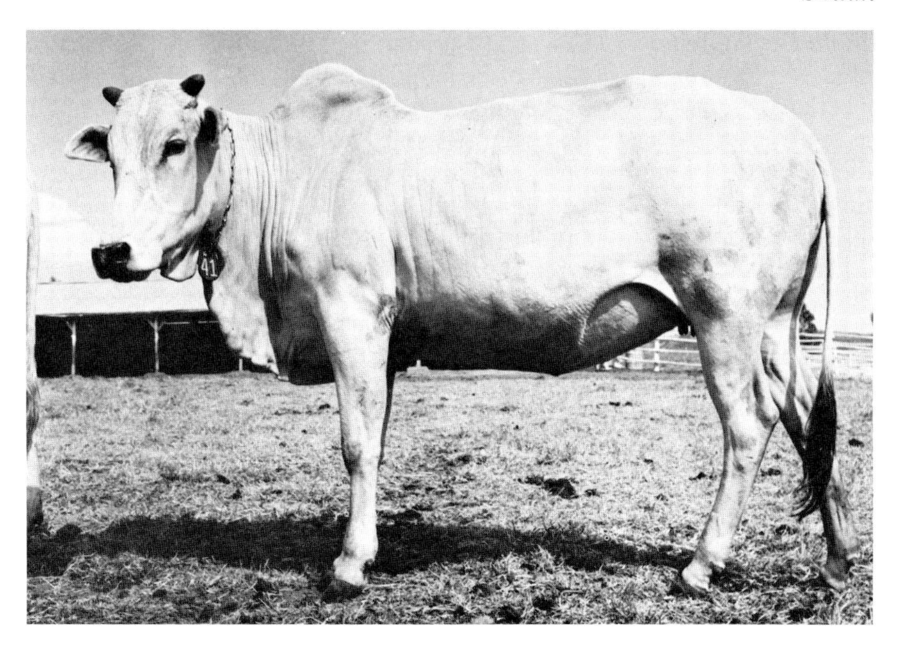

Kenana cow. Courtesy J. H. R. Bisschop

cattle: on good feed cows weigh up to 1,000 pounds, bulls 1,250 pounds. The color is solid, rather light grey, grading to very dark, nearly black, on the shoulders and neck of the males. The breed has a fair beef conformation except for the sloping rump and is a fair milk producer. The dewlap and sheath are pendulous and quite large.

The Kenana is found east of the conflux of the Blue and White Niles at Khartoum and south to the Ethiopian border.

Butana bull.

Butana.—Another result of Shorthorn Zebu interbreeding with the Sanga, the Butana usually has a higher milk yield than, but not as good a beef conformation as, the Kenana. Selected native cows on reasonable feed give as much as 4,000 pounds of milk a lactation. Cows weigh as much as 1,000 pounds; bulls may weigh 1,200 pounds. Short, rather thin horns, upturned and often poorly formed, characterize the animals, which are a solid color ranging from medium to dark red.

The Butana is seen along the Nile and on the plains to the west.

Butana cow.

Baggara.—Prevalent in the western part of Sudan, this type of cattle, hardly well enough differentiated to be called a breed, is another Shorthorn Zebu-Sanga admixture. It is not as well established as the Kenana and the Butana, however. The color varies from white to black, sometimes with colored markings. The horns are of medium size, quite thick at the skull and upturned.

Dinka.—Considered to be of pure Sanga ancestry, this breed is found in southern Sudan. It is a small, rough animal with typically large upswept horns and somewhat resembles the Ankole cattle of Uganda.

Baggara bulls trailed to Omdurman from western Sudan.

Dinka cow in a village in southern Sudan.

Exotic Breeds.—Sudan contains very few exotic cattle. A small number of Friesian, Swiss Brown, and Ayrshire cattle have been imported by private owners to upgrade the native cattle by crossing. Several of the British beef breeds and Zebu bulls from India and Pakistan have been used in experimental breeding programs at government stations, but nothing of noteworthy significance has resulted from these programs.

MANAGEMENT PRACTICES

The grazing region extends in a wide belt across the country between the northern desert and the southern forest, a distance of nearly 1,000 miles from east to west and 500 miles from north to south. Rainfall varies from 10 inches in the north to 35 inches before the humid areas in the south are reached. From here the cattle population falls off rapidly because of unfavorable climatic conditions and, in the extreme southwest, as a result of tsetse fly infestation. In the northeast there is another small area where the tsetse persists.

Nomads or seminomads seasonally traveling with their herds in search of water and grass are the cattle owners. Sheep and goats invariably are included in their operations. Water is frequently more of a problem than grass, and the Western idea that a bovine animal should water every day is far from attainment in the dry season. Long migrations are seasonally made to the areas where stockwater is available—away from the rivers in the rainy season for better grass and back again when the rains cease. Herds of 2,000 head are gathered by the members of a tribe for these seasonal movements, which result in severe overgrazing along the routes traveled, leaving large areas of grassland unutilized because of the distance from water. The extent of usage of the natural feed can be judged from the fact that the cattle population on approximately 500,000 square miles of fair to excellent grazing lands probably does not exceed 7,000,000 head. This is more than thirty acres for each animal unit after making due allowance for the sheep, goat, and camel populations. Under proper management between eight and twelve acres an animal unit would be good usage. The necessary facilities for water and control of grazing to accomplish this would require substantial development, but there is no question about the potential that exists for cattle raising in Sudan.

Although the nomad is basically a good cattleman, some of his practices are unusual when viewed in the light of Western standards. Cattle numbers are maintained and increased if possible as a matter of prestige. There is no breeding season, and usually bulls are left entire and run with the herd. During the dry season intervals of three days may elapse between stock waterings. The unusual affection of the Dinka tribesmen for their cattle is exemplified by the "song" bull: a young man of any stature in the tribe is allotted one of the better young bulls, which is then castrated, the only time this practice is resorted to. The steer becomes his constant companion, is cared for as a household pet, and is even taken along when the young man goes courting.

In the south, where the rainfall is heavier, cattle can be maintained the year round in the vicinity of the villages. Government experiment stations maintain limited numbers of cattle, but the nomad continually traveling with his livestock accounts for most of the cattle in Sudan.

MARKETING

Livestock markets are located in the larger towns. Cattle for slaughter are bought by the meatshop proprietor from traders who buy in the grazing area and trail their animals to market, often a distance of hundreds of miles, which can require several months. This trailing to market is a profession followed by herdsmen who work at it continuously. Cattle raised in the vicinity of a town may be bought direct from the owner. Large purchases often go straight to the abattoir in Omdurman. Prices are low: 700- to 900-pound bulls (there is no trade in steers) were bringing from $40.00 to $50.00 in 1965. As in all other Moslem countries, mutton is the preferred meat and retails for 27 cents a pound, compared to 18 cents for beef. Very little attention is paid to the quality of the live animal and none to the cuts of beef.

ABATTOIRS

The Central Abattoir, in the old town of Omdurman, across the Nile River from Khartoum, is the largest slaughterhouse in Sudan. The average day's run is 175 bulls, 500 sheep, and usually a few camels. This establishment is quite modern and sanitary. Equipment is good, including mechanical hoists, conveyors, and power saws. Slaughter is in accordance with the Moslem law: the animal is brought onto the killing floor, thrown, and faced East toward Mecca before its throat is cut by a holy man. Operations start at 8:30 P.M. Carcasses go direct into waiting trucks and then to the shops, the day's run being on the road by 6:00 A.M. This procedure is essential where temperatures of 120 degrees are common in the summer. Refrigeration was not available and custom demands that meat be consumed when freshly slaughtered. (Refrigerated storage at the abattoir was planned for 1965 for the high-priced trade.) The plant is operated by the municipality on a custom basis. The cattle kill consists entirely of bulls; it is prohibited by law to kill females without a veterinarian's certification that the animal is incapable of reproduction, but old cows are undoubtedly surreptitiously utilized back in the country for meat by their owners.

CATTLE DISEASES

Many of the diseases with which cattle in sub-Sahara are plagued are found in Sudan. There has been practically no introduction of exotic breeds into the herds of the native cattle owners, who hold most of the cattle in the country. The degree of tolerance which has been developed in the indigenous cattle is the major factor in holding losses from tick-borne and infectious diseases to a reasonably low figure.

This tolerance applies only to a limited extent for trypanosomiasis, carried by the tsetse fly in certain areas. Sudanese cattle are highly susceptible to this disease, except for a dwarf type of cattle in the Nubian Hills, in northern Sudan, and another, apparently unrelated dwarf type in the tropical region next to the Congo border. Both of these areas are tsetse infested but the cattle there have developed a substantial tolerance to the disease.

In recent years the Ministry of Animal Resources has promoted disease-control measures, and these have been effective in reducing death losses. It is claimed that the mortality rate of mature cattle has decreased from 12 to 2 per cent in the past ten years. While this is probably an optimistic estimate, there is no doubt that losses from disease have been materially reduced by dipping and prophylactic treatment. In addition to trypanosomiasis, which is confined to the relatively unimportant cattle areas mentioned, contagious pleuropneumonia and rinderpest are the most prevalent diseases. Outbreaks of foot-and-mouth disease, rabies, and anthrax also occur. Brucellosis is undoubtedly present, but cattle operations are conducted in such a manner that the degree to which it affects production is not known. The inherited tolerance to many diseases of the indigenous stock, coupled with the fact that exotic cattle have not been extensively introduced, enhances the effectiveness of the preventive measures taken.

GOVERNMENT AND CATTLE

Several noteworthy cattle projects have been initiated by the Kuku Livestock and Poultry Scheme, a United States AID activity under the auspices of the Ministry of Animal Resources of Sudan. One of the plans is a program to induce the seminomads in the Khartoum area to take up dairying and lead a settled life. Four thousand acres have been placed under a gravity-ditch irrigation system supplied by motor-driven pumps lifting water only twenty-five feet from the Nile. Two central milking stations were in operation in 1965, each with a 300-cow capacity, and a third was under construction. A modern dairy plant pasteurizes the milk and markets bottled milk, butter, and cheese.

The base unit for a co-operator in the scheme is a ten-acre plot of irrigated land and 8 cows, and an individual is permitted to hold two units. The co-operators, seminomads who formerly made very little use of their cattle, are now milking a total of 600 cows twice a day in the two milk parlors. Each owner's cows are milked separately and the milk weighed. This is the basis on which he is paid. Milking is by hand but sanitary standards are well maintained. The milk delivered to the pasteurizing plant is said usually to have a bacteria count that would qualify it as Grade A according to United States standards.

The benefits of grain-feeding cattle are also being demonstrated at Kuku. Sudan is one of the few African countries which has the potential to grain-feed cattle. With a density of only 14 cattle to the square mile, ample water for irrigation from the Blue and White Niles and, a fair rainfall over much of the country, the productive ability of the land has only been scratched. At Kuku young Kenana bulls purchased from native owners have gained as much as three pounds a day on a 100-day feed test, which shows a good growth potential.

Nearly finished Kenana bull, two and one-half years old, 1,100 pounds.

The most basic program at Kuku is that designed for improvement of the Butana cow for milk production. A small herd of the best cows the owners could be induced to sell was obtained and a selection program initiated for milk-producing ability. Chances of developing a good milk-producing type from the Butana appear to be excellent: native cows have been found which, without any selection, give as much as twenty-five pounds of milk a day.

Programs such as these are sometimes disappointing in their terminal results. There is usually a retrogression to the old customs when control is placed back in African hands. The outcome of these endeavors in Sudan remains to be seen.

The Ministry of Animal Resources maintains a number of experiment-station activities both as demonstration centers for better husbandry methods and for improvement by selective breeding. The breeding station seventy miles from Juba, in southern Sudan, is working with the Dinka and related animals. On the Blue Nile, south of Wad Medani, the Um Benein Livestock Improvement Station has a herd of 500 Kenana cattle on which selection is being made to develop a better milk-producing type. Calves are taken from their dams at birth and hand-fed so that individual milk production may be determined.

OUTLOOK FOR CATTLE

All of the physical elements for the development of a great cattle country are present in Sudan, but the human drive for accomplishment is lacking. Two breeds of good indigenous cattle—the Kenana, susceptible to development as a beef type, and the Butana for a milk or milk-beef, dual-purpose animal—are grown there. The hot climate, while limiting the possibilities for the introduction of exotic cattle except the Zebu, is withstood well by both the Kenana and the Butana. Properly developed, the grazing areas could support several times the present cattle population. Grain feeding, since the potential productive capacity of the arable land exceeds the need of the population for years to come, is also a possibility.

Having traveled for ages the seasonal routes from water to grass, the nomad is reluctant to change his way of life. The instability of government precludes any large-scale investment for development of a sizable cattle industry. For the foreseeable future, cattle raising in Sudan will probably continue much as it has in the past.

Tanzania

Land area (sq. mi.):	363,000
Population (1966):	10,500,000
Density (per sq. mi.):	29
Agricultural (90%) (1964):	9,475,000
Per capita income (1964):	$61
Cattle population (1966):	10,000,000
Offtake (11%) (1966):	1,100,000
Year visited:	1965

THE REPUBLIC OF TANZANIA is slightly smaller than Texas and Arizona combined. It was formed by merging the Republic of Tanganyika and the Sultanate of Zanzibar, which occupied the island of that name. One thousand square miles in area, Zanzibar lies off the northeastern coast of Tanzania. Located just south of the equator, the Tanzanian mainland is largely high plateau except for two narrow strips of low land, a 500-mile stretch along the Indian Ocean in the east and the lakes of the Great Rift Valley in the west. The northern border is with Kenya and Uganda, the southern with Mozambique, Malawi, and Zambia. The Indian Ocean lies to the east and the Congo (Kinshasa), Rwanda, and Burundi are on the west. On the coast and in the Rift Valley the climate is tropical, but the plateau land, generally 4,000 feet in elevation, varies from semiarid to regions of 30-inch rainfall and because of the altitude has a warm but equable climate. In the northwest and below Mount Kilimanjaro in the northeast, the grassland and brush areas of the plateau rise up to high mountain slopes which enjoy heavier rainfall and are good crop lands.

Ninety-eight per cent of the population is African, mainly Bantu tribes, and about 1 per cent is Arab. Indians, somewhat less than 1 per cent of the people, handle most of the small commerce. In 1962 the total number of Europeans was approximately 21,000.

Zanzibar ruled the northeast coast before the Portuguese obtained control in the sixteenth century. Except for some British explorations, little attention was paid to the area until the latter part of the nineteenth century, when Germany declared the land a protectorate and initiated some agricultural development. What then became known as German East Africa was occupied by the British during World War I. After the war the area was mandated by the League of Nations to Britain under the name of Tanganyika. Independence was attained in 1961. In 1964 Tanzania was formed by the union of Tanganyika with its small communistic neighbor, the Sultanate of Zanzibar, which had become independent in 1963.

In addition to all the usual African endemic diseases, the cattle of Tanzania are plagued with the most widespread tsetse fly infestation of any country on the continent. The fact that the government may become communistic or that the tsetse fly confines cattle herds to badly overgrazed areas are matters of no concern to the 90 per cent of the population who enjoy either a pastoral nomadic or a rudimentary agricultural way of life.

Cattle of the nomadic tribes are run in those drier grass and bush areas that are reasonably free of the tsetse fly, and the agricultural tribes farm their plots in the heavier rainfall regions. Including large estates, there are fewer than 1,000 European farms, and they account for less than 1 per cent of the agricultural land.

Except for the heavily populated urban area of Dar es Salaam and a few small towns, the African population lives in the bush. They all raise cattle—the nomadic and seminomadic tribes and even the primitive agriculturists who in recent years have begun to keep a few head.

CATTLE BREEDS

The Ankole cattle of the neighboring countries of Uganda and Rwanda are seen in northwestern Tanzania. An inferior type of Boran is found in the north, below the Kenya border.

Most of the cattle of the country, probably over 90 per cent, cannot be classified as a breed or even as a type. Locally they are called Small Zebu or Tanganyika Zebu. They derive from the Shorthorn Zebu, possibly influenced by the cattle the Arabs had, for they were strongly entrenched in the region when the Portuguese arrived on the east coast. Mature Tanganyika Zebu bulls that reach the Tanganyika Packers' abattoir in Dar es Salaam average 560 pounds liveweight. Color can be anything and horns follow no pattern. There is no conformation similarity, but there usually is a hump, large and varying widely in shape. The Tanganyika Zebu is slow maturing, partly because of poor nutrition. Heifers calve at three and one-half to four years of age and reach maturity at six or seven years. Milk yield is only a few pints a day.

Because of the care they give their animals, the Masai, who travel the

Tanganyika Zebus.

country along the eastern side of the high plateau, tend to have better cattle than the other tribes. There is no distinguishing difference, however, between the cattle of the Masai tribesmen and the other animals except in condition and size.

Small dairies on the outskirts of Dar es Salaam contain a few cows of the milk breeds. Most of the animals are Friesians, with a scattering of Jerseys. There are also a few dairy cattle on the European farms.

MANAGEMENT PRACTICES

With the exception of the limited number on the European farms, the cattle of the country are run by nomadic and seminomadic tribes on the vast grasslands. Large areas are completely unutilized because of tsetse fly infestation. Overgrazing is common, occasioned by the routes traveled to find grass and water. Mature animals weigh from 500 to 700 pounds and could weigh more on better nutrition. African-owned cattle are used primarily to supply milk, which is taken at the time of night corralling.

Being better cattlemen than the other tribes, the Masai have located in the grazing areas of heavier rainfall. They see that their herds are on the best grass available and that they have more frequent access to water. The Masai's cattle, although of the same type, outweigh the Tanganyika Zebu of the other tribes by 100 to 150 pounds.

The traditional cattle-raising tribes utilize the milk and, in the case of the Masai, some blood as well. Beyond this, very little use is made of their herds, although the ratio of the cattle to the human population is one of the highest in Africa. A few head are sold if there is a need for cash. Now and then an animal is slaughtered, and anything that dies of natural causes is eaten.

Tanganyika Zebus belonging to Masai tribesmen. Photographed in northern Tanzania. Courtesy A. F. Dinewall

The native farmer clears and crops for a few years a small plot sufficient for his needs. When the fertility of the thin soil is exhausted, he moves to another plot. Cattle are sometimes used for draft, although the hand hoe is used for most cultivation. The cow or two that a farmer keeps may be milked for household use, and occasionally he may sell an animal. The cows are maintained by pasturing them on whatever grass can be found in the vicinity of the cultivated plots.

Apparently little if any attention has been given to selection for any desired characteristics by the cattle-raising tribes in the area that is now Tanzania—not even to the commonly desired attributes that the African cattle owner likes in his cattle, such as color or size and shape of horns. The primary objective is quantity, to be maintained as a matter of prestige. Title to cattle is transferred only when the headman or his sons desire a new wife or when the money economy encroaches, causing a want that requires cash.

Feeding cattle for market is unthought of in Tanzania, but two European-managed operations, the Mkata and the Ruvu farms of the National Development Corporation, involve the growing out of native cattle on grass. Bulls at two to three years of age, weighing about 350 pounds, are bought in the northern part of the country at a laid-in cost of $20 a head. They are grown out on native grass for eighteen to twenty months to a finished weight of from 550 to 600 pounds, gaining approximately four-tenths of a pound a day. This procedure doubles the value of the animal, which usually sells for $40 or slightly more a head after improving from poor condition as received to what would be low-good by USDA standards. The management program is simple

Tanganyika Zebu bulls, three years old, 350 pounds, as received at Ruvu Farm. These animals were purchased to be grown out on grass.

and effective. The bulls, as received in lots of a few hundred head, are allowed a week to recover from the effects of shipping before being sorted for disease and size, sprayed, castrated, and vaccinated for anthrax; other prophylactic treatment may be given, depending on the area in which the animals originated.

The Ruvu Farm sells 7,000 to 8,000 head annually. All cattle are corralled at night for protection from theft and predatory animals. Stealing and death losses, under good management, can be kept at less than 2 per cent. Such operations have now been nationalized; but, as sometimes happens in East Africa, the new regime has learned that the European know-how is necessary to make such an enterprise profitable and, therefore, hires a European manager.

Nearly finished Tanganyika Zebu steers, four and one-half years old, 550 pounds. Photographed at Ruvu Farm.

Spray run at Ruvu Farm. It has a capacity of 750 head an hour.

There is very little dairying in Tanzania. A small farmer on the edge of town, usually an Indian, may have 10 to 15 cows, which average about ten pounds of milk a day. These animals sometimes show admixture of European milk breeds, Friesian or Jersey, but often are simply the Tanganyika Zebu. If bottled and distributed by the owner, the milk brings 12 cents a pint; if sold wholesale, the price is 9 cents. In Arusha, just below the Kenya border, there is a pasturizing plant, which serves this area; and the National Development Corporation is planning a processing plant at Dar es Salaam to combine raw milk produced in the area with powdered and skimmed milk.

The one large dairy, the Kiraura Farm, across the river from Dar es Salaam, was taken over by the principal trade union from the former

Indian owner. Two hundred-nine cows form the herd, mostly Friesian stock obtained from Kenya at the time of the distressed condition of the dairy owners there. Pasturing is during the day since the immediate area is infested by tsetse flies and the cattle cannot be in the vicinity of the brush in the morning or late afternoon. A supplement consisting of brewers' mash and unpressed cottonseed is fed ad lib when milking. Milk production averages ten pounds a day. The operation is a typical example of what happens when an effort is made to maintain the European breeds of cattle under tropical conditions: the cows are in poor condition and production is quite low for the quality of the feed they are on.

MARKETING

Wherever the Europeans' sphere of influence declines in East Africa, the Asian (usually the East Indian) tends to take over the day-to-day operations of commerce, invariably at a rather high cost to the African who is unoriented to a money economy. This situation does not, however, apply to the cattle trade in Tanzania. Cattle for slaughter are hard to obtain, and Arab or Somali traders (usually Tanzanian citizens) dominate the market. These traders are established throughout the area in which native cattle are grown and undoubtedly buy more cattle than anyone else could. The trader buys a few head here and there from the native Africans and also is a large buyer at the government-controlled auctions in the bush. He then consolidates his purchases and either sells to the European-operated abattoir in Dar es Salaam or to local butchers in the towns which have municipal abattoirs where slaughtering is carried out on a custom basis.

Most African-owned cattle pass through government auctions held in the grazing areas. Prices in 1965 averaged approximately $30.00 a head. Official 1966 figures showed that 394,000 head of slaughter cattle went through the regular outlets; yet about 1,100,000 hides were sold, indicating an offtake of 11 per cent. Clandestine slaughter causes this discrepancy, which has never been explained.

ABATTOIRS

The two principal slaughtering facilities at Dar es Salaam offer a startling contrast in the preparation of meat for human consumption. The municipal slaughter slab kills 150 to 200 head a day, mostly bulls but including a few steers. Town butchers purchase the animals for slaughter from traders and hold the cattle in small corrals near the abattoir. A herdsman takes them out for such grazing as can be found nearby and returns them at night. Animals are designated for slaughter as required by the owner. The killing occurs under a shed roof over a concrete floor in accordance with Moslem-approved religious practices (see "Abattoirs" under "Sudan") but without heed to the most elementary sanitary procedures. A veterinarian inspector casually examines the liver and lungs. Carcasses go to waiting trucks for delivery in town and are sold to consumers within a few hours after slaughter—an essential procedure in the hot climate since refrigeration is not available. Tanganyika Packers, Limited has been asked to take over the processing of cattle and leave the slaughter slab for the handling of sheep and goats.

On the other side of town is an abattoir which was originally set up by Liebig's Extract of Meat Company, Ltd., as Tanganyika Packers, Limited. A majority of the stock is owned by the Tanzanian government, but the plant is operated by British personnel under the direction of Liebig's. The usual high standard of the British abattoir is meticulously maintained. All of the carcass except the tenderloin is used in canned meats or extracts. The cattle supply is obtained from Arab and Somali traders and African co-operatives at the government auctions, shipped by rail to Dar es Salaam, and kept in a large holding area adjoining the plant. This method equalizes the work load and also enables the poorest animals to make some improvement in condition before being killed. All animals are typical Tanganyika Zebus, mostly bulls which average 560 pounds liveweight and would grade a USDA standard. Cattle are paid for on a cold-dressed-weight (CDW) basis, the current price being equivalent to 7 cents a pound liveweight.

CATTLE DISEASES

All the plagues that African cattle are subject to take their toll; yet disease loss is probably no greater in Tanzania than in some neighboring countries that have better-organized control measures because indigenous cattle account for practically the entire cattle population and have developed a degree of tolerance for the infectious and parasitic diseases that exotic cattle do not have.

Rinderpest still occurs but there have been no serious outbreaks recently. Tsetse fly infestation, however, covers three-fourths of the potential grazing land and has made these areas uninhabitable for cattle as well as man. Even the indigenous animals of the region have not developed a significant tolerance to trypanosomiasis. In recent years because of uncontrolled wild life, the fly has encroached on clean areas. Calving rates are low and death losses high, but the over-all cattle population continues to gradually increase since the offtake is low.

GOVERNMENT AND CATTLE

The operation of the experiment station established by the British at Tanga, on the northern coast, is being continued by the Agricultural Department of the National Development Corporation. Efforts were made to establish a fixed type of improved milk animal by crossing Boran bulls and Boran crosses with European milk breeds on Tanganyika Zebu cows. The Boran bulls used were imported from Kenya. Also at Kongwa, West Kilimanjaro, and Nachingwea stations, good stock from Kenya has been introduced to upgrade the local cattle. At the Kongwa farm there is a small abattoir which supplies carcasses of superior quality to Dar es Salaam.

The National Development Corporation program involving cattle improvement is dependent on securing funds from the World Bank. Bettering existing farms and establishing two new ones are in their plans and will cost an estimated $3,600,000. Attempts are being made to localize the cattle raising of the nomadic Masai tribesmen, who are considered somewhat a menace to more settled cattle raisers because the Masai consider the acquisition of other people's cattle a legitimate enterprise.

OUTLOOK FOR CATTLE

Any substantial improvement in stock raising in Tanzania is difficult to foresee, although the country has the potential for a vastly expanded cattle industry. Much of the land area which to a large extent is unutilized or poorly utilized could be developed for productive grazing. Because of inadequate rainfall for cropping, poor soils, and the terrain, three-fourths of the land could be better kept in grass than put to other use. This change would require taking out the brush, eliminating the tsetse fly in infested areas, and providing for adequate stock water. To prevent tsetse reinfestation, the game would probably have to be killed or segregated in areas enclosed by a perimeter free of all wild life. Reservoirs and wells would be necessary for water. The National Development Corporation is taking small steps along these lines, but to finance such projects in a cultural atmosphere which sees no reason for them is simply poor economics.

The most rapid increase in cattle usage could be effected if the owners of the 10,000,000 head in the national herd could be induced to sell off stock at a reasonable age. Even this simple step is a long-range problem which only time can solve. Here the question arises as to what reason there is for an expanded cattle industry. Tanzania is not over-populated, with a density less than the average of Texas and Arizona, with which she has been compared in area. Why try to create strange and unknown wants among a people who enjoy life and a wide range of freedom in their present cattle economy? Their philosophy that a man owning cattle always has a good, tangible asset that in time of emergency will get him what he wants may have genuine validity. The civilized world has known many catastrophes from runaway inflation in its money economy.

Uganda

Land area (sq. mi.):	91,000
Population (1966):	7,550,000
Density (per sq. mi.):	83
Agricultural (90%) (1964):	6,795,000
Per capita income (1964):	$69
Cattle population (1966):	3,627,000
Offtake (1965):	70,000
Year visited:	1965

UGANDA is the garden spot of Africa. Located in central East Africa, bisected by the equator but at elevations between 3,000 and 4,000 feet, the climate is pleasant though warm. The growing season is twelve months long and the rainfall pattern is excellent, decreasing from 60 inches along Lake Victoria to 20 inches in the north, with a semiarid strip in the extreme northeast. Not quite as large in land area as Mississippi and Louisiana, Uganda has a population one-fifth greater. The eastern border is with Kenya, the southern with Tanzania and Rwanda; Congo (Kinshasa) is on the west and Sudan is on the north. Uganda has its own "fertile crescent"—exceeding in productivity that of Mesopotamia—a productive belt lying inland for fifty miles along the northern shores of Lake Victoria, the headwaters of the White Nile.

When they entered the country only 100 years ago, the British found a well-organized tribal society, frequently at war among its elements. Unlike some of its neighbors, Uganda was developed by British administration, rather than by colonization. An independent, federated state since 1962, it is a political enigma, consisting of four kingdoms each with a king and his ministers, nine districts self-administered to a degree, and one territory. Political power is primarily in control of the Baganda tribe, which comprises a large majority of the population in Baganda Kingdom, but what are basically tribal conflicts arise from time to time and reach sizable proportions.

Cotton, coffee, and sugar provide the substantial trade balance enjoyed by the agrarian economy. The stock-raising seminomads are on the fringe of the settled communities and enter into the economic picture to only a limited extent in the number of cattle, goats, and some sheep that are marketed. Throughout the country the trade-minded Asian has worked his way into the handling, almost exclusively, of commercial enterprises at the retail level and tends to exploit the African quite generally even in cattle trading.

The Bantu-speaking peoples, which include the Baganda, are the largest tribal group; Nilotic and Nilo-Hamitic peoples make up the rest. About 1 per cent of the population is Indian. The European element, mostly British, numbered about 11,000 in 1965 and was gradually decreasing.

CATTLE BREEDS

Only one distinctive breed, the Ankole, is seen in Uganda. Most cattle are nondescript, the result of countless generations of interbreeding the different cattle types in the general area. The cattle of the country show more influence of the Boran than of other African breeds; but the individuals in which this influence is noted do not present the conformation of, and are smaller than, the true Boran.

Karamojong.—Named for the tribe that runs them, Karamojong cattle

Karamojong bull showing Boran influence.

Karamojong cow in good condition.

constitute most of the commercial cattle. They are mixed animals, neither a breed nor a type, although many individuals show much Boran influence. For generations the Karamojong tribe has been adept in acquiring other people's cattle, which probably accounts for the diversity of the animals in their hands. Color, conformation, and size vary widely.

Unless drought conditions have prevailed or overgrazing has occurred in the area adjacent to available water, these cattle are usually in fair condition. The Karamojong people are among the best African stockmen.

Ankole.—One of the most distinctive breeds of cattle in all Africa is the Ankole, named for the tribe which runs them. These animals, seen in southwestern Uganda along the Congo and Rwanda borders, are Sanga descendants. There are said to be more than 400,000 of these cattle in Uganda and another 800,000 in Burundi and in Rwanda. They are surviving in spite of more than the usual handicaps to cattle raising in Africa—extreme overgrazing, stockwater shortage, and infectious disease. Although tolerant of many African cattle diseases, the Ankole cattle, along with their owners, are particularly susceptible to tubercu-

Ankole headman and cattle.

losis. Probably 60 per cent of the cattle and 25 per cent of the people are so affected.

Large, uprising, and outswept horns are the startling characteristic of the Ankole cattle, which in the past have apparently been subject to selection for shape and size of horns. They frequently measure five feet in length, six inches in diameter at the base of the skull, and as much as six feet between tips. This same breed in Rwanda, known there as the Watusi, has even larger horns, which are somewhat lyre-shaped and have been obtained by selective breeding and by training the horns of young animals. These cattle are considered sacred by the tribesmen.

In good condition a mature Ankole cow may weigh 800 pounds, a bull 900 pounds. These weights are only attained at seasons when good grass is available in areas not overgrazed. Average weights are lower because of inadequate nutrition. The Ankole is a humped animal, but the hump is hardly discernible in the female and not very large in the male. Many black animals are seen, but color can be white, red, or varied. The conformation is uniform but very poor, possibly because of generations being grown on deficient nutrition. The breed is slow in maturing, not attaining full growth until six years old. Milk is taken regularly and furnishes an important part of the diet of the owners. No use is made of the Ankole for draft purposes.

Ankole with exceptionally large horns.

Nondescript Cattle.—The cattle in the cultivated areas are locally referred to as Shorthorn Zebus but bear no uniform resemblance to cattle of that name in other countries. As seen in the countryside, they are mixed, nondescript, and without any common characteristics. They are extremely varied in color, many are humped, and some show no hump at all or only a vestige of one. Horns are usually small, and a few animals are naturally polled. A cow weighs from 500 to 700 pounds; a bull rarely weighs as much as 900 pounds.

Exotic Breeds.—The few exotic dairy cattle, located on the European-managed estates, probably do not number 1,000 head. Beef breeds have

Nondescript cattle. Photographed in western Uganda.

not been introduced commercially for upgrading, but imported semen is used in a government project referred to later.

MANAGEMENT PRACTICES

Holding cattle for prestige and lobola is universal in Uganda. Three-fourths of the cattle—from a few to perhaps 20 or 30 head to an owner—are held by the African crop farmers in the area north of Lake Victoria.

Cultivated landholdings are in the hands of the small African farmers except for a few large private estates which raise tea and sugar cane and are responsible for only a small fraction of the total agricultural production.

Oxen are used for farm draft, and the cows for milk to be consumed by the owner's family. As the agricultural community slowly changes to a money economy, more cattle are being sold for slaughter.

In the Karamoja District, which lies along the Kenya border to the east, cattle raising is the sole occupation of the seminomadic people. This area has the lowest rainfall in Uganda, precipitation decreasing from 25 or 30 inches in the southern part to semiarid conditions in the extreme north. Stockwater is frequently more of a problem than grass, and animals are moved long distances for water. The situation in recent years has been alleviated to some extent by government-drilled bore holes with hand-operated pumps.

The Karamojong are one of the most primitive peoples in Africa. The male native wears nothing but a narrow skin or strip of cloth, thrown

Karamojong chief supervising watering of his cattle at one of the government bore holes near Moroto.

over one shoulder and carried as a cape; the female wears only a short skirt, tucked in at the waist. The male herder carries a long spear and a tiny, hand-carved wooden stool to sit on. Like the Masai, the Karamojong traditionally consume milk and blood from their cattle; but the use of blood is now decreasing and grain is becoming a part of the diet.

The average Karamojong owns 20 head of cattle, but a few tribesmen may keep several hundred each and a major chief may grow as many as 1,000. All cattle are confined at night within a perimeter barrier strongly built of logs. Inside this are thornbush pens for separating the cattle of different owners and the cows from their calves. The living huts are also located inside this enclosure. Cattle are slaughtered as a rule only on ceremonial occasions, but an animal that has died of natural causes is utilized as food. Cattle stealing is simply a way of life and is part of a young man's training for entrance into manhood. In the more remote areas organized raiding, particularly across the border into Kenya, is common. There is no stigma attached to these raids. Such undertakings in the tribal culture are equivalent to a European sporting event. If ineffectively protected, herdsmen are killed so their stock can be taken. Herding is done by men, not by boys.

In Ankole, in southwestern Uganda, the tribesmen run the Ankole breed exclusively. These people are seminomadic and have fixed abodes in the areas to which they periodically return. Cattle raising is their only occupation. The average ownership is about 50 head, but important chiefs may have herds of several hundred. The cattle owner's hut and night corral of thornbush are utilized as long as there is sufficient grazing in the area; when grazing has been exhausted, the area is abandoned and a move is made to better grass. At night the cattle owner brings his herd in and the calves, which have been held in the corral or run separately if old enough, are allowed to suck for about one minute. The cows are then milked and the calves turned back to them. When they finally cease their efforts to obtain nourishment, the calves are again separated from their mothers until morning, at which time the same procedure is repeated before the cows go to pasture. The deficient nourishment this practice occasions results in a heavy death loss in young calves—as much as 70 per cent. Milk production does not average two pints a day, although exceptional individuals have given up to eight pints. The lactation period is short.

Cows normally calve at four years of age and then at intervals of two to three years. Inadequate nutrition, the resulting poor condition, and disease are largely responsible for this low productivity. Areas convenient to water are badly overgrazed—1 head to two acres or less in areas where the normal growth is such that five or six acres a head would be required for adequate nutrition. Grazing lands are communal and no control is exercised over the number of cattle an owner can run. When first seen by the Europeans, Ankole tribesmen followed the custom of killing male calves at birth unless they were wanted for breeding. The custom is prevalent to some extent today.

Tsetse fly infestation exists in parts of Ankole. Losses from trypanosomiasis are sometimes so high that the cattle are eventually moved out of large areas. Ankole cattle are quite tolerant to tick-borne diseases.

Because of transportation difficulties, very few Ankole cattle reach the Kampala market. Generally they serve no economic use other than for the milk consumed by the owner and his retainers. Even for an individual owning several hundred cattle there is no surplus of milk. Requirements rise with the number of cattle—more wives, more retainers and herdsmen, and their families.

MARKETING

In an effort to effect an orderly flow of cattle to market, auctions have been established by the government in the Karamoja District. Native cattle are trailed to them on the specified sale days. Bidders are butchers desiring animals for slaughter; traders, usually Indians, who buy for resale to butchers; and representatives of Uganda Meat Packers, Limited, which operates an abattoir in Kampala. Most commercial cattle are obtained at these auctions. Traders working through the cultivated areas also obtain slaughter animals by direct purchase from the small farmers.

Cattle bought by the head at the government auctions sold for an average price of 7.7 cents a pound in 1965. Retail meat price was fixed by the government at 21 cents a pound with practically no differentiation for specific cuts. Average liveweight of the cattle slaughtered at Uganda Meat Packers, the majority of which come from Karamoja, is 580 pounds and the dressing percentage is 51 per cent. The total number of cattle offered for sale was approximately 3,000 head a month, representing an offtake slightly over 1 per cent. Other sales such as those from the small farms are not included in this number but are relatively small.

ABATTOIRS

The municipal slaughterhouse in Kampala is run by the city on a custom basis. Local butchers buy cattle a few head at a time and slaughter them as necessary to fulfill their requirements. Carcasses are inspected by British veterinarians. From 6 to 8 per cent of the animals slaughtered are condemned because of measles (*Cysticercus Bovis* cysts). About 20 per cent of the animals are so infected, but many which show only a few cysts are passed. Tuberculosis infection runs 4 to 7 per cent, but not all of these animals require condemnation. Slaughter facilities are primitive, and the approved Moslem method for killing is used (see "Abattoirs" under "Sudan"). Processing is accomplished with practically no attention to sanitation. All meat is sold warm the day it is killed.

Built by the British and maintained and operated at their usual high standard, a modern abattoir in Kampala was taken over by the Uganda Development Corporation, a government organization, and is now operated as Uganda Meat Packers, Limited. Management and supervision are still European. As far as sanitation is concerned, carcasses could go direct to Smithfield's in London. The plant has a capacity of 200 head daily and is optimistically being enlarged. Stock for slaughter, however, was in such short supply in 1965 that the average kill was only 2,000 a month. This shortage is so acute that slaughter cattle were being shipped across Lake Victoria by boat from Tanzania. In spite of this situation another slaughterhouse is being planned by Uganda Meat Packers at Soroti, the railhead for the Karamoja district, in anticipation of more Karamajong cattle being brought to market.

Twenty-five per cent of the output of the Uganda Meat Packers' abattoir goes to the local trade in Kampala, the remainder for export as canned, frozen, or chilled beef—the latter for Cyprus and Arab countries not requiring rigid inspection. Killing methods are approved by the diplomatic representatives from the Moslem countries, who quite exceptionally permit the use of electric shock in the killing box. The throat of the animal is cut when the carcass is on the rail. This authorization of killing procedure is extremely liberal compared to the usual Moslem requirements.

CATTLE DISEASES

Of the endemic African cattle diseases, the trypanosomiasis of the tsetse fly takes the greatest toll, with tuberculosis second. Tsetse fly infestation has completely driven the cattle from large areas. A Tsetse Control Board, set up by the British in 1947, is still functioning. Some progress has been made in controlling the spread of the insect, and there has been a limited reclamation of infested areas for grazing. In parts of Ankole, however, death losses from trypanosomiasis run as high as 20 per cent, while in others cattle raising has been abandoned. Although game parks have been established, the still limited control of wild life is the major cause for spread of the tsetse fly.

Tuberculosis is widespread; up to 7 per cent of the cattle slaughtered at the Municipal Slaughterhouse in Kampala are found to be infected. The Ankole breed is particularly susceptible. The only satisfactory method of combating the disease is by slaughter, and this is unthinkable in Uganda: the African cattle owners simply would not submit to it.

Some spray runs have been provided in the tribal areas. Their use is not looked upon with much favor, and they are too widely scattered to provide effective control of parasites. The indigenous cattle, however, have acquired some measure of tolerance to most of the tick-borne diseases.

Cysticercus Bovis is prevalent because of the close association of cattle and people. Although it does not harm the cattle, the debilitation it causes in the human population is bound to be extensive.

GOVERNMENT AND CATTLE

Government activities concerning cattle are largely confined to such routine work as inspection at the slaughterhouses, quarantines on cattle movements within the country, efforts to increase the marketing of cattle by establishing tribal auctions, provision of spray runs and for stock-water in the tribal areas, and some attempted control of the tsetse fly. All of this work is largely supervised by the European personnel who were engaged in the work before independence. The needs are considerably in excess of both the funds available and the people necessary for administration.

In 1963 a program was initiated by United States AID in cooperation with the Department of Veterinary Services to demonstrate good cattle management methods in the country. An area in northwestern Ankole was cleared of the tsetse fly by the usual method of brush clearing and spraying, after which the game was killed off by government hunters. Test groups of 20 cattle were then held for six months in small fenced areas of the reclaimed land to determine if there was any reinfestation. The whole area is to be fenced into 3,200-acre units which will be cross-fenced for controlled grazing. These units are designed to maintain individual herds of 400 cattle and are to be allotted to African cattle growers who will agree to follow approved management practices for disease control, breeding, grazing, and marketing under competent supervision and control.

On one corner of the area allotted to the scheme, a 13,000-acre unit has been set aside to establish management practices best adapted to the environment. Breeding experiments are in progress to determine the results of crossing three breeds on the indigenous cattle. Random mixed lots of Boran cows imported from Kenya, native Zebu cows, and Ankole cows—150 head to a lot with all three breeds represented—are to be used. Frozen semen will be used to breed one lot to Red Poll bulls and a second to Angus bulls; the third lot is to be bred naturally to Boran bulls. If Africans can be induced to handle the 3,200-acre units in the prescribed manner, it will be a unique and progressive step in their cattle raising.

OUTLOOK FOR CATTLE

Uganda has a sizeable potential for beef cattle production. Cattle in relation to human population is 50 head to 100 inhabitants; yet the offtake for commercial slaughter is estimated at about 2 per cent. The cattle population is less than that of Mississippi and Louisiana, with which Uganda has been compared in land area. The natural grazing areas, effectively managed and cleaned of tsetse flies, would permit much larger cattle numbers to be run. In addition to such a possible increase in the cattle population, a greater crop production obtainable by modern farming methods from the arable lands could even provide grain in excess of human requirements and the animals could be finished to heavier slaughter weights. The major requirements for the attainment of such objectives are, first, a change in the cultural practice of holding cattle for prestige and utilizing only a small milk yield; second, effective disease control, particularly elimination of the tsetse fly; and third, the acceptance by the pastoral African of good management methods.

Zambia

Land area (sq. mi.):	291,000
Population (1966):	3,780,000
Density (per sq. mi.):	13
Agricultural (77%) (1950):	2,810,000
Per capita income (1964):	$170
Cattle population (1966):	1,300,000
Offtake (5.8%) (1964):	75,000
Year visited:	1965

THE REPUBLIC OF ZAMBIA, FORMERLY Northern Rhodesia, was granted independence in 1964. Occupying an extension to the southwest of the high plateau area of East Africa, this landlocked country is bounded by the Republic of the Congo (Kinshasa) and Tanzania to the north; Malawi to the east; Mozambique, Rhodesia, and South-West Africa to the south; and Angola to the west. Lying between latitudes 9 to 19 degrees south of the equator, the climate is alleviated by the altitude and is more temperate than tropical. The elevation is between 3,000 and 4,000 feet. Zambia is in a 25- to 60-inch rainfall belt. In size its area is comparable to that of Texas and the part of Louisiana lying west of the Red River. The drainage is principally to the Zambezi River, which divides Zambia from Rhodesia. The country enjoys a major economic advantage over most of her newly independent sister states—a good income and source of foreign exchange from the large copper and other mineral deposits.

Ninety-nine per cent of the population is African, almost entirely members of the Bantu tribes, and 1 per cent is European. The latter originated from settlers who came from either what is now Rhodesia or from South Africa or are connected with the operation of the large copper development and commerce generally.

The Rhodesias for the most part were settled by emigrants moving northward from South Africa before and after the turn of the century. What was Southern Rhodesia, just across the border from South Africa, was colonized first. After her agricultural possibilities had been sampled, settlers moved into what used to be Northern Rhodesia, which became a British protectorate in 1924. Nyasaland was the third part in the Federation of Rhodesia and Nyasaland but was only lightly populated during the colonizing period. Following the dissolution of the federation in 1963, Nyasaland became the independent state of Malawi, and Northern Rhodesia became independent Zambia. Southern Rhodesia, now known as Rhodesia, in 1965 took, and is now sustaining, a new approach to independence in Africa—a unilateral declaration in the pattern of the American colonies in 1776. Two-thirds of the land in the colonized area that was once Northern and Southern Rhodesia went to Zambia, along with one-third of the human population and one-fourth of the cattle. Zambia is more arid and less developed than her neighbor to the south.

CATTLE BREEDS

Three well-recognized breeds of indigenous cattle, not including the Africander, are found in Zambia. There have been no noteworthy introductions of exotic cattle breeds unless the Africander, which is indigenous to South Africa but has been brought to the Rhodesias in recent times, is considered one.

Angoni.—The Angoni traces back to the Shorthorn Zebu. It is a fairly

Angoni cow and calf in better than average condition. Photographed at Mazabuka Experiment Station.

compact animal of medium size; a mature cow in good condition weighs an average of 900 pounds, and a bull 1,100 pounds. A dark-red color predominates but is by no means distinguishing since other colors, particularly black or black with white markings, are common. Although the horns are usually short and stubby, they vary widely in size and shape. The hump is prominent on the female as well as on the male.

Barotse.—This Sanga derivative, locally referred to as the Longhorned Sanga, has a fair beef-type conformation. If in good condition, a mature cow may weigh up to 1,000 pounds and a bull 1,300 pounds. The horns are rather large and upthrust, with a wide spread. Color varies widely; dark red is perhaps most prominent, but blacks, browns, or any of these colors with white markings are common. The hump is small and usually not noticeable on the female.

Barotse bull.

Barotse cow.

Tonga.—Also of Sanga origin, the Tonga outnumbers both the Angoni and the Barotse. Locally it is called the Shorthorn Sanga. While the horns are of fair size and similarly shaped, they are not as large as those of the Barotse. Colors are not uniform—black, red, browns, and tans, all with white markings and often a white topline and bottomline, are all found. The hump is quite small, especially on the female.

Tonga cows.

Africander.—This is the principal breed of the European grower and is often the result of upgrading from a base of the indigenous breeds bred back continuously to Africander bulls. The Africander was brought to the area that is now Zambia from South Africa through Rhodesia during the colonization period. It is as well acclimated as the native cattle because of the similarity of the two environments.

MANAGEMENT PRACTICES

The cattle of the African tribesmen are run in the traditional manner, usually herded by boys during the day and corralled at night. The grazing land varies from open veld to thick bush cover and forest. Tsetse fly infestation makes large areas unusable for the grazing of livestock. Stockwater is often a limiting factor which results in local overgrazing. To provide protection against the ever-present ticks in the reserve areas, the previous government provided spray runs. They are still maintained but are not used systematically; as a result, death losses are considerably higher in the African-owned herds than in those of the Europeans.

African owners run more than 80 per cent of the total cattle population of 1,300,000. Their stock is of the indigenous breeds, with a minor admixture of Africander blood in grazing areas near those of European ranchers. Milk is taken when the herds are corralled at night. Oxen are used extensively as draft animals. The offtake for slaughter is small, probably less than 3 per cent; but it is slowly increasing through the efforts of the government to induce the African to improve the management and marketing of his cattle.

In 1967 the European ranchers were running approximately 200,000 head of Africander cattle and, with an average offtake of 15 per cent, were supplying well over half the animals slaughtered. For the most part these properties are well managed, fenced, and cross-fenced for grazing control and have good watering facilities. Seasonal calving is practiced because of the concentration of rain in the one wet season. Many growers use a three-month breeding period so their animals will be able to start calving at the beginning of the rains. Heifers are bred as two-year-olds if about 700 pounds in weight; otherwise, they are kept until the next season. Three per cent bulls are normally used, and calf crops of 65 per cent are common. The Africander is a rather slow breeder, and this calving percentage is said to represent an average operation. Steer calves at eight or nine months will wean at a little more than 400 pounds. These animals are often run separately on grass to slaughter age. They weigh 450 pounds at the end of the first winter, 650 to 700 pounds at eighteen months, and 950 to 1,000 pounds when they go to the abattoir at the end of their third summer at thirty months of age. Dipping or spraying for external parasites is done at five- to seven-day intervals during the wet season and at longer intervals in the dry period, varying with the locality.

MARKETING

All cattle for slaughter are processed in registered abattoirs or those of the Cold Storage Board, a statutory agency established by the British and continued by the new Zambian government. The board sets a floor for all cattle prices in the country, fixing minimum prices to be paid for each grade of cattle on the cold-dressed-weight (CDW) basis a year in advance. These prices vary seasonally in an effort to encourage more uniform marketing throughout the year. The board is firmly committed to buy on this scale regardless of the volume of offerings. Prices in 1965 varied from 25 cents a pound for prime grade to 20 cents for standard, which most of the finished European growers' steers grade; then from 12.5 to 17 cents for the two lowest grades, into which most of the native cattle fall. The grading system was recently simplified to include only four grades, compared to seven in Rhodesia. They can be roughly compared to United States grades as follows:

Zambia Grade	*USDA Grade*
Prime	Low-choice
Standard	Good to low-good
Commercial	Low-standard
Utility	Low-canner

The European grower usually sells direct to the board, scheduling his shipments several months in advance. He can thus plan his program for the year and know exactly the minimum price he will get when he ships, provided he prejudges correctly how his cattle will grade. If a drought, a disease outbreak, or any other unforeseen cause forces him to market cattle in a hurry, he has to arrange shipping dates as best he can with the board, and he takes whatever price was prescheduled for the day on which the animals are killed. Summer prices, December through February, are customarily set about 20 per cent higher than fall prices, April through June.

In an effort to get more of the traditional cattle of the African owner on the market, the Cold Storage Commission of the Federation of Rhodesia and Nyasaland established cattle auctions in strategic locations where the native herds are run. Continued by the Zambian Cold Storage Board, the sales are held on dates specified in advance. Native cattle are driven in the morning of the sale and held in corrals, frequently only a single bull belonging to one owner and rarely more than 3 or 4. Each animal offered is run on the scale and the price that will be paid for it is called out. This is based on the board's price scale effective on the date of the sale. The grade of each animal is determined by an experienced government cattle grader, and the liveweight converted to the CDW equivalent by a table of average fixed dressing percentages. This, less a reduction to cover shipping costs and condemnation losses, is the price offered. The owner either accepts or rejects this amount for his animal as he hears it called. If he accepts, he is paid cash forthwith and the animal is inoculated and placed in the shipping corral; if he rejects the price offered, he collects his animal and drives it back to his village. The procedure illustrates the fundamental difficulty involved in getting the African to dispose of his cattle. Often when the owner hears the price for his animal called, he decides that he values the beast more than the money and takes it back home. In times of scarcity of animals for slaughter, efforts have been made to get more animals put on the market by raising the price. This generally results in fewer animals being offered. The African owner sells only because he has a specific need for a certain amount of cash—taxes, some clothing, perhaps a bicycle. If he can get his cash needs by selling 3 head, he sells 3; if the price is high enough so he can get the amounts he wants from 2, he will sell only 2. Although he is money-conscious to this extent, he sells only his poorest animals no matter what his needs may be. The ultimate success of the auction plan hinges on developing additional cash needs of the African.

Native auction at Mwenda, held by the Cold Storage Board.

CATTLE DISEASES

All the cattle diseases of southern Africa are endemic in Zambia. Control measures before independence were much the same as those described under "Rhodesia" and are now conducted by the Veterinary Services of the Ministry of Agriculture.

GOVERNMENT AND CATTLE

The part that government played in cattle protection and development in the country before independence has been continued under the Veterinary Services of the Ministry of Agriculture. Some of the senior

European personnel still continue in government service. Central Research Station, at Mazabuka, has 80,000 acres of good veld, 8,000 acres of which have been cross-fenced into 125-acre paddocks. Boran bulls from Kenya and Bhagnari bulls from Pakistan have been crossed on native Angoli cows. Nothing outstanding was demonstrated, and the present management feels the most productive course is to improve the indigenous breeds of the country by systematic selection within the breed. Feeding tests on the indigenous breeds have shown that for the same age and nutrition these breeds will approach within 100 pounds of the finished weight of the Africander and that the dressing percentage of the indigenous animals will be 2 per cent higher. These results on animals which have had no selection for growth or early maturity in their background would indicate a reasonable possibility of developing a good beef type from them.

OUTLOOK FOR CATTLE

Development of the veld would permit a severalfold increase in the present cattle population of 1,300,000, which with only 4.4 head to the square mile is one of the lowest densities in Africa. There are large areas in Zambia which if not actually virgin cattle country could be converted to the equivalent of such. The low density of the human population, only thirteen persons to the square mile, is the reason for the small number of native herds. Tsetse fly infestation prevents cattle grazing in approximately 40 per cent of the country. No adequate control measures have been inaugurated. A start was made toward control of the tsetse when the former government established game parks to isolate the wild life—the first step in eradicating the insect. The United States government is assisting in research on the sterile-male technique, which has been successful in the control of other insect pests. In other parts of Africa where this has been tried, the method was considered to have little effect because of the small number of larvae, usually only two to six, produced by the female.

Present cattle prices would support a commercial cattle industry. Controlled grazing would limit bush encroachment and could increase production nearly everywhere that native cattle are now being run. In the background, however, is the African who merely wants to keep his cattle and would not market any more than he now does no matter how many he owned.

Not only does Zambia present this favorable stage for an expanded cattle industry, but in her mining income she has the financial capability to develop cattle ranching without foreign aid. European management, however, is essential to the development of large-scale cattle operations. Provision of the conditions necessary to obtain this is a step which most newly independent countries of Africa have not seen fit to take. Zambia may be different in this regard, for the government has conducted economic feasibility studies on a number of nationally owned estates with the intention of contracting with foreign cattle operators to supply the management for large-scale ranching operations.

Algeria

Land area (sq. mi.):	920,000
Arable land area (sq. mi.):	26,000
Population (1964):	11,300,000
Density (per sq. mi.):	12
Density (arable area):	435
Agricultural (71%) (1963):	8,050,000
Per capita income:	$225
Cattle population (1963):	610,000
Year visited:	1965

THE DEMOCRATIC AND POPULAR REPUBLIC OF ALGERIA occupies most of the northwestern corner of Africa. The Mediterranean Sea on the north, Morocco on the west, and Tunisia on the east form the borders of the inhabited area. The vast desert extends south to the Mauritania, Mali, and Niger borders and east to Libya. The inhabited area along the coast lies within the same latitudes as North Carolina; the desert is mostly within the latitudes of Mexico.

What is now Algeria has been under the control during historic times of many ancient and modern world powers. Romans, Carthaginians, Byzantines, and Arabs, followed by Spaniards and Turks, all had their day, with the Berbers at intervals regaining control of their own country. Algeria gained virtual independence in the eighteenth century, but France began to take over in 1830 and proceeded to a very effective colonization of the coastal area and finally to complete subjugation of the native Moslem population when the south was brought under control at the beginning of the twentieth century. Nationalist movements began to gain force between the two world wars and ended in open rebellion in 1954. Then followed seven years of war, until independence was attained in 1962. Mass exodus of the French and the socialistic policies of the inexperienced governments which have been in power have led to a rapid deterioration of the economy.

In areal extent Algeria is the second largest country in Africa—920,000 square miles or three and one-half times the size of Texas. The productive area, including sparse grazing and forest lands, is only one-fifth of this, and the arable land is but 26,000 square miles, half the size of North Carolina. The Atlas Mountains rise from the cultivated coastal plain and valleys, then fall off to the vast stretches of the Sahara Desert to the south. With oases included, the desert carries one sheep or goat to half a square mile—after making allowance for what the camel consumes.

Cattle in the cultivated area are largely descendants of the European milk or dual-purpose breeds introduced by the French. Agriculture has suffered perhaps more severely than the other activities of private enterprise in the takeover by the socialistic system, which has followed Yugoslavic procedures. Management of the dairy operations, which usually were only one part of a farm enterprise, rapidly deteriorated when all the European-owned farms were sequestered and collectivized. In areas with ample moisture from either rainfall or irrigation, good wheat land is seen unplanted and even uncultivated. Citrus and olive groves are frequently unpruned and uncared for; machinery lies idle in the fields. Forty per cent of the tractors in the country were inoperable in the spring of 1965 because of lack of both replacement parts and mechanics to install them. This situation idles the other farm machinery dependent on the tractors.

Severe erosion is seen in the hilly areas where the mountains grade into the coastal plain, the result of centuries of overgrazing and removal of trees.

When the Frenchman left or was removed after independence, a vacuum in direction and know-how resulted. This occurrence sharply contrasts with the case of Africa south of the Sahara, where the newly created governments were satisfied with placing their nationalists in the top posts and leaving the lower echelons of management largely in the hands of the incumbent, trained civil servants of the former governments.

After a small bonus for the farm worker is declared at the end of the year, the returns from agriculture go into the state coffers. No provision is made for financing the farm operation during the next growing season: thus the agricultural plant is bled white. After having been decimated during the revolution against France, the cattle of the country suffer the penalties of this situation and the ravages of uncontrolled disease. Cattle numbers in 1965 were less than two-thirds of the population in 1954, when war broke out.

CATTLE BREEDS

Brown Atlas cattle, which have been in the area for many centuries, are the only indigenous cattle in Algeria. The French introduced their milk and dual-purpose breeds as their colonization proceeded.

Brown Atlas.—Found throughout the area lying west and north of the mountains, thus including nearly all the nondesert areas of Algeria, this breed is thought to have existed there since before the time of the Roman occupation during the first few centuries of the Christian era. Brown Atlas were the cattle of the Berbers, the earliest known inhabitants of what is now Morocco. The simplest hypothesis of their origin is that the breed was derived from the short-horned humpless cattle which were carried by the migrations out of Egypt and along the coastal plains of Africa during the second millennium before Christ. There has been some mixture with the blood of the European breeds in the area where the coastal plain meets the hills; but the small, rugged Brown Atlas is still in the hands of the villager in the mountainous back country and extends to the region where cultivation begins.

Mature Brown Atlas cow, 700 pounds.

Extremely hardy as the result of countless generations receiving little care and poor nutrition, the true Brown Atlas is a small, nonhumped animal; a mature cow weighs 750 pounds if in fair flesh at the end of the rainy season, a bull about 900 pounds. The horns are small and may turn either up or down. Color varies from tan to brown and occasionally to light grey. It is apparent that the French breeds have had some influence on the animals seen around the more heavily populated areas; but the small size, fine bones, and rather compact conformation of the original Brown Atlas have been largely retained. The Brown Atlas is an extremely poor milk producer with a short lactation period; but, dur-

Montbéliard bull.

Tarentaise cow.

ing the rainy season when grass is available, it fleshes well and is a fair meat animal.

Exotic Breeds.—At the turn of the century, when the French control of Algeria was consolidated, herds of European breeds of cattle were established as the agriculture of the country was built up. The French Friesian, Montbéliard, and Tarentaise breeds predominated. Primarily dual-purpose, though with more emphasis on milking ability than on beef production, these breeds were maintained in unusually pure strains on the well-managed French farms and did well in the Algerian climate. Although subject to hotter and drier summers than France, Algeria has more moderate winters and, overall, is almost as hospitable an environment for the northern breeds of cattle as Europe.

Young French Friesian bull.

MANAGEMENT PRACTICES

Village-dwelling farmers, each usually having only a few head of cattle, maintain most of the Brown Atlas and related types. Attended by boy herdsmen during the day, these animals are driven into corrals in the villages at night for milking and safety. They are sold as necessary to meet cash needs, but the cultural element of holding cattle as a matter of prestige persists. As utilized today, these cattle are primarily draft and milk animals. Mules, burros, and some horses and camels are also used for farm work and transportation, the number of these animals exceeding the total cattle population. In the villages cows are milked for household use and the surplus is sold. In the vicinity of the larger towns and even in the populated areas, small herds are maintained for milk production exclusively, only males and a few unwanted females being sold for slaughter.

On the collectivized farms of the former French colons, dairy herds of the European breeds are still maintained; but their management was such that in 1965 the future of these herds was unpredictable.

Some years ago an attempt was made to start an artificial insemination center, the Dépôt des Reproducteurs, at Blida, twenty-five miles southwest of Algiers. The facilities of a remount station of the French cavalry were utilized for this purpose. Results were practically nil because of lack of technical direction; in 1963, however, a French veterinarian was put in charge and some progress was made in the next few years, but the program was abandoned when he left.

MARKETING

Cattle are bought and sold in the livestock section of the souk, a historic institution of the Arab countries of North Africa. It is an open-air market held on the same day each week in a given area, the specified day varying from one town to the next. All articles of commerce are offered in trade; in the larger towns with sizable abattoirs, the livestock section is an active market for cattle, sheep, goats, and often camels. In Algiers several hundred head of cattle, twice as many sheep and goats, and

Cattle section of the Algiers souk.

always a few camels are sold on souk days. At Tlemcen, located in a livestock area, the souk is held on Monday and is mainly for cattle; a week's supply of animals for the adjoining abattoir is bought and held in a holding yard.

Animals are brought to the souks by fellahs, small owners with 1 or 2 head of cattle, and by an occasional large operator with a dozen or more. Sales are by individual bargaining between the owner and a butcher or a commission man who is connected with an abattoir. All cattle sold are marketed off the grass without preparation. Grain feeding to fatten cattle is an unknown practice, and good pasture is so scarce that even a short feeding period on grass before marketing is exceptional. Most animals sold would grade standard, with only a few low-good by USDA standards.

ABATTOIRS

Abattoirs, always municipally operated, in slaughtering follow the Moslem practice of throwing the animal, facing the head toward Mecca, and cutting the throat. To facilitate skinning, the carcass is usually inflated—with compressed air in the larger plants such as in Algiers, by hand pump in Tlemcen. Carcasses are hung for skinning and eviscerating. In Algiers, where the facilities are above average, they are transported by overhead rail. Refrigerated storage for chilling is available, although most meat is sold the day it is killed. Slaughter starts at nine o'clock in the evening and is completed by six o'clock the following morning. Meat is delivered to the shops by ten o'clock. Only a few shops, those catering to the high-class trade, have refrigeration facilities.

CATTLE DISEASES

When the disease-control measures of the French were relaxed after independence, severe outbreaks of foot-and-mouth disease occurred. Other cattle diseases became rampant as the supply of pharmaceuticals vanished and was not replenished. The cattle population in 1965 was still decreasing from such causes, and it appeared that for survival they would have to depend on naturally developed resistance. In this respect, the well acclimated Brown Atlas will certainly surpass what remains of the exotic breeds.

GOVERNMENT AND CATTLE

With the abandonment of the artificial insemination center at Blida, the only government programs involving cattle were two range and pasture management projects sponsored by the European Economic Community. These activities were initiated in 1964. The government is too involved in maintaining itself in power and in attempts to create an international image to devote either effort or money to agricultural pursuits.

OUTLOOK FOR CATTLE

The need is acute for the development of a healthy cattle sector in the agricultural economy of Algeria. The limited milk supply is entirely inadequate and of inferior quality; most of it is unpasteurized and produced under deplorable conditions. Milk yields are so low that most of the feed consumed goes for maintenance of the milking animal itself and very little is actually converted to milk. This situation could be corrected more readily in Algeria than in many other countries of Africa because it has been demonstrated in the past that the good European milk breeds adapt readily to the climate.

There is equal need for more beef. Much of the meat supply is now imported from Europe as veal, frozen or chilled, or as cattle on the hoof. While lamb and mutton would continue to be the preferred meats of the predominantly Moslem population, more beef would be consumed if it were available. Overgrazing on the nonirrigated areas prevents such growth of grasses as the available moisture can support. Controlled grazing and seeding of the right types of grasses could materially increase the carrying capacity. Systematic selection within the Brown Atlas breed could result in the development of a good meat, possibly even a meat-milk, type, adapted to the conditions under which the animals are run. The breed is ideally adapted to the country and has demonstrated its hardiness and ability to thrive if it only gets sufficient rough forage. An effort was made by the French to preserve the breed pure by segregation of a small herd. This plan has been abandoned, and the development of an improved type by selection was never attempted.

Good management—controlled grazing, better watering facilities, and introduction of improved grasses—would substantially increase the number of cattle that could be handled in the available grazing areas. Such developments in any magnitude are not attainable under present conditions. The cloud which the political situation casts over the economy of the whole country also covers her cattle. Their future can only be expected to deteriorate until this condition changes.

Libya

Land area (sq. mi.):	679,000
Arable land area (sq. mi.):	11,800
Population (1966):	1,675,000
Density (per sq. mi.) (arable area):	142
Agricultural (72%) (1960):	1,200,000
Per capita income (1964):	$435
Cattle population (1964):	147,000
Offtake (5.5%) (1962):	8,000
Year visited:	1965

THE LIBYAN ARAB REPUBLIC, in land area, is fourth among African countries; but only 3,000 square miles of land, in a narrow belt of oases along the Mediterranean coast, are under irrigated cultivation. Beyond the coastal plain the land rises to a ring of low hills called the Jebel, 30,000 square miles in extent, with a maximum rainfall of 10 inches, then falls off to the desert—more than 646,000 square miles of it—with only an occasional oasis. Partly protected from the hot desert winds, the irrigated area can be compared with the Río Grande Valley in southwestern Texas. The Jebel is a poor grazing area beyond which the Libyan Desert stretches south for 1,000 miles. The Mediterranean Sea lies to the north, Sudan to the southeast, Niger and Chad to the south, and Tunisia and Algeria to the west.

The United Kingdom of Libya was declared independent of Italy in 1951 and was then ruled by King Idris for eighteen years. The country existed at a subsistence level for the first half of King Idris' reign, during which foreign aid was required to feed the people. The discovery of large petroleum reserves in 1958 saw a booming economy develop which still persists. A coup d'état in 1969 abolished the monarchy and constitution. A provisional constitution is the current basis for government by the Revolutionary Command Council, a group of young military officers headed by an appointed president.

CATTLE BREEDS

The cattle of the country are referred to simply as Libyan Cattle, as distinguished from the European milk breeds of the dairies around the cities. They are generally nondescript as a result of miscellaneous inter-breeding with exotic cattle.

Libyan Brown Atlas.—Small, brown cows, usually not exceeding 600

Libyan Brown Atlas cows on average grazing.

A Libyan Brown Atlas on good pasture.

pounds in weight and having characteristics very similar to those of the Brown Atlas of the countries to the west, are seen rather frequently in Libya. Locally these animals are not distinguished from the nondescript types and are called, along with the latter, Libyan Cattle. The animals have sufficient uniformity to justify classification as a breed, however, and must have had the same ancestors as the Brown Atlas common to Morocco, Algeria, and Tunisia; they are, therefore, here designated as Libyan Brown Atlas.

The first of the two examples shown on the opposite page belongs to a seminomadic owner. It is representative of the average cow. The second picture is of a cow belonging to a village dweller who has managed to maintain her on fair pasture.

Exotic Breeds.—The Italians, during their period of occupancy from 1912 until they, along with the German army, were driven out in 1943, made considerable progress in developing the agriculture of the country. This included establishing a fair but small dairying industry near the cities. These dairy herds have been continued by the Libyan owners who have maintained the individual breeds. Many of these dairies are well managed by their present owners. New breeding stock is imported from Europe to maintain the quality of individual herds, and husbandry practices generally are better than those commonly found in North Africa. The Friesian predominates, with the Jersey probably ranking second among the exotic breeds.

Friesian cows which formerly belonged to an Italian dairy.

Pantelleria.—When they were developing the dairy industry in Libya, the Italians introduced this little-known breed. These animals probably originated from the same cattle as did the Libyan Brown Atlas, the progenitors of the Pantelleria being carried to the island of that name from Africa in very early times. After many generations and some improvement by selection for milk productivity, Pantelleria cattle were brought back to Africa by the latter-day Italians.[1]

Pantelleria sire. Photographed at Sidi Mesri Experiment Station.

MANAGEMENT PRACTICES

With the exception of the dairy herds, the livestock of the country is run by nomads and seminomads. Sheep and goats predominate in a ratio of 20 to 1 head of cattle. The Jebel is naturally a better sheep than cattle country; the sheep and goats tend to keep the continual brush encroachment on the overgrazed land in check. Stockwater is the limiting factor in all Libyan livestock raising; large areas are underutilized because of lack of water, while others are overgrazed where water is available. Almost all watercourses are dry except in time of flood, and a continuous supply of water depends on underground cisterns. These cisterns are formed by the excavation of an underground reservoir—sometimes with a number of galleys running to it—beneath the layer of the hard

[1] R. S. Temple, personal communication.

limestone that is found close to the surface in many areas. The floor and wall surfaces must be plastered to prevent loss by seepage. The limestone layer under which the cisterns are built prevents evaporation loss in the dry, hot climate. Surface water is conserved by a system of ditches which gather the water for storage. Cisterns of this type were constructed by the Romans and have been maintained over the centuries by removing accumulated silt and replastering the sides and floor as necessary. Some of these very old cisterns are still in use. Under a United States AID program, in recent years many abandoned cisterns were rehabilitated and new ones constructed, all of which added appreciably to the carrying capacity of the land. Droughts, which are not infrequent, are decimating in their effect on livestock. When these dry spells occur, there is not sufficient runoff to fill the cisterns and livestock perish immediately.

Common practice of many seminomad stock owners is to scratch grain in a small plot of ground in the spring, leave with their sheep and goats for several months in a continuous search for grass and water, and return in the fall to harvest what may have happened to grow.

Under these conditions natural selection in the cattle of the country has produced a particularly hardy animal with the ability to withstand long periods at low nutrition levels, the limited water supply, and the hot, dry climate. Because of poor nutrition, calving rates are low, probably less than 50 per cent, and the mortality rate is high in young calves. Milk is utilized when it can be obtained by the owners. Cattle are held to a considerable extent as a mark of prestige in the Arab world of North Africa but not to the degree prevailing in Africa south of the Sahara. The Libyan cattle owner sells his animals as he needs cash but does his best to keep the size of his herd up.

Dairies in the irrigated areas near the cities have been continued in the rather modern type of operation introduced by the Italian farmers fifty years ago, albeit with somewhat less meticulous husbandry practices. Management, however, is better than might be expected—good feeding is the rule, records are often kept of individual cow production, and sanitation is fair. Such dairies are usually maintained as a diversification of the general farm operation which will involve cultivation of olives, grapes, peanuts, potatoes, and various other crops.

Cows are fed in the barn—green chop for much of the year, hay in the dry season. Alfalfa and berseem, an annual resembling alfalfa, yield ten cuttings a year on irrigated land. Grain supplements are generally used. Herds vary in size from 10 to 100 head. Loafing sheds and yards are beginning to be used, but most herds are kept in stalls or stanchions the year round. Labor is a serious problem, the farmer having to compete with the high wages of the oil industry. To counter this, some farmers employ a share-cropping system under which the worker receives a percentage of the crop in return for his labor.

An artificial insemination center has been established at the Sidi Mesri Experiment Station, which supplies thirty-nine outlying stations. Semen is kept for five days under ordinary refrigeration. The speculum method is employed for impregnation. The main objective is to upgrade the Libyan cow to a higher milk-producing level. Cultural objections to the practice and the lack of transportation and communications handicap the program. A 40 per cent conception rate is obtained on the first insemination.

MARKETING

Cattle are sold in regular livestock markets around the larger towns but are only a small part of the trade, which is largely in sheep and goats, along with a sizable number of camels. Most animals are brought in on the hoof, although trucks are beginning to be used. Sales are by the head and individual bargaining takes place for each transaction. Prices are phenomenally high—in 1965 a 900-pound cow in only fair condition brought $350. The general run of cattle is what has been referred to as Libyan and Libyan Brown Atlas, with a sprinkling of discarded dairy animals.

Tripoli cattle market.

ABATTOIRS

The municipal abattoir, operated by local authority, is usually adjacent to the market area. Slaughter is under roof; killing is in the prescribed

Moslem manner discussed under "Abattoirs" in "Sudan." Carcasses are hung on hooks to eviscerate and skin. Sanitation is rudimentary. The average liveweight of cattle processed is 600 pounds, the dressing percentage about 45 per cent. Carcasses are dispatched to the butchershops the day the animals are killed, and the meat is sold before nightfall.

CATTLE DISEASES

Imported cattle are rigidly controlled to prevent the introduction of contagious diseases. A three-month quarantine period is required on animals brought in, and they must be certified as clean of foot-and-mouth disease, tuberculosis, and brucellosis before they can be maintained in Libya.

GOVERNMENT AND CATTLE

The government is actively engaged in cattle development, mainly in increasing milk production. The experiment station at Sidi Mesri, near Tripoli, maintains a herd of 200 milk cows. Crossing experiments using Zebu breeds on exotic milk types are being conducted in an effort to obtain a good milk-producing animal that can better withstand the heat stress of the climate.

In 1965 a trial running dairy cattle on irrigated pasture was underway at Sidi Mesri. This was a definite innovation from the practice of bringing all feed to cattle in a barn.

OUTLOOK FOR CATTLE

Milk-type animals seem to hold the future of cattle in Libya. In the heavily populated, irrigated coastal plain, all of which is intensively farmed, there obviously is no room for a beef animal. In the Jebel, controlled grazing, the introduction of good types of dry-land grasses, and more stockwater cisterns would permit an increase in the cattle population. It is, however, better sheep than cattle country; in addition, Libyans, most of whom are Moslem, much prefer mutton to beef. The high level of the economy will support a healthy dairy industry, which can be expected to grow.

Morocco

Land area (sq. mi.):	174,500
Population (1964):	13,300,000
Density (per sq. mi.):	76
Agricultural (85%) (1945):	11,400,000
Per capita income (1964):	$160
Cattle population (1965):	2,900,000
Offtake (10%) (1964):	300,000
Year visited:	1965

THE KINGDOM OF MOROCCO lies in the northwest corner of Africa in the same latitudes as that part of Texas north of San Antonio. In area it is equal to New Mexico and the panhandles of Texas and Oklahoma. The western coastline, on the Atlantic Ocean, is separated by the Strait of Gibraltar from the northern coast, on the Mediterranean Sea. Algeria is to the east and south, and Spanish Sahara lies on the extreme south.

Agriculturally Morocco falls into much the same general pattern as the other countries of the Maghrib, the area in northwestern Africa occupied by the Atlas Mountains and the plains lying between them and the Atlantic and Mediterranean coasts. This region has sufficient rainfall to support forests and permit grazing and cultivation. To the east and south of the mountains, there is only desert—traveled by nomads with their camels, sheep, and goats—and an occasional oasis. Morocco, however, has a smaller proportion of desert than the other North African countries since more than half the land is in forests or is utilizable for grazing and crops.

Most of the people are descendants of Berbers who inhabited the area long before the Christian Era and of Arabs who invaded the country in the eighth and eleventh centuries. There are more than 150,000 Europeans, a large number of whom are French.

Historically Morocco has followed much the same pattern as Algeria —held at various times by most of the ancient Mediterranean powers and, late in the nineteenth century, coming under the influence of France, eventually becoming her protectorate in 1912. Independence was attained in 1956; two kings—father, then son—have since maintained a stability in government. There is a feeling of law and order in the air. Private initiative is seen in agricultural and livestock developments as well as in other sectors of the economy. The Ministry of Agriculture is undertaking progressive programs involving cattle improvement. The fact that these things are happening is significant, even though most of the cattle population, the small native herds of the villager, has not yet been affected.

CATTLE BREEDS

The majority of the cattle in Morocco are of the Brown Atlas breed, which is indigenous to all the area in North Africa which drains seaward from the Atlas Mountains. These animals are the same as the Brown Atlas described under "Algeria." A related type, locally known as a Oulmès or the Zaers, for these regions in Morocco, is described as a somewhat larger and more fleshy animal than the Brown Atlas. Basically these animals are probably of the same stock, such differences as are noted being the result of management practices, better nutrition, and, possibly, a selection for color.

To some extent cattle ownership is still a matter of prestige with the small native owner, who milks his cows for family use and reluctantly sells them for slaughter when there is a need for cash.

Brown Atlas cow kept on good nutrition by a European grower.

During their period of influence the French introduced some of their milk and dual-purpose breeds, principally the Montbéliard, the Tarentaise, and the French Friesian. These breeds still predominate in the dairy herds and are usually maintained in a pure state for milk production. However, miscellaneous interbreeding with the local Brown Atlas of small farmers and also crossing by some of the large farmers in planned efforts to upgrade the native cattle also occur. The practice of

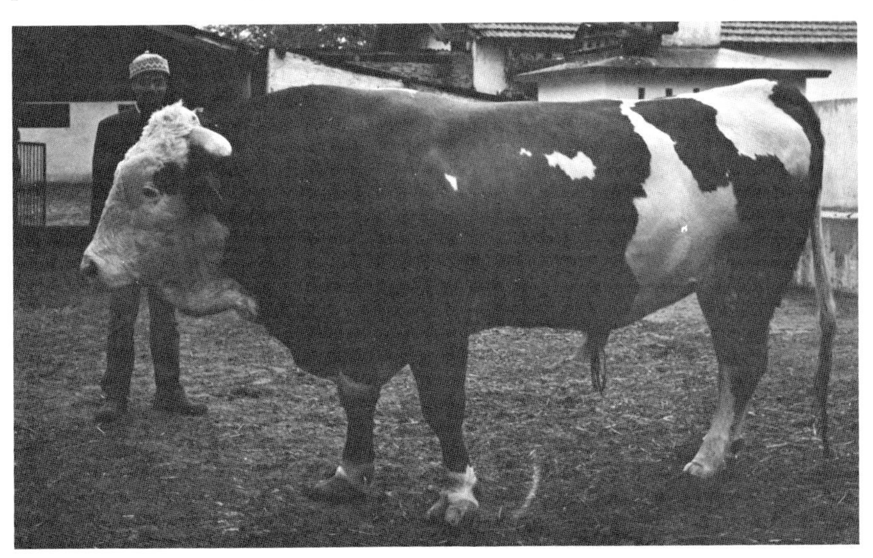

Montbéliard bull of a European grower.

Mature Brown Atlas cow (left) and her two-year-old daughter by a Montbéliard bull.

crossing European bulls on Brown Atlas cows is growing, mainly to increase the number of milk animals. The first cross, which has the advantage of the hybrid vigor, shows a marked improvement both in size and in milk yield.

MANAGEMENT PRACTICES

Cattle owned by the small farmers, particularly in the hilly and rolling country lying between the cultivated areas and the Atlas Mountains, are generally pastured throughout the year and supplemented in the dry season with the minimum of straw necessary to maintain life. Much of the land on which such cattle are run is badly overgrazed; but during the period of lush growth in the rainy season, from late winter until the beginning of summer, feed is generally adequate and animals reach a fair condition.

During the dry period, from late summer to early winter, available forage, largely in the stubble fields, is soon cleaned up and the condition of the cattle rapidly deteriorates. No hay is put up except on the large farms. Straw is stacked when available to the small livestock owner and fed sparingly during this period. It is stacked loose and is covered with a mud plaster for protection from rain. Frequently, the small stack is utilized on the self-feeder principle, an animal being tethered so that it can barely reach the end of the stack that has been opened. As the straw is consumed, the animal is moved closer so that there is no waste and consumption can be held to a minimum.

Small owners in the country frequently put their cattle out under the management of herdsmen who receive for their services a varying percentage of the weight increase, as much as 50 per cent, or a share of the calf crop. All cattle in Morocco, except on the large farms and the small dairies near the larger towns, are grazed on unfenced areas under the constant attendance of herdsmen and are corralled at night. Common practice is to separate calves from their dams at one month of age. They are then permitted to suck twice a day, both before and after each milking—first, to prepare the cow for milking and last, to be sure she is properly stripped. Bulls are left entire except those designated as draft animals; the latter usually are castrated when two years old.

Around the cities there are dairies following modern management practices and having as many as 100 cows of the European breeds. Cows, young stock, and bulls on feed for slaughter are maintained in barns the year round, either tied or in stanchions. Cows may or may not be turned out in a paddock for an hour or so during the day for exercise.

In the dry season they are fed straw or hay and a grain supplement. The roughage is baled or stacked and fed liberally. Green chop is hauled in during the growing season. Some of the better-managed dairies have pit silos, and good-quality silage is fed along with the dry roughage. A grain supplement, usually consisting of oats and barley together with a large, limalike bean called *feve* or a small, round legume called *feverole* to increase the protein content, is given to the cows in milk and young stock being fattened for market. Most of the milk supply comes from small dairies milking from a few head to 25 or 30, the size of the operation being geared to the quantity of feed the small farmer can produce. The management of these dairies is often quite casual and sanitation practically unknown.

Bulls, except those kept for breeding, and occasionally a few heifers are fed to weights of about 1,000 pounds at two years of age and the best reach a grade of low-choice by USDA standards. Bulls for slaughter are not castrated.

Milk production is often only one phase of the large-farm enterprises. In addition to the forage and grain required for the stock, various combinations of citrus, grapes, wheat, and occasionally tobacco are raised. Often olive groves and cork trees are also grown. The owner invariably lives in town.

MARKETING

Livestock in Morocco is sold in the souks. These are markets held on a designated day each week, a region often being known by the day of the week on which its souk is held. All articles of commerce, both new and used, and many services including barbers, teeth extractors, and "healers" are available in the larger souks. The livestock sector is one of the major activities in the souks of areas in which cattle and sheep are raised or in towns in which there are municipal abattoirs. Cattle, sheep, goats, and a few camels are sold at these markets. Sales of cattle are both for slaughter and for animals to go back to the country for milking or growing out to heavier weights but without the feeding of grain. Transactions on animals for slaughter are between the farmer-owner and the wholesale butchers, who have the animals processed in the municipal abattoirs and then sell the dressed meat to retail shops.

For the past few years government price controls on meat have led to chaotic variation in retail prices. Beef to the consumer nearly doubled in price in the five years before 1965, increasing from 32.5 to 60 cents a pound; and the price of mutton increased from 42 to 70 cents a pound.

Government-supported beef imports from France have not remedied the basic cause, a shortage of animals to supply the demand. It is probable that the cattle population is decreasing as a result of this situation.

ABATTOIRS

Fairly modern abattoirs, operated by the municipality, are located in the cities. Some have chilling facilities, although most meat is sold the same day or, at most, the day following that on which it is slaughtered. Killing is in the orthodox Moslem manner described under "Abattoirs" in the section on Sudan. Inspection by veterinarians and condemnation of infected carcasses are the practices in the municipally-operated plants.

In the small village slaughtering is done in the open, often at the local souk. The meat is sold in booths screened off from the killing area. Although the carcasses are not inspected, the meat is handled as cleanly as such conditions will permit and is offered for sale in a very presentable form.

GOVERNMENT AND CATTLE

In 1964 a government program was launched to improve the dairy cattle and the management methods. A government-operated farm outside Fez is raising purebred Tarentaise bulls, which are supplied to nearby villages to improve the native stock of the small cattle owner. Thirteen dairy centers with 100 cows each are to be established in the Meknes area; eleven will be stocked with imported Tarentaise cows and two with Friesian cows to be furnished by West Germany. These operations will serve as demonstration and vocational training centers in dairy management and will also supply breeding stock to farmers or to other government farms. Twelve artificial insemination stations with two bulls each are to be set up in the same general area for the improvement of the stock of nearby small farmers. A modern dairy plant planned for Meknes will process the milk produced at the dairy centers and from all farms in the area which wish to co-operate.

OUTLOOK FOR CATTLE

Government activities indicate a progressive attitude toward cattle raising for milk production. What the impact will be on the industry remains to be seen. Little, if anything, is being planned to improve the lot of the native cattle growing in the hills and comprising the largest part of the cattle population. The potential of the Brown Atlas cattle has never been determined either as a beef animal or as a milk producer. These hardy beasts with a proven ability to maintain themselves under adverse condi-

Bulls decorated for sale at a village souk.

tions have good fleshing qualities if given reasonable nutrition. This breed warrants, but is not getting, some attention in the cattle development programs and is probably doomed to eventually being bred out by the European breeds.

Private enterprise is playing an increasing part in both milk and beef production. The examples of the large farms running sizable dairy herds and feeding out the bulls for slaughter, all under reasonably good management, will certainly have an impact on the cattle economy. Occasionally a small farmer will employ modern methods. Exceptional instances are seen of small dairies milking up to 20 cows under conditions that would warrant a grade A rating in the United States and accomplishing this on a holding of eight to ten acres.

Tunisia

Land area (sq. mi.):	63,400
Population (1964):	4,565,000
Density (per sq. mi.):	72
Agricultural (75%) (1966):	3,920,000
Per capita income (1964):	$168
Cattle population (1965):	565,000
Buffaloes (1965):	370,000
Year visited:	1965

THE REPUBLIC OF TUNISIA extends southward from the northernmost tip of Africa into the Sahara Desert and is bordered by Algeria on the west, the Libyan Desert on the southeast, and the Mediterranean Sea on the east and north. Compared to many other African countries, Tunisia is small in area, being only about half the size of New Mexico and lying between the same general latitudes. The southern half of the country is desert. Annual rainfall varies from less than 4 inches where the desert borders the arable area to a maximum of 24 inches in the northwest corner. The economy depends primarily on agriculture; other than cereals the principal crops are citrus fruits, grapes, olives, and dates.

Historically Tunisia has followed much the same pattern as Morocco and Algeria, finally falling into the French orbit during the latter part of the nineteenth century. Independence was attained in 1956 and orderly government has since followed. During their seventy-five years of control, the French developed the country a great deal and left their mark on the culture and customs. Farming is accomplished in a husbandlike manner. Crops are sown when they should be, vineyards are free from weeds, and citrus and olive groves are well pruned. Although tractors are coming into use, animal draft power is still quite generally employed in farming and farm transportation—cattle, buffaloes, camels, and horses all being so used. The people work. Thousands of men and women in the country who would otherwise be unemployed are kept busy on large reforestation projects in hilly areas unsuited to cultivation. Tunisia is the best run of the Moslem states in North Africa.

CATTLE BREEDS

Brown Atlas cattle like those found throughout North Africa and discussed under "Cattle Breeds" of Algeria are the indigenous cattle of Tunisia. There has been considerable admixture with the European milk breeds that the French introduced, and many of the cattle seen in the countryside are a nondescript type resulting from this interbreeding.

Brown Atlas cow. Photographed near Kasserine Pass.

Brown Atlas ox team. Photographed in the Aïn Draham area.

In some areas, however, particularly from Thala to Aïn Draham in the western part of the country, the true Brown Atlas is seen.

Near the larger towns such as Tunis and Béja, there are herds of European dairy cattle, often with Brown Atlas crosses among them. Friesian, Montbéliard, Tarentaise, Jersey, and Swiss Brown breeds were imported by the French colonists during the period of French influence, which lasted from 1882 to 1956. Many of these animals are still maintained pure, but there has been some crossing with the Brown Atlas, particularly in the smaller herds. The Friesian constitutes a large majority of the milk cattle.

Mixed dairy herd. Photographed near Béja.

MANAGEMENT PRACTICES

In the low rainfall areas of the cultivated lands which border on the desert, camels supply most of the draft for farming. Forage is scarce and very little roughage is stored for the dry, hot summer season. As rainfall increases toward the north and is sufficient for winter wheat, the straw is saved for feed and some hay is put up. On the large farms, which were taken over from the French colons and are now operated by the government, mechanized farming is the general practice. Pasturing throughout the year is common.

Artificial insemination has not been used, but in 1965 a government center was being established at the Sidi-Thabet Experiment Station for the preparation of semen and for the training of technicians. Plans call for several substations to serve cows in the dairying areas, with the use of imported Friesian, Montbéliard, and Tarentaise bulls as sires. The station's own herd of these breeds is being bred artificially.

CATTLE DISEASES

Brucellosis is widespread and is considered the most serious cattle disease. It is hoped that the Cape Bon project for disease control, dis-

Tarentaise bull. Photographed at Sidi-Thabet Artificial Insemination Center.

cussed under "Government and Cattle," will pilot the way to its control. Outbreaks of foot-and-mouth disease occur but have not caused serious losses. Internal parasites are also a problem.

GOVERNMENT AND CATTLE

Efforts to increase cattle numbers and improve management methods are being made by the government. The programs to accomplish these ends have been largely instigated by French and United States advisors and are being actively pursued.

In trying to permanently locate the Bedouin, who for centuries has traveled the desert with his sheep and camels, the government has built settlements near some of the larger cities. A family unit consists of a small house and a shed and yard designed for a few head of cows. Pasturing is provided on nearby common grazing lands. The object is to increase the dairy herd of the country, but so far the Bedouin has been slow to accept this new way of life. He is still inclined to keep his sheep herds and travel the desert, returning only occasionally to his home near the city.

A major program designed to control disease and illustrate the advantages of artificial insemination is planned for the Cape Bon Peninsula, a well-developed agricultural area of 600 square miles, isolated by the Mediterranean Sea on three sides and having a main road separating it from the rest of the mainland. A rigid quarantine will be maintained along this road; cattle will be allowed to cross the restricted area to market but no cattle will be allowed to come into the area. Artificial insemination services will be free, and immunization treatment for the common cattle diseases will be provided. The program should serve as a demonstration of what can be accomplished by controlled breeding and by disease-prevention practices and should also provide a means of determining the best breeds to utilize in upgrading.

OUTLOOK FOR CATTLE

Tunisia is not a cattle country but has a definite need for draft and milk animals. Sheep and goats outnumber cattle 7 to 1. The population is more than 90 per cent Moslem and has a strong preference for mutton over beef. Under such circumstances there is little place for beef cattle. The European milk breeds, however, will probably increase and be developed. They are adaptable to the climate and do well if given reasonable protection from the endemic diseases. The Brown Atlas will continue to be used for draft until it is finally bred out.

United Arab Republic

Land area (sq. mi.):	386,900
Inhabited area (sq. mi.):	13,600
Population (1966):	30,055,000
Density (per sq. mi.) (inhabited area):	2,246
Agricultural (65%) (1964):	19,500,000
Per capita income (1961):	$154
Cattle population (1964):	1,590,000
Buffaloes (1964):	1,590,000
Year visited:	1965

THE UNITED ARAB REPUBLIC (EGYPT) occupies the northeast corner of Africa, with Libya on the west, Sudan on the south, the Red Sea and a short frontier with Israel on the east, and the Mediterranean Sea on the north. It lies within the same latitudes as the peninsula of Lower California.

Egypt had risen to its zenith and had fallen to Persian conquerors 500 years before the Christian Era. All of the ancient Mediterranean powers then had their day as conquerors of the Nile and were followed by the Saracens, Mamelukes, and Turks. The area was contested by the French and the British during the latter part of the nineteenth century, when the European powers were vying to increase their African holdings. The British prevailed, and Egypt was under the control of Britain until 1922, when a treaty providing for its independence was concluded. The next year Egypt became a kingdom until 1952 when a military coup resulted in the establishment of a republic. Union with Syria six years later formed the United Arab Republic; but after three years, in 1961, Syria withdrew from the union and Egypt continued alone, still under the name of the United Arab Republic.

The Egypt best known to the world is the Nile Valley, a narrow strip of 12,000 square miles of intensively cultivated, irrigated land following the meandering river southward from the Mediterranean coast for nearly 1,000 miles to the Sudan border. Here more than 30,000,000 people strive to exist in an area one-fourth the size of Louisiana.

Aswân High Dam will add 1,400 square miles of cultivated land in the next few years, and the Tahreer Province scheme may eventually add another 1,500. Nevertheless, 368,000 square miles of desert area remain—one of the least productive of African desert regions. This area is three times the size of Nevada and is perpetually traversed by 50,000 nomads. By the time the new lands are under water, the population increase can be expected to consume the production from them.

Since the days of the Pharaohs, the base of the Egyptian economy has been the use of cattle as draft animals; for some centuries past this has also included buffaloes. Although in relation to the human population there are few of these animals—only 5 cattle and 5 buffaloes for every 100 persons—the agriculture of the country depends almost entirely on them for cultivation. Without them, cotton, the main foreign exchange earner, would disappear because it would be impossible under the present economy to replace the draft animals with tractors.

Individual land ownership is limited by law to 100 feddan (103.8 acres) and the number of farms of more than 25 acres is negligible. The average Egyptian farmer cultivates less than 2.5 acres, many less than 0.5 acre. His production levels are low as viewed by Western standards but he comes close to averaging two crops a year on the land he tills. He has a buffalo and a cow, quite frequently 2 of one or the other. They supply the draft for tillage and, as is often necessary, the power for irrigation, as well as milk for household use and sale. The Egyptian farmer uses a plow identical with that used in the time of the Pharaohs, a metal-shod stick connected by a pole to the crude yoke of the draft team.

Rainfall, when there is any, amounts to only 1 or 2 inches annually in most of the cultivated areas, although at the mouth of the Nile Delta it increases to 8 inches. South of Cairo there has been no rain for centuries. Although there are some deep wells in the Tahreer Province scheme, most agriculture depends on irrigation from the Nile, either directly by overflow or ditch or from wells supplied by the underground flow. The Pharaonic system of scratching seed in the ground after the flood waters had deposited their layer of silt and receded was used until fairly recent times, but now less than one-sixth of the cultivated land is so handled. Storage and diversion dams feeding canals are employed to get more land under water, and the cattle-powered Persian water wheel is still an important element in the irrigation system. The more uniform supply of water which results from the storage systems permits a two-crop program, but the fertilizing effect of the floodwater silt is lost.

The legend of the fertile soil of the Nile Valley is fast becoming a myth. Soil nutrients are removed faster than they are replaced. All plant growth above ground is completely consumed either as food for man or beast or as fuel; as a result, the humus content of the soil is negligible. The average corn yield was only 47 bushels an acre in 1965; the cotton yield, 1.25 bales. Some fertilizer is used but not enough to replace the annual mineral loss from cropping.

CATTLE BREEDS

Although in the days of the Pharaohs, Egypt harbored both the longhorn humpless and the shorthorn humpless cattle that contributed largely to many of the present-day African cattle breeds, little trace of descendants of these types is in evidence today.

The single exception is the small, light-boned brown cow seen in the region of the lower Nile; locally called Baladi cattle, they bear a marked resemblance to the Brown Atlas of the countries to the west. Usually undernourished, they are utilized to the extreme limit of their ability to work and produce milk and offspring. Baladi cattle have acquired a high degree of tolerance to the hot, dry climate and to endemic diseases and have the ability to exist on poor-quality feed. With reasonable nutrition, available only on the government experiment farms, a mature cow may weigh 800 pounds and a bull a maximum of 1,000 pounds. The

Baladi cow on good feed. Photographed at Sakha Experiment Station.

Baladi cows of a small farmer.

rather small animals have bodies that are quite narrow and poorly muscled. Although most are solid dark brown, some are varying shades of brownish red and there is even a light-tan animal occasionally. The horns are small in diameter, short, upturned, and stubby, with a forward thrust.

Throughout most of Egypt the local cattle are nondescript animals, sometimes humped, but without any common characteristics. They are primarily used for draft. These nondescript animals are also sometimes referred to as Baladi cows, but they seldom bear much resemblance to the distinct breed of cattle described above.

Nondescript cattle common throughout Egypt.

The buffalo, brought to Egypt by the Arabs during the first centuries of the Christian Era, now accounts for nearly half of the total cattle population.

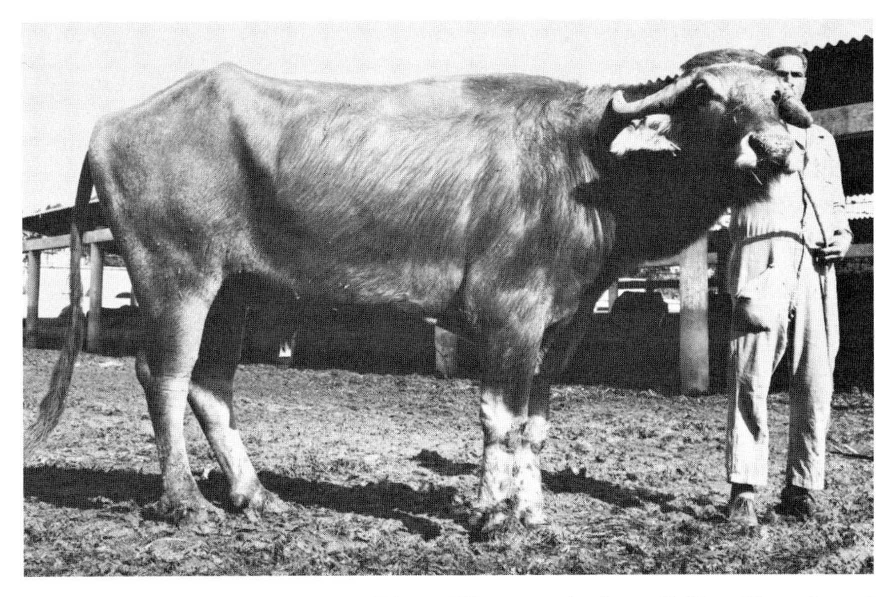

Murrah buffalo in good condition. Photographed at Sakha Experiment Station.

MANAGEMENT PRACTICES

Among cattle, both cows and oxen are utilized for draft and are more important than buffaloes for this purpose. Cows are milked but are not nearly as important a source of supply as buffaloes because they do not produce as much milk on the nutrition level at which they are maintained and they are required to raise both male and female calves to weaning age. Buffaloes are also prized for the high butterfat content of their milk. Even though the cattle are maintained on the same routine as the buffalo cow which raises a female calf, they are invariably in poorer condition—an evidence of their inability to utilize the available coarse feed as well as buffaloes. Only 28 per cent of the country's milk production is from cattle. The few dairy herds of European milk breeds around the large cities furnish only a small fraction of the milk supply, and there are large imports of milk, milk substitutes, butter, and cheese.

When using buffaloes for draft, the Egyptian works only the cow; the male, even if castrated, is considered too difficult to handle.

(Throughout Asia castrated males as well as cows are worked.) The common practice is to sell male buffalo calves at approximately forty days of age for slaughter. In addition to being used for draft, the cows are milked to the limit of their ability to produce.

Forage of any kind for livestock is costly and difficult to obtain. Stubble pastures rented to seminomadic herdsmen are gleaned to such an extent that one acre suffices for 300 sheep or goats or 50 head of cattle for one day. When cattle are grazed on crop residues, they are under the constant supervision of herdsmen. If they are not grazed, fodder is brought to them from wherever it can be cut.

Berseem is grown to the limit possible under a farmer's cropping practices and government regulations. An annual clover quite similar in appearance to alfalfa, it provides excellent cattle feed for the limited time it is available; but unfortunately it also hosts the disastrous cottonleaf worm. Requirements for salable crops or those for human consumption are so pressing that the cultivation of berseem is limited. In an effort to obtain maximum cotton production, it has been made illegal to irrigate forage crops after May 10 of each year.

MARKETING

Livestock markets in Egypt are principally devoted to trading between farmers. Cattle for slaughter are sold by the farmer directly to the government for processing in the government abattoirs from which carcasses go warm to government stores. In 1965 the fixed price paid for live cattle was equivalent to 8.5 to 10 cents a pound liveweight, although all sales were by the head. There is a constant shortage of meat, particularly of beef and poultry, at the retail level. Meatless days, as many as three a week, are decreed as considered necessary by the government; and usually the housewife must queue at the shops to make a purchase. The choice cuts of beef are siphoned off for the military and a few of the luxury hotels. What remains is sold retail for about 60 cents a pound. Dairy products are in even shorter supply. To obtain butter, a European often resorts to the black market.

CATTLE DISEASES

External parasites are mitigated by the hot, dry climate, and there is no incidence of the tick-borne diseases prevalent over much of Africa. Brucellosis is not considered serious, although in the government experiment stations the practice is to calfhood vaccinate with Strain 19. There is some testing for tuberculosis with slaughter of infected animals, but no general program to eradicate the disease. Outbreaks of foot-and-mouth disease occur and are treated locally by vaccination of all animals in the affected area. Quarantine regulations are enforced on imported animals, which must be negative for tuberculosis, brucellosis, and foot-and-mouth disease before being released.

GOVERNMENT AND CATTLE

Experiment stations well in line with Western standards are operated by the Animal Production Department of the Ministry of Agriculture. Its broad objectives are to determine and develop types of buffaloes and cattle best adapted to the conditions of the Nile Valley and then to produce bulls for distribution in the country to upgrade the native stock. A herd of 1,300 buffaloes and 800 cows is maintained at the Sakha Experiment Station for this work. The average farmer with only 1 or 2 animals obviously cannot afford to keep a bull, and the local sires available are invariably of inferior quality. Good bulls at central village breeding stations could well improve the farmer's stock.

Buffalo cows yielding 4,000 pounds of milk a lactation, as compared with 1,500 pounds for the better-producing animals in the country, have been developed at the experiment station in Sakha.

An artificial insemination center has been established in each of the three zones in the Tahreer Province scheme. The stud herd in the Rowad Zone consists of 10 imported Friesian and 2 Red Danish bulls. Fresh semen is supplied by each center to substations, and the speculum method is employed for inseminating. Results have been mediocre until 1965, only 40 per cent conception being obtained on the first insemination and 80 per cent for as many as five breedings.

Efforts of the experiment stations in cattle development have been directed mainly toward upgrading the local cattle for milk production by the use of bulls of the European milk breeds. Elaborate experiments have been conducted using Friesian, Guernsey, Jersey, Swiss Brown, Simmental, and Red Danish sires to determine their relative advantages. These tests have indicated that the Friesian cross is the most desirable from a milk production standpoint and because of a greater tolerance to the heat stress of the hot, dry climate. Nothing appears to have been done to improve the milk-producing ability of the Baladi cow by systematic selection within the breed. Although the feeding of grain to cattle for beef production would appear to be the wildest fantasy in a land without sufficient cereals to feed its people, experimental work is being done to determine the best cross on the Baladi cow for fleshing quality. Friesian, Simmental, and Hereford bulls have been used.

OUTLOOK FOR CATTLE

Although the work at the experiment stations follows well-established practices in this field, the applicability of the results to improvement of Egyptian cattle is open to question. The large majority of the cattle and buffalo populations is in the hands of the small farmer. Most of what his land will produce must go for subsistence of the family, either as grain to be consumed directly or as cotton to be sold to buy food. This necessarily can leave only a meager ration for the cattle, which are essential for cultivation. The primary need of the Egyptian cow is simply something more to eat. If improvement in milking ability is sought, assurance should first be obtained that it will not be at the cost of a lower draft capability.

Cattle in Egypt undoubtedly will continue as they have for centuries —at a bare subsistence level. A decrease in number would be beneficial, but this is economically impossible since agriculture depends on the small farmer with his minimal holding and inability to finance a tractor.

Angola

Land area (sq. mi.):	481,400
Population (1964):	5,100,000
Density (per sq. mi.):	11
Agricultural (85%) (1966):	4,350,000
Cattle population (1966):	1,700,000
Offtake (4%) (1966):	70,000
Year visited:	1966

ANGOLA, a Portuguese province of thirteen districts, is a rectangular-shaped area lying between Congo (Kinshasa) on the north, South-West Africa and Botswana on the south, and Zambia on the east. It has a 1,000-mile coast on the Atlantic Ocean in the west. From a narrow, dry coastal plain, the land rises to a broad plateau 4,000 to 6,000 feet in elevation. The region has an annual rainfall of 35 inches and a very equable climate. Near the eastern border the plateau slopes to the Zambezi River drainage. In the south and west there is some desert and semiarid country. Although slightly larger in area than Texas, Oklahoma, Arkansas, and Kansas combined, Angola has less than one-fourth their population.

Having landed in 1482, the Portuguese were established in Angola before Columbus reached the Americas. They have ruled the area continuously except for the few years in the middle of the seventeenth century when the Dutch took and held the seaports until being forced out. The present population is 98 per cent African, largely of the Bantu tribes, but with a few Bushmen in the southeast. The remainder of the population is predominantly Portuguese, and immigration is actively encouraged.

Nationalist and anti-Portuguese Africans led a rebellion in 1961, but it was confined to an area below the Congo border in the north. While suppressed primarily by white Portuguese troops, Africans from areas of Angola where there was no fighting served, and still serve, in the army units. The movement has disintegrated, largely because of conflicting tribal and personal interests among the leaders. Some insurgency persists in some economically unimportant areas, but in 1966 it was well contained. Metropole Portugal, as the mother country is called, may border on police-state methods in keeping the situation in Angola in hand, but life proceeds in an orderly manner. Price controls are general and are detrimental to some extent to an expansion of the economy, particularly in cattle raising.

Luanda, the capital, is one of the cleanest cities in Africa, and the smaller towns throughout the country reach comparable standards. The Portuguese policy of assimilation of the African is a long-range program. For fifty years, until 1961, as an *assimilado* an African who cared to renounce his tribal customs and could meet certain educational and financial qualifications could obtain equal citizenship with the Portuguese. In 1961, however, all African residents of Angola were granted full citizenship. Time will tell the ultimate effect of this action on multiracial problems; but the Portuguese, who initiated the policy, have had nearly 500 years of close contact with the African tribesmen, far longer than that of any other Europeans.

Agriculture is the backbone of the economy and coffee is the principal foreign exchange earner. The country is self-supporting in oil and has mineral resources that have only been tapped. Cattle, 80 per cent of which are owned by Africans, being largely unutilized are only a minor element in the economy. With its varied resources, which are to a large extent undeveloped, Angola has room for expansion along many lines and stock raising is not the least of them.

CATTLE BREEDS

Cattle in Angola can be classified into three groups: indigenous types, which are mostly in the hands of the tribesmen; exotic breeds imported from South Africa; and mixtures of the two. The Africander, being native to South Africa, is much better acclimated than the other exotic breeds, which are of European origin although they came to Angola from South Africa. Herds of the exotic breeds are maintained largely to produce bulls for use in upgrading native cattle, although some of the milk breeds are maintained pure in the dairies.

Barotse.—In the eastern part of the country, bordering Zambia, Barotse cattle are maintained by tribes which are not cognizant of national boundaries and graze their herds in a wide area that lies in both countries. The animals are a Sanga type, as described in the "Cattle Breeds" section of the chapter on Zambia under the heading "Barotse."

Porto Amboim.—Locally called Porto Amboim cattle, these animals, similar to the Barotse but smaller in size and with a less pronounced hump, are seen in widely separated parts of the western plateau area as well as in scattered locations along the coast. Mature cows weigh approximately 800 pounds, bulls as much as 1,000 pounds. Color varies widely—black, black and white, red, and tan. The black-and-white pattern seen below is common.

Porto Amboim cow. Photographed south of Sá da Bandeira, in southern Angola.

Native Cattle.—In some parts of the central plateau, a rangy type of cattle with long horns, obviously of Sanga derivation, is locally referred to as Native Cattle. The color is varied—red, tan, black, and mottled black and white. Mature cows may weigh 700 pounds. Oxen, which are often used for draft, are maintained in better condition and may weigh more than 900 pounds. Although a local distinction is made between Native Cattle and the Porto Amboim, both probably trace to the same origins and one is as native as the other.

Native cattle ox team. Photographed in the Cela area.

Native cow. Photographed in the Cela area.

Barra do Cuanza.—The Barra do Cuanza, grown only by European farmers, is found on the coastal plains south of Luanda. Its derivation is not known.

The Barra do Cuanza has a fair beef conformation, only a slight vestige of a neck hump, and very large horns with a wide upturn. Cows on good nutrition weigh as much as 900 pounds, bulls 1,100 pounds. Ranging from red to a light tan bordering on a cream color, the hair is

Light-tan Barra do Cuanza bull, Pecuaria do Barra do Cuanza.

Barra do Cuanza cattle from a red herd, Pecuaria do Barra do Cuanza.

short and of a very uniform solid color. Some growers run separate herds of the red and the light-tan animals.

Exotic Breeds.—Exotic beef breeds have been introduced by both the government and individual growers to improve the fleshing characteristic of the native cattle. The Africander is generally preferred; but other breeds such as the Santa Gertrudis, the Simmental, the Charolais, and the Red Poll are used.

Purebred herd of Africanders used by a large grower to supply bulls for breeding to native cattle.

Purebred herd of Simmental.

Swiss Brown, Red Danish, and Friesian cattle, the latter predominating, are the milk breeds used by the European dairy operators. All these breeds are imported primarily from South-West Africa.

Friesian cows of a small dairy. Photographed in the Cela area.

MANAGEMENT PRACTICES

African tribes own a majority of the cattle of the country and run them in the traditional African manner—grazing on the public domain in areas of better rainfall and in nomadic search for grass and water in the drier regions. The African's impelling instinct to maintain or increase his cattle numbers prevails. Except for providing milk for household use, this livestock is little utilized and is disposed of only when there is some pressing need for cash; even then, only the older animals are sold.

Indigenous cattle obtained from the tribesmen are the usual base stock of the large European growers. Many owners are now upgrading their animals with exotic breeds, of which the Africander is preferred. The Veterinary Services in their experiment stations are also promoting this practice as well as the introduction of European milk breeds.

Although some are even larger, the fazendas of the European growers usually range from 10,000 up to 100,000 acres. These holdings are fenced and cross-fenced for pasture control. In the 16- to 20-inch rainfall areas of the coastal plain and in the south, the stocking rate is eighteen to twenty-five acres a head; on the high plateau, where the precipitation varies from 35 to 60 inches depending on elevation, five to ten acres a head is common. Pastures are rotated for good usage and some are saved to furnish matured growth for the dry season, usually May to October, which corresponds to winter in the Northern Hemisphere.

Breeding practices vary, but the breeding season is usually long in order to permit the use of a minimum number of bulls. Five to six months breeding is common where there is good rainfall, and often bulls are left with the herd the year round in the warmer, drier areas. A good operation gets a 65 per cent calf crop, with a death loss of about 2 per cent. In a representative operation in which native cows are being upgraded, native calves, weaning at eight months, weigh approximately 200 pounds, Africanders 385, Simmentals 425, and Africander-native first crosses 265. On the ranch in the lower rainfall area where these comparisons were made, a native cow required twelve and one-half acres to carry her through the year and the exotic breeds required about twenty-five acres.

Dipping for external parasites every seven days in the wet season and every ten to fourteen days in the dry season is essential. Compulsory prophylactic treatment is required for pleuropneumonia, blackleg, and anthrax. A few growers calfhood vaccinate for brucellosis, but this practice is not general in spite of widespread evidence of the disease.

Calves are weaned when about nine months old, castrated, and grown out on grass. Africander-cross steers from the high plateau region go to market when they are about two and one-half to three years old and weigh 900 pounds. In the drier areas slaughter animals are usually grown out to weigh approximately 1,000 pounds at three and one-half to four

Two-year-old steers nearly finished on grass. At left, a second cross of an Africander bull on a Native Cattle cow; at right, a first cross.

years old. No grain feeding is practiced. In addition to breeding and growing out cattle for market, some operators purchase young Native Cattle from the African cattle owners and pasture them along with their own raised stock until ready to market. Kept on good pasture these Native Cattle weigh 900 pounds at four years of age.

Dairying is confined mainly to the western side of the high plateau area. Near Nova Lisboa and the recently colonized Cela area, herds of the European breeds of dairy cattle are seen. Cela is a community of small villages, now seventeen in number, each the center of a group of farms averaging 142 acres in area. It was financed and built by the government to encourage European-managed agriculture. Native labor cannot be employed by the holders of these farms—a highly unusual provision in Africa. The farms are primarily dairy operations; milk not consumed locally is used in butter and cheese. The price paid the producer for milk is under government control; in 1966 the base price was 2.85 cents a pound for that used for manufacturing purposes and 3 cents for first-class milk which is pasteurized and bottled for household use.

Red Danish, Friesian, and Swiss Brown are the principal breeds introduced on the dairy farms. Herds average fewer than 50 cows each and are of good European stock. Bull calves are castrated when weaned and kept on pasture until sold at an average age of eighteen months, weighing 550 to 650 pounds. Milk production is rather low, considering the quality of the cows and the relatively good pastures they are on. It averages approximately twenty-five pounds a day for the straight European breeds, primarily because of the sparse feeding of concentrates since many farmers rely on pasturing exclusively. Brucellosis and venereal diseases are serious problems on dairy farms, and adequate control measures are lacking.

Artificial insemination is employed to a considerable extent in breeding dairy cattle. At the Centro de Inseminacao Artificial, serving the Cela area, three excellent Red Danish bulls and three Swiss Brown bulls of equal quality are maintained. A modern laboratory processes the semen, which is held under ordinary refrigeration for a maximum of three days. Currently the center is breeding 7,000 cows annually after only three years in operation. The over-all conception rate is 75 per cent, 45 per cent on the first breeding. It is felt that these low rates are the result of the prevalence of brucellosis and venereal diseases. The facility is operated by the Veterinary Services and service is free to the farmer, showing the extent of the government's endeavor to encourage dairying. A similar but larger center is also maintained in the Nova Lisboa area.

MARKETING

In Luanda all cattle are marketed through the guild, a co-operative organization of retail butchers which buys direct from the grower, has the animal processed in the municipal slaughterhouse, and distributes the dressed carcass and offals to the shops. In the rest of the country, marketing practices follow no fixed pattern. Traders operating independently or through government markets enter into the transactions involving the sale of cattle for slaughter by Africans. Large European growers sell direct to butchers, who have the processing done in the municipal plants. Two abattoirs, one in Sá da Bandeira and another in Moçâmedes, are operated by SOFRIO, a private company, and buy direct from growers as well as from traders and also process cattle which the company raises. Over the country live cattle prices ranged from 8 to 9 cents a pound in 1966. The government recognizes three grades of slaughter cattle on which it pays a premium: the "Special" grade, an animal less than three years old in good flesh, brings a premium equivalent to 1.5 cents a pound liveweight; No. 1 grade, 1 cent; and No. 2 grade, .75 cent. Most of the cattle slaughtered in Luanda are worn-out work animals and do not command a premium.

The cattle population for 1966 was estimated at 1,700,000 head by the Veterinary Services and is expected to increase to 2,000,000 head by 1970. The number slaughtered in 1963 is given as 77,000 by the United States Department of Agriculture, and the Veterinary Services place it at 70,000 for 1966. These low numbers represent an offtake of only 4 per cent.

ABATTOIRS

Slaughtering, with the exception of the private plants referred to under "Marketing," is handled by municipal abattoirs. The largest of these is the Matadouro Municipal, in the center of Luanda. Built thirty years ago to handle 40 head of cattle a day, it is currently handling an average of 120. Cattle are brought to the plant both on the hoof and by truck from the holding stockade of the guild on the outskirts of the city. From a walled yard next to the main building, the animal is dragged to the killing floor by a rope over its head and is pulled by ten Africans to a rhythmic chant which times a series of jerks until the animal is in the desired position. The head is then held by the horns (by the ears and neck if there are no horns) while the killer with an accurate thrust sticks a chopa into the neck at the base of the brain. A chopa is a short, heart-shaped dagger generally used for killing cattle in Portuguese abattoirs. The animal then falls, is immediately bled, and is dragged by several pairs of hands to the dressing floor, where it is hoisted by the forelegs, again by hand, for skinning and eviscerating. There are no cooling facilities. The morning they are killed, the dressed carcasses, with hearts and livers attached, are displayed on hooks and are sent to shop owners. Inspection is by a veterinarian of the Veterinary Services. Sanitation is all that could be expected in the overcrowded facility.

Abattoirs in the country, which kill only a few head of cattle on the days they operate, each consist of a small building provided with water, trenches in the floor for removal of waste, and a small holding enclosure. They are operated by the municipalities and inspected by a veterinarian.

The abattoirs of SOFRIO at Sá da Bandeira and Moçâmedes are engaged principally in the preparation of frozen meat for export to Portugal, although there is some shipment to Luanda. Their operations are rather limited in scope, and processing is not modernized; sanitation is excellent, however, and there are good freezing and refrigerated storage facilities. The abattoirs are a good example of what could be done to utilize the potential for beef production in the country.

CATTLE DISEASES

Persistent efforts are made by the Veterinary Services to prevent and control disease. A well-operated laboratory in Nova Lisboa, including a newly established veterinary college, produces most of the vaccines used in the country. Compulsory vaccinations are required for pleuropneumonia every ten months and for blackleg and anthrax every twelve months.

Foot-and-mouth disease is endemic, and when outbreaks occur the area involved is quarantined; there is no vaccination, however, for serum is not available. There is considerable controversy among the veterinary services about the cause of the spread of the disease between the Huila district of southern Angola and Ovamboland in South-West Africa, which lies across the border.

Brucellosis is recognized as being prevalent in most of the country, but in 1966 there were no effective control measures. Calfhood vaccination when requested by growers is given gratis, as are all other preventive and control treatments. It is not unusual in a well-managed herd for 2 per cent of the cows to be eliminated on an annual blood test for brucellosis. Heartwater and enterotoxemia are prevalent in many areas. Anaplasmosis is also common and causes sizable losses. Rabies, carried mainly by the dogs of the Africans, is fairly prevalent. Although it is unusual to do so in Africa, many natives keep dogs as pets but take no effective preventative measures against the spread of rabies among them.

Tsetse fly infestation, which sometimes occurs in the northeast and along the Congo River, is said to be reasonably well controlled by treatment of infected animals every four months. The high plateau area, being cooler, is practically free of the fly.

There is no quarantine control on the cattle smuggled across the border from South-West Africa.

GOVERNMENT AND CATTLE

Efforts of the government to foster a healthy beef cattle and dairy industry are most contradictory. Disease-control measures; preparation of vaccines; experiment-station work, which includes supplying good breeding stock; and artificial insemination centers are well administered and should produce good results. The fixed price for milk, however, is too low for the dairy operator to be able to feed his cows the grain that would enable them to reach their milking-potential. The fixed price for retail meat, even with the subsidies provided the grower for marketing younger animals, can only be lived with by the large operator with extensive holdings. This situation stems from the highly centralized control in Lisbon of all phases of the economic life of Angola.

OUTLOOK FOR CATTLE

One of the world's best remaining cattle-raising regions with its potential greatly underutilized is the belt that stretches across Africa from the Atlantic to the Indian Ocean between the 5th and 25th southern parallels. This area, 2,000,000 square miles in extent, includes Angola, Zambia, Mozambique Botswana, Rhodesia, and parts of the Congo (Kinshasa), Tanzania, and South-West Africa, in all of which there is only an estimated 16,000,000 head of cattle (8 to the square mile.) While some of this area is desert and marginal grazing land and parts are subject to tsetse fly infestation, none of these handicaps are of such magnitude that they detract materially from the ultimate capability. Angola occupies one-fifth of this undeveloped region but has only one-fifteenth of the cattle on some of the better grazing lands.

The stability of the government today cannot be questioned, in spite of the pressures it is under. Yet government regulations and bureaucracy are deterrents to an expanding cattle industry. The effect of price controls has been mentioned. There is also the problem of inadequately staffed government offices not only in the departments directly related to cattle raising but also in the department which grants land concessions. Granting concessions in cattle-raising areas in the south is often long delayed because of the understandable requirement that proper provision be made for the welfare of the nomadic African cattle raisers who have been living on the land. Some South African cattle investors have been waiting more than three years for land concessions.

Even more fundamental, however, is the large outlay of capital that would be necessary for a sizable expansion in stock raising; and this Portugal does not have. The government is interested in attracting such capital, and at least one major American cattle producer is known to be investigating the possibilities.

Angola has a large potential for cattle raising which will be utilized some day—the growing world-wide shortage of beef will force it. If outside capital is to be the source of such development, however, the road will first have to be paved by the government in Lisbon because of its highly centralized control of all that occurs in Angola.

Mozambique

Land area (sq. mi.):	303,000
Population (1964):	6,900,000
Density (per sq. mi.):	23
Agricultural (75%) (1964):	5,200,000
Cattle population (1966):	1,130,000
Year visited:	1965

THE PORTUGUESE PROVINCE OF MOZAMBIQUE is a beautiful tropical area stretching along the Indian Ocean for 1,300 miles on the southeast coat of Africa. It is a land with a future, when the Portuguese can find the funds to develop it. In area Mozambique is equal to Texas and Louisiana combined. Most of the population lives south of the Save River in an area 60,000 square miles in extent, between the river and the South African border. The northern borders with Tanzania and Zambia are separated by the lower tip of Malawi, which extends down into Mozambique for 250 miles. The western border touches Zambia and extends along Rhodesia and South Africa and ends with Swaziland in the extreme South. On the east the shores of the Indian Ocean stretch the full length of the country from north to south. Mozambique lies within approximately the same latitudes in the Southern Hemisphere as those occupied by Mexico and Central America in the Northern Hemisphere. Except for a narrow strip along the coast, the land north of the Save is largely game country, probably the best in Africa. It is heavily forested except for local areas of brush-free savanna. All the region but these savanna areas is infested with the tsetse fly and is uninhabitable for man or cattle. When population pressure requires, the tsetse can be eliminated and the region made productive. The population density even in the heavily populated area is fewer than 100 persons to the square mile, three-fourths that of Louisiana.

The Portuguese were the first Europeans in Mozambique, landing there in 1498 as they established their trade routes to the East Indies.

They found Arab slave traders already entrenched but proceeded to establish their own settlements. Of the many places in Africa where the Portuguese touched first in their world-wide exploration, Mozambique is the only one to have stayed under Portugal's continuous control until today. The Dutch attempted to gain a foothold on the coast in the seventeenth century but were soon driven off.

Bantu-speaking tribes account for nearly 98 per cent of the population. Most of the European element is Portuguese and now numbers well over 100,000, probably 1.5 per cent of the total. Indians and Chinese, who are mostly engaged in trade, form minor segments of the population. Mixing of the races has been on a relatively small scale—in 1960 the number of persons of mixed blood was estimated at 36,000 or .5 per cent of the total population.

CATTLE BREEDS

The two principal breeds are indigenous to the general area, the Landim to Mozambique itself and the Africander to South Africa. Both are excellently adapted. A few European growers have introduced exotic beef breeds for crossing experiments, but little has been accomplished by these efforts. The number of crossed animals in the country is inconsequential.

Landim bull.

Landim.—Apparently differentiated out of the original Sanga type of East Africa by native selection, this breed was eventually brought by tribal migrations into the area that is now Mozambique. It is the same animal as the Nguni found across the border in Natal, a province in South Africa.

Polled Landim bull.

For an African breed the Landim is rather large. In conformation individuals are quite uniform, well muscled, and of medium bone with parallel toplines and bottomlines. Color varies widely from nearly all white to black, brown, tan, and almost any combination of these. White with mottled black markings and a white topline and bottomline is common. The hump is quite small, in the female often unnoticeable and in the male not much larger than the seasonal sexual swelling of a nonhumped breed. The horns of the cow are rather thin, wide, upswept, and sometimes lyre-shaped; on the bull the horns are shorter and thicker. There is a strong tendency to be polled.

The Landim is hardy in the subtropical environment and fleshes well on grass. Under African care a mature cow rarely weighs 800 pounds, but with proper nutrition and reasonable management a good cow may weigh as much as 950 pounds. The breed is exceptionally fertile when kept on good feed, has a high degree of tolerance to most of the endemic

Landim cow.

diseases, and is susceptible to development as either a good milk or beef type.

Africander.—This breed was brought into Mozambique from South Africa by Portuguese settlers and predominates in the herds of European growers. Indigenous to a similar, though somewhat less tropical, environment, it does well. Many of the European herds were built from a Landim base through the continued use of Africander bulls.

Africander cows with calves. These animals belong to a European grower.

One of the flaws of the Africander as a beef animal as it has been developed in South Africa is the sloping rump. This distinctive characteristic of most humped cattle was probably accentuated by the original selection criterion—to produce a good draft animal—and remains a rigid requirement of the Africander Breed Society. At Chibanza Xinavane, a progressive European ranch, selection has been made for thirty years to obtain a type with a more rounded rump. This program has resulted in an animal with a better beef conformation than that of the traditional Africander. A number of the Africander herds in the country exhibit this tendency toward a more rounded rump (probably resulting from the use of sires acquired from Chibanza Xinavane or Mazeminhama, the two best Africander breeders in southern Mozambique). It would not be surprising to see these cattle recognized some day as a new type of the breed.

Africander bull showing the improved rump developed at Chibanza Xinavane.

Friesian.—The principal milk breed is the Friesian. The dairy herds are largely the result of upgrading Africander or Landim cattle, many of which are now excellent Friesian representatives. Starting with a base of well-acclimated indigenous cows, the growers developed a nearly pure Friesian, disease and heat-stress tolerant, that does well in the tropical climate. Average milk yields in a good herd run as high as 7,000 pounds a lactation—a production which none of the European milk breeds could accomplish in the tropical climate unless maintained under practically clinical conditions.

A fifteen-sixteenths Friesian cow bred up from an Africander base.

MANAGEMENT PRACTICES

Two-thirds of the cattle population is owned by Africans. They pasture their animals during the day, kraal them at night for protection, and maintain them largely as a prestige symbol. Their animals are principally the Landim breed or variations of it resulting from admixture with other African types as caused by tribal migrations. Although good Landim animals are seen in some native herds, African management, low levels of nutrition, and indiscriminate breeding have resulted in the general run of stock being inferior to the potential of the breed. This fact is obvious from the results of the selection program that the Veterinary Service has followed for years to improve the Landim.

European owners account for 400,000 head of the total cattle population of 1,130,000. Some owners have been afflicted with the crossing complex. Limited numbers of both European breeds and American-developed breeds with Zebu influence have been introduced at various times, and the practice is continuing to a limited extent. Sussex, Hereford, Friesian, Jersey, Swiss Brown, and American Brahman animals have entered the picture. The European beef breeds, after the hybrid vigor of the first cross is lost, do not appear to accomplish much in the improvement of the local cattle. All European-imported breeds require special care in nutrition and in prophylactic treatment for the tick-borne African cattle diseases.

Some farms of European growers cover 25,000 acres, and some are even larger. They are often stocked with several thousand head of cattle, usually of the Africander breed. Many follow modern management practices, using fenced and properly rotated pastures and good provision for stockwater. Frequently two breeding seasons are scheduled: February through April and August through October. Some growers, however, leave bulls in the herd the year round.

Calves are castrated when approximately six months of age and weaned at eight months. In a commercial herd steers are marketed off the grass at three and one-half to four years of age and usually average more than 1,000 pounds. Exceptionally well managed farms ship Africander steers at two and one-half to three years of age, weighing 1,100 pounds. This weight is attained by feeding the animals a small amount of protein supplement during the dry season, when only matured grass is available in the pastures.

Disease-control measures are rigidly adhered to. Spraying for external parasites once a week during the wet season and every two weeks during the dry season is a common practice. Vaccination for anthrax, blackleg, and brucellosis on heifer calves is practically universal. Under such a prophylactic program, disease losses are held to less than 5 per cent.

Grain feeding in preparation for market is unknown, but a step forward is being made through the expanding practice of growing out slaughter animals on grass, using native cattle. Such livestock is purchased when the African owners can be induced to sell. Steers at one to one and one-half years of age are preferred, but heifers and other stock are purchased if yearling steers are not available. They are run on grass through the dry seasons for two to two and one-half years and marketed when about three and one-half years old, weighing 800 to 1,000 pounds and grading mostly in the Improved Type (discussed under "Abattoirs"). Because he practices better management, the European owner gets considerably better weights among his cattle than does the African grower.

Recognized practices are followed by the well-operated dairies. Milking machines are used, often in conjunction with milk parlors. Cattle are often pastured when the grass is green and are kept in a loafing yard with an adjacent shed at other times. Sanitation is good in the larger dairies. The Friesian cross is the preferred dairy animal. It is said to do better under the climatic conditions than either Swiss Brown or Jersey crosses.

MARKETING

Cattle for slaughter are bought direct from the owners by the municipal abattoir in Lourenço Marques or from the traders on the cold-dressed-weight (CDW) basis. In all sales involving cattle, both buyer and seller are legally required to register the transaction. While this requirement cannot be enforced on all transfers of African-owned cattle, it effectively controls the movement of most commercial cattle.

ABATTOIRS

Operated by the municipality, the principal abattoir is located in Lourenço Marques. The thirty-year-old plant is well run under reasonably sanitary conditions except for the handling of the digestive tract. This phase of the operation is let out to Africans on a contract basis and is conducted in a very primitive manner. Inspection otherwise is good and is handled by employees of the Veterinary Services. Except for a small volume of chilled and frozen beef that is shipped to the north and for ships' stores, all meat is delivered warm for sale in the local shops. The average daily slaughter is 140 cattle, which includes 20 veals—animals under one year of age—and 100 head of small animals, mostly sheep and hogs. Most of the cattle killed are Landims or related Landim types by cross with Africanders.

Carcasses are graded and paid for in accordance with government-established prices. These grades and prices as revised in 1966 were as follows:

Improved Types	*Price per Pound, CDW*	*Common Types*	*Price per Pound, CDW*
Extra I	$.29	First	$.18
Extra II	.26	Second	.16
Extra III	.255	Third	.14
Extra IV	.20	Fourth	.125

Butchershops buy dressed carcasses at a fixed price of 17 cents a pound. When cattle are graded as Improved Type, the difference is paid by the Fundo de Fomento Pecuario (Cattle Breeding Protective Fund). If the cattle are graded as common type, except for first grade, the difference goes into the Fundo. For all grades there is a deduction of 2.3 cents a pound for the Fundo.

In 1965, Mozambique had been subjected to a two-year drought period, and the cattle being killed were mostly of the lower grades. There is practically no distinction for cuts, and in the retail market sirloin or ground meat sell for the same price, about 28 cents a pound. Butcher-shops buy dressed carcasses at the same price the abattoir pays the producer. Cattle are killed by the chopa, a short dagger with a heart-shaped point, which is expertly wielded to sever the spinal cord at the base of the brain.

CATTLE DISEASES

All the endemic diseases of African cattle are encountered, but the control measures of the Veterinary Services are quite effective in holding losses down. Legal requirements include vaccination for anthrax once a year for all cattle, for blackleg each year for three years, and for brucellosis at four to nine months of age for heifer calves; testing for tuberculosis by inoculation of tuberculin in the lower eyelid and slaughter of reactors; blood-testing for brucellosis when animals are transferred between the north and south areas; and periodic dipping of all cattle for which spray races are supplied in the settlement areas.

Treatment for foot-and-mouth disease is now compulsory only on outbreaks. Rinderpest has been eliminated. Most of the European growers dip or spray for external parasites at one- to two-week intervals. Anaplasmosis is prevalent but most cattle acquire an early immunity to it. Some lumpy skin disease is showing up. This can be controlled by immunization measures, but the incidence has not been large enough to require treatment.

GOVERNMENT AND CATTLE

Government control and development of livestock is vested in the Department of Veterinary Services. The long-range advantage of improving indigenous cattle by selection within a breed adapted to the tropical climate and the disease-infested environment has been recognized by the department for many years. Herds of pure native Landim cattle are maintained at two experiment stations engaged in this program. Selection within the breed has produced a Landim cow with an average mature weight of 930 pounds, compared to 775 pounds for native-raised cattle. Hereford, Africander, Friesian, Swiss Brown, and Jersey bulls are used in crossing experiments on Landim cows at the Estação Zootécnica Central at Chobela. These tests have shown that the pure Landim has a higher birth rate and lower mortality than either the pure Africander or the pure Hereford, the latter being a very poor third. Work on the dairy breeds has shown that the Friesian has done better than the Swiss Brown or the Jersey in milk production and has proven more tolerant to disease.

In a wide area around the station at Chobela, a large majority of the cattle show distinctive Landim characteristics but are inferior in conformation as well as size to Chobela animals. It is impossible to estimate how much of this difference has resulted from selection and how much from better management practices, but certainly selection has contributed to the improvement.

Before the establishment of the higher price schedule for dressed carcasses in 1966, the grower was suffering severely from the low level of the government-fixed prices. This new schedule, with its built-in incentive for improvement in quality, should materially improve conditions in the industry.

OUTLOOK FOR CATTLE

Mozambique has the potential in areas now unutilized for at least a tenfold increase in her cattle population. In the more populated region south of the Save River, the number of cattle is one-fourth less than that in Louisiana, with which it has been compared in area and which is admittedly not cattle country. By control of grazing and provision for adequate stockwater in the native grazing areas, the number of cattle here could readily be doubled. There is ample know-how available from the European grower, who manages his herd advantageously. Furthermore, the political climate is such that he has no fear of the future, and, with the several generations of development behind his holdings, he has no intention of leaving. All Africa, along with much of the rest of the world, is short of beef and there is no basic reason why Mozambique could not enter the market of those areas which do not have rigid quarantine restrictions. This would require sizable investments in abattoirs and in freezing and chilling facilities.

Rhodesia

Land area (sq. mi.):	150,000
Population (1966):	4,330,000
Density (per sq. mi.):	29
Agricultural (82%) (1962):	3,530,000
Per capita income (1963):	$204
Cattle population (1964):	3,855,000
Offtake (9.5%) (1964):	365,000
Years visited:	1965, 1966

RHODESIA is the name under which the former British colony of Southern Rhodesia declared its independence in 1965. The country lies between the 15th and 22nd parallels, in southern Africa. Landlocked and roughly circular in shape, it is enclosed on the west by Botswana, on the south by Botswana and South Africa, on the north by Zambia and Mozambique, and on the east by Mozambique.

Native Africans, consisting almost entirely of the Bantu-speaking peoples, account for 94 per cent of the total population; Europeans, predominantly British but some of Dutch descent, for nearly 6 per cent. Asians and colored persons (in the South African sense) number about 20,000.

Because of the elevation, Rhodesia enjoys an equable climate and has an annual rainfall varying from 15 to 35 inches. In area it compares with the western half of Iowa, all of Nebraska, and the eastern half of Wyoming. The human population is slightly less than that part of the United States, and the cattle population is one-third as large. The west, where the rainfall varies from 15 to 25 inches, is good stock-raising country. The east, with as much as 35 inches, is something of a corn belt, although tobacco is the major cash crop.

Until past the middle of the nineteenth century, the region that is now Rhodesia was the seat of a flourishing Arab slave trade. The British, under the leadership of Cecil Rhodes, ended this practice and settlers began moving in at the turn of the century. These immigrants were largely South Africans of British descent looking for greener pastures. As colonization proceeded in the general area and to the north, Southern

Rhodesia became the guiding spirit in the Federation of Rhodesia and Nyasaland. Following the dissolution of the federation, Southern Rhodesia took a new approach to sovereignty in Africa by declaring unilateral independence in November, 1965, thus becoming Rhodesia. In the heat of the political turmoil which followed this action, the outcome cannot be fairly forecast. Time is on Rhodesia's side, and in 1969 the country remains one of the three African states that enjoy a modern, Western-type of day-to-day life in both town and country.

CATTLE BREEDS

European ownership accounts for one-third of the total cattle population, a larger proportion than in most other African countries. This situation, coupled with the fact that the colonists brought their cattle with them when they came to Rhodesia, has resulted in considerable dilution of the indigenous breeds except in the more remote areas.

Some native breeds that were in the hands of the tribesmen when the colonists came to Rhodesia have been developed into excellent beef types by selection for fleshing characteristics. Breed societies for these improved indigenous breeds have been established and require exceptionally high standards for registration.

Mashona.—Natural selection over the centuries produced in this indigenous breed a hardy animal tolerant of the disease and the parasites of the dry areas where it was run. Improvement of the breed was started in 1941, when F. S. B. Willoughby obtained some of the best cows and a few of the best bulls which he could induce the chiefs of the Mashona tribe to part with. Selection for a beef conformation over the next twenty-five years resulted in the excellent animal grown today. For registration the Mashona Breed Society requires that:

1. All animals meet the conformation standards established.
2. Cows have calved in the ratio of twice in three years and are shown with 2 offspring meeting the breed conformation standards.
3. Bulls are shown with an entire season's get, of which 90 per cent and a minimum of 18 calves must meet the breed conformation standards.

There is a strong tendency in the Mashona breed to be polled, and some entire herds show that characteristic; horns are not discriminated against, however. Dehorning may be, and is, practiced. Horns when present are small, growing upward and out from the head. Solid dark-red or solid black animals predominate.

The differentiation of this remarkable and distinctive animal in such a short period of time (twenty-five years) shows that definite genetic types had been developed in Africa. The Mashona is of Sanga origin. The improved type has an excellent beef conformation and a very well-rounded rump. Although a rather small animal, three-year-old steers off grass weigh 1,000 pounds and, with a little supplemental feed in the dry season will reach 1,100 pounds at two and one-half years of age.

Mashona bull. Photographed at Ellerton Farm.

Mashona cow. Photographed at Ellerton Farm.

Tuli.—Another indigenous animal that has been developed by selection within the breed, the Tuli is also of Sanga descent and is native to the low veld in southwestern Rhodesia in a 15-inch rainfall area. In 1942 selection was begun at the government breeding station in Matabeleland on a herd of 20 native cows and 1 bull.

One of the original foundation cows of the Tuli breed. Courtesy Tuli Cattle Society.

In addition to weaning and growth weights, the progeny tests included carcass evaluation on the hoof. In 1951 the average weaning weight on all male and female calves at the station was 350 pounds. In 1959 the heifer calves averaged 480 pounds, the bulls 560 pounds. Part of this increase resulted from better pasturing practices, but genetic improvement was certainly a major factor.

A remarkable animal has resulted from this program. The improved Tuli is medium boned, with a good beef conformation; cows weigh as much as 1,000 pounds and bulls 1,500 pounds. Steers off good sorghum-grass pasture at two and one-half years of age weigh about 1,050 pounds and dress 58 per cent.

Light golden brown, although preferred, is not a fixed color, for light grey is also common. Tuli Cattle Society standards allow any solid color except black. There is a definite polled tendency: nearly one-fourth of the animals in the average herd do not have horns. When horns are present, they are short, thin, curved inward, and sometimes loose. The hump is small, not much larger than the normal sexual swelling in the male and frequently not discernible in the female.

Tuli bull in a commercial herd.

Tuli cow in a commercial herd.

A Tuli Breed Society has been organized along the same lines as the Mashona Society and has similar registration requirements.

Nguni.—This indigenous breed has also been developed as a beef animal by selection within the breed. It is the same breed as the Nguni of South Africa and the Landim of Mozambique. The Matabele tribesmen, who were good cattlemen, had selected a particular side color pattern which was distinctive in their herds. These cattle were known as Nkone, the tribal name for the color pattern, and were of the Nguni breed. European growers starting with a base of Nkone cattle obtained from the Africans have developed an excellent beef animal, which when finished on grain produces a 500-pound carcass at two years of age. Performance-tested bulls are advertised as having a feed conversion of 4.49 total digestible nutrients for one pound of gain. Mature cows average 900 pounds, with individuals weighing as much as 1,100, and bulls average 1,400 pounds, with some reaching a weight of 1,800 pounds. The breed society recently changed the name of the improved breed to Manguni.

Angoni.—Representatives of this breed, which is described in the chapter on "Zambia," are seen in the northern part of the country in native herds.

Africander.—Cattle of this breed were brought into Rhodesia from

Purebred Africander bull.

Purebred Africander cow.

South Africa before 1900, by the first settlers; the descendants of those animals now account for the majority of the herds of the European growers. In addition to the commercial herds there are many purebred studs. The same animal is seen in South Africa.

Exotic Breeds.—A number of British beef breeds have been established in the high veld. Foremost among these are the Hereford, the Aberdeen

A good Friesian herd. Photographed at Glenara Estates.

Angus, the Shorthorn, and the Sussex. Under good management and careful prophylactic treatment against endemic diseases they do well. Some herds are equal to anything in Britain or the United States.

Excellent dairies are found near the population centers in which the Friesian predominates, although there are good examples of other European milk breeds. All these animals require more meticulous care than the native cattle. There are 125,000 dairy cattle of all breeds.

MANAGEMENT PRACTICES

Cattle run by the Africans are handled in the traditional manner of settled tribes, herding during the day and kraaling at night. Such animals are usually owned by village dwellers. Open grazing areas are set aside for their exclusive use and are well provided by the government with plunge dipping tanks and stockwater facilities. There has been considerable admixture with European breeds, and the native cattle have frequently taken on a nondescript appearance. Number is considered more important than quality. Milk is taken for household use; cattle are being used with greater frequency for draft, but are only used for meat on ceremonial occasions. The lobola custom is universal among the cattle-owning tribesmen.

Operations of the European growers run from a few hundred head on a few thousand acres to several thousand animals on a spread as large as 300,000 acres. The average farm is about 5,000 acres. The commonly accepted stocking rate is 1 head to 15 acres in the western stock-growing country, which is ample in a year of average rainfall but runs into severe feed shortages in drought periods. In the low veld in the southeast, where the rainfall is lower, 25 to 30 acres a unit is considered good practice. All European stock raising is under fence and is cross-fenced in paddocks for grazing control. On the large ranches, pastures of several thousand acres suffice for this purpose. The old practice of saving matured grass for pasture during the dry season is giving way to that of putting up hay for feed during the winter months.

On some of the better ranches, the breeding season has been shortened to three months, mid-December to mid-March. Four per cent bulls are commonly used. A 65 per cent calf crop is generally considered satisfactory but a well-managed operation gets between 80 and 90 per cent. Calves are grown out on grass to finish 900-pound steers at two and one-half to three and one-half years of age, depending on the growth of grass available. All Rhodesia has recently suffered several bad drought years, with 1965 being one of the worst.

Most operators dip every five days for tick control except during the winter, when the interval is lengthened to two weeks. Heifer calves are vaccinated for brucellosis at four to eight months of age, and all calves are vaccinated for blackleg.

Western and southern Rhodesia, with 15 to 25 inches of rain, are well adapted to cattle raising under range conditions. In the north and east, centering around Salisbury, excellent farming country enjoys more than 35 inches of rain in normal years. Here a regular corn-belt fattening operation is developing. In these operations steers weighing 800 to 900 pounds off the grass are put on a full feed of corn and either cottonseed or groundnut meal for about three months and finish at 1,200 to 1,300 pounds. Such fattening operations, however, were limited in 1965, and most of the cattle of the European owners go to the abattoirs off grass at four to four and one-half years of age. The majority are of the Africander breed, although there are representatives of many exotic breeds and their crosses.

In the cultivated areas of Rhodesia, good purebred establishments raise breeding stock of the European breeds as well as the Africander. Purebred bull prices to local growers generally run from $500 to $800, with tops as high as $2,000.

Many well-operated, hygienic, modern dairies, both large and small, are located near the population centers. The Friesian cow predominates, accounting for 80 per cent of the dairy animals; the Jersey is next, followed by representatives of other dairy breeds.

At Lilfordia Estates, outside Salisbury, there is an interesting contradiction to the often-heard opinion that the African buffalo cannot be domesticated. When Kariba Dam was completed, many wild animals in the inundated area took refuge on islands in the reservoir as it filled. "Operation Noah" was instigated by the Rhodesian Game Department to ferry this stock to shore. Lilfordia obtained 12 female buffalo calves and 1 bull, which were put with a group of Africander cattle and have since run with them in fenced pastures. In 1965 these buffaloes were three years old and apparently were as docile as the Africander cows. While it can be said that these buffaloes have only been tamed and not domesticated, they have submitted to confinement and man's presence in considerable contrast to the behavior of their wild counterparts. The African buffalo is a different branch of the Bovini tribe from the long-domesticated Asian animal and has never been useful to man.

Artificial insemination has received little attention. A few large breeders have been interested to the extent of running trial programs on

Three-year-old African buffaloes running with a herd of Africander cattle. Photographed at Lilfordia Estates.

a sizable scale. The government has displayed little interest in artificial insemination. The large pastures used and the exclusive handling of cattle by herders on foot add to the difficulty of the practice. The Veterinary Services and the experiment stations have lost interest because of the lack of funds; and, without renewed government sponsorship, it is doubtful whether artificial insemination will be of much importance for some time to come.

MARKETING

Cattle marketing in Rhodesia is a revelation in price stabilization. Although it would have to be called a form of subsidy in that the Cold Storage Commission, which controls the mechanism, is a government agency, in practice Rhodesian marketing seems to avoid the basic evils of outright subsidy. The commission operates the major abattoirs, killing about 60 per cent of the cattle slaughtered, and is the country's only legal exporter of meat. It sets the minimum prices it will pay for dressed carcasses a year in advance. All cattle are bought on this basis except those purchased direct from Africans at the auctions in the tribal areas. Each owner's animals are identified as received at the abattoir, and he is paid on the cold-dressed weight (CDW) and the grade of the cattle he shipped. The price paid on the spot for animals purchased from the Africans is arrived at by an arbitrary conversion to dressed weight for the visually appraised grade as each animal is weighed. The established CDW prices on the day of sale then apply.

Privately operated abattoirs must at least equal, and usually have to exceed, the commission's minimum price to get cattle. The commission can also pay a higher price than its established minimum if necessary to obtain the desired runs for its abattoirs. If the prices established result in a net loss for the abattoir at the end of the year, the government pays the difference. The stability that the program gives to prices at all levels in the cattle industry is a highly satisfactory feature. Growers, feeders, and private packers complain about the prices fixed, but it is doubtful whether any of them would voluntarily change to a free-market system.

There are seven established grades of dressed beef. Chiller grade 1 (rolled "super"), of which there are very few, must show a good finish and must not have more than four permanent teeth. This is the only grade which cannot be met without some supplemental feeding. As seen in the Bulawayo abattoir, it is equivalent to a USDA low-choice. There is not much difference between the next two grades—Chiller grade 2 (choice) and Chiller grade 3 (prime)—the requirement for Chiller grade 2 being that it is a somewhat younger animal. Both would grade good by USDA standards. The next three grades—G.A.Q. (good average quality), F.A.Q. (fair average quality), and X quality—fall in the standard range. The last grade, Inferior, is a poor canner. The bulk of the carcasses grade Chiller grade 3. This requires a good, grass-fattened three- to four-year-old animal. In 1965 prices varied for Chiller grade 2 from 22 cents a pound at the beginning of fall (April–May) to 25 cents in the summer (December–February). These prices on a CDW basis are equivalent to 11.5 to 13.5 cents a pound liveweight. Prices scheduled for the highest grade were 15 per cent above these, and for F.A.Q. grade were 25 per cent lower.

Because of the shortage of beef in 1965, the Cold Storage Commission adopted a novel plan, the Grazier and Feeder Scheme, to bring heavier cattle to market. Under this scheme the commission buys distress cattle on the hoof and sells them to an approved operator at the price for which they were purchased, agreeing to buy them back at a specified time at the established price then in effect. Credit, carrying interest rates as high as 11 per cent, is supplied the buyer; no capital is required other than the outlay necessary for feed. The operator is assured of a minimum price when he ships. It was evident that the scheme was doing something to get heavier cattle on the market, for more than $10,000,000 had been invested in 225,000 head in the hands of cooperators in the spring of 1965. Practically all of these cattle would otherwise have been slaughtered at the time they were purchased at much lighter weights by the commission.

As in all other parts of Africa, the problem of getting the African to market his cattle persists in Rhodesia. Only one-fifth of the cattle slaughtered are from the native herds, which comprise two-thirds of the total cattle in the country. Even this low offtake is high for African-owned cattle and results from the program of the Cold Storage Commission to get native cattle on the market. Cattle sales organized by the commission are held in the tribal areas on fixed dates, along the same lines as in Zambia, where they were set up before the dissolution of the Federation of Rhodesia and Nyasaland. This convenient opportunity for the African to sell some of his cattle is probably more successful in Rhodesia because there civilization has enclosed the African further and he has more need for money.

ABATTOIRS

Rhodesian abattoirs adhere to the usual high standards found wherever the British have initiated meat processing. The plant of the Cold Storage Commission at Bulawayo is as modern in facilities and exemplary in sanitation as that to be found any place in the world. The British killing box and captive bolt are used. From kill to carcass on the rail requires forty-five minutes. The chilling and freezing facilities are thoroughly modern. Before the embargo that followed independence, chilled meat shipped by rail to Cape Town then by ship to England was on the table in twenty-five days. There is complete by-product recovery. Except for the native cattle which the commission buys direct for cash at the auctions in the tribal areas, all cattle are processed and the producer is then paid on a CDW basis.

CATTLE DISEASES

In the past Rhodesian cattle have suffered severely from the endemic diseases of Africa. The rinderpest epidemic that nearly wiped out both cattle and game in all of southern Africa in 1898 left fewer than 25,000 head of both native and European-owned cattle. This disease was eliminated by extreme measures at that time, and the country has now been free of it for many years. Such outbreaks of decimating diseases as now occur are handled in a manner that prevents serious loss.

By constant vigilance and immediate action at the first sign of an outbreak, the Veterinary Services of the Ministry of Agriculture maintain effective control. Compulsory spraying or dipping and prophylactic treatment are common practices even for the tribesmen's cattle, for which the government provides dipping tanks. Both control of the wild life by slaughtering the species that carry infection and the installation of hundreds of miles of gameproof fences are major control measures.

Foot-and-mouth disease is kept under control by vaccination and quarantine when outbreaks occur. It has been proved that cattle in an infected area rather quickly develop a high degree of tolerance. Outbreaks in recent years have occurred only occasionally; there were none in 1965. Tuberculosis is present but is held in check by slaughter. Brucellosis is definitely present and, although free vaccination is available, it is not widely used except in well-managed European herds. Spot testing has indicated an incidence of more than 10 per cent, and this disease must be the cause of greater losses than are appreciated by the growers.

Currently the tsetse fly is the most serious threat to Rhodesian stock. Gameproof fences are maintained to isolate an unpopulated belt between the cattle areas and the game parks. Instead of now indiscriminately slaughtering all game, as was formerly done where intermixture of livestock and game is unavoidable, only known hosts of the fly are being killed off. This practice has been accomplished by collecting engorged flies which had fed in heavily infected areas in which all types of game were present. The blood from the fly was then examined for trypanosomes and the animal it had fed on determined from the blood grouping. This procedure led to the discovery that only two types of wild hog and two of the many varieties of antelope were the principal hosts of the tsetse fly. Killing off these species produced effective control within a comparatively short time.

This control measure of systematically hunting these particular wild hogs and antelopes was in active progress when Rhodesia declared her independence. Just south of the Zambian border, over a narrow belt 300 miles long, 1,000 African hunters, under the direction of 10 white supervisors, were hunting down these hosts of the tsetse fly with rifles and ammunition. This area is populated almost entirely by tribesmen. With the declaration of independence, as a precautionary measure the guns were taken from the hunters for a short time but were soon returned and the game hunting proceeded. There were no defections.

While both prophylactic and curative drugs effective against the type of trypanosomiasis prevalent in Rhodesia are available, elimination of the tsetse itself in areas where cattle are run or game is to be preserved is considered the only satisfactory control measure. It has been found that the trypanosome, the carrier of trypanosomiasis in the blood stream, very quickly becomes resistant to the drugs employed against it.

GOVERNMENT AND CATTLE

Government activities in marketing and disease prevention have been discussed under these headings. The Livestock Experimental Stations are the third major contribution of government to stock raising.

There is an active interest in bettering the quality and in increasing the quantity of beef production in Rhodesia. Much of the investigational work at the Matopos Research Station, near Bulawayo, has been directed toward obtaining an improved slaughter animal. Native cattle have been systematically crossed with European breeds and comparisons made with the Africander. Results have led to the conclusion that the indigenous cattle and the Africander are the most economical beef producers under the general conditions encountered in Rhodesia. Extensive tests carried through several generations at the station have shown that under the natural conditions which exist, the indigenous cattle will outdo the exotic breeds and their crosses on native cattle except for the first cross, with its hybrid vigor. The most popular British beef breeds—the Hereford, the Aberdeen Angus, the Sussex, and the Shorthorn—with good management and under carefully controlled conditions reach market weights earlier than the indigenous breeds, however. Because of the altitude, the climate of Rhodesia is equable enough to enable these British breeds to adapt to it, but they require a more rigid control of endemic disease and parasites and better nutrition than do the indigenous cattle.

Feeding experiments have shown an economic advantage in using some supplemental feed during the dry season. The usual practice is to carry cattle through the dry season (corresponding to winter in more temperate climates) on standing, matured grass which is low in protein. Supplemental feeding in 1965 was a revolutionary idea in an area in which cattle have always fed only on pasture. It was shown that market weights could be reached a year earlier by moderately increasing the nutritive level during the dry season.

An interesting result of these investigations was the differences found in the teeth of young exotic animals and of young indigenous cattle. At eleven months of age the milk teeth of a Nguni-type calf were larger and more firmly set and had a definitely better biting surface than those of a comparable Hereford calf. At two years of age the Nguni had a fairly usable set of teeth except for the first two teeth, which were gone. The Hereford still had a full mouth of small, widely spread stubs. At twenty-seven months both had two good front teeth and the remainder of the Nguni's milk teeth were still usable but the Hereford's milk teeth were completely gone. At forty-one months the Nguni had six well-formed teeth with good wearing surfaces, but the Hereford's were definitely inferior.

OUTLOOK FOR CATTLE

Judged by Western standards, Rhodesia is further advanced in the production of beef cattle than any other African country. More progress has

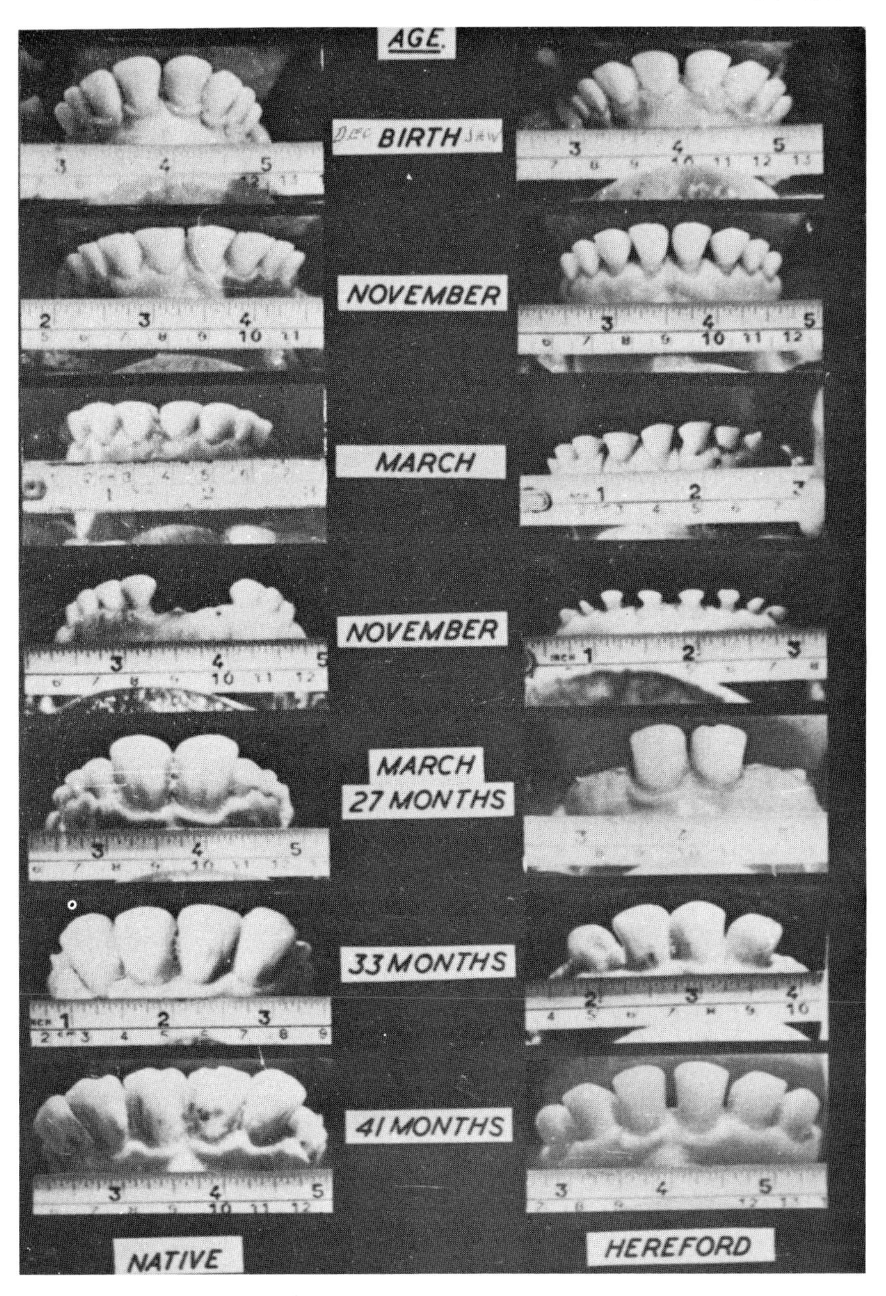

Teeth of Nguni and Hereford stock at different ages. Matapos Research Station. Courtesy J. D. G. Steenkamp

been made in getting the African to market his cattle as a result of the auctions established by the Cold Storage Commission. Grain feeding to finish cattle is rarely practical on the rest of the continent because what is grown is needed for human consumption. Finishing cattle on grain in Rhodesia is still practiced on a small scale but is increasing because of the pressure which, by controlling prices, the Cold Storage Commission can put on the grower to market younger cattle at better weights.

The future of Rhodesia's cattle operations can be visualized by considering western Nebraska and eastern Wyoming as a feeder-raising area and eastern Nebraska and western Iowa as the feeding belt where corn is grown. Climatic conditions would give about the same carrying capacity for cattle and grain yields as that part of the United States, which roughly equals Rhodesia in land area. Winter temperatures are much higher in Rhodesia, which is to her distinct advantage. The cattle population of this section of the United States, however, is nearly three times that of Rhodesia. The productivity of the Rhodesian soil permits the feeding of grain to livestock without depriving the human population, which is less than that in the part of the United States used for comparison. The major differences in cattle-raising conditions between the American and the Rhodesian regions are the milder winters in Rhodesia on one hand and, on the other, her far greater incidence of decimating cattle diseases. It has been demonstrated that the latter can be kept under control, however.

The largest potential for increased beef production is still a reasonable offtake from the herds of the African tribesmen. The European runs 42 per cent of the cattle in the country with an offtake of 16 per cent. The African has some 2,200,000 head, from which the offtake is less than 5 per cent. This is a high ratio for African-owned cattle and is increasing. If it reaches 10 per cent, the number of animals slaughtered annually would increase by one-third.

Physical capabilities, however, may not be the determining factor in the future of Rhodesia's cattle industry. In spite of the modern aspects of life in both town and country and the typically British atmosphere of law and order, in early 1965 the Rhodesian future could not be said to be assured. Very substantial progress had been made by 1969 in that the status quo had been quite successfully maintained in spite of the controversies with Britain and the United Nations. The political uncertainties involved, however, pose a different problem from those which hang over much of Africa. The conflict in Rhodesia is not between black men and white men but between the current British government and the former colonists. The African of Rhodesia enters the picture only as a political red herring. It is unthinkable that some middle way out of this situation will not be found.

Healthy progress in the cattle industry can be expected if there is no radical change in the political situation and Rhodesia's normal trade channels are again opened. If government falls into the hands of the small African minority now seeking it, however, the story will be sadly different.

South Africa

Land area (sq. mi.):	471,820
Population (1966):	18,300,000
Density (per sq. mi.):	39
Agricultural (44%) (1960):	8,040,000
Per capita income:	$463
Cattle population (1965):	12,000,000
Offtake (12.8%) (1965):	1,530,000
Year visited:	1965

THE REPUBLIC OF SOUTH AFRICA occupies the southern end of the continent, lying between the 22nd and 35th southern parallels, roughly the same latitudes south that encompass north Mexico and Texas in the Northern Hemisphere. South-West Africa, Botswana, and Rhodesia border South Africa on the north; the South Atlantic and Indian oceans surround the country on the west, south, and east. In the northeast there is a short frontier with Mozambique.

The high plateau region of East Africa extends across South Africa from north to south and falls off to narrow coastal plains on the Atlantic coastline in the extreme south and rises to the Drakensberg (Dragon) Mountains on the east. West of this escarpment the main drainage is to the west by the Orange River, while east it is by short, fast rivers to the Indian Ocean. The Limpopo River flows through northern Transvaal around the Drakensberg Mountains to the Indian Ocean.

The climate on the coastal plains is subtropical. On the high plateau it is more temperate, with some frost in the winter at the higher elevations. Precipitation increases from a semiarid area in the west to 30 inches over much of the plateau area and 40 inches and above on the narrow eastern coastal plain. The land area is slightly larger than that of Texas, Oklahoma, and New Mexico combined.

The Portuguese were the first on the coast, and although they quickly established themselves in what is now Angola on the west and Mozambique on the east, they used the Cape of Good Hope as only an occasional port of call and made no attempt to settle any part of what is now South Africa. In 1652 the Dutch established a supply station, the first permanent white settlement, on the Cape, used by their ships traveling to India and by the middle of the seventeenth century had started settlement in a small way around the Cape area. These early settlers, by force of circumstances, developed into a rugged people, holding off the Hottentots pretty much on their own, receiving only spasmodic help of any kind from their homeland. By the end of the eighteenth century there were approximately 20,000 of these settlers. The British took the Cape in 1795, moved out in 1803, but came back in force three years later to stay. Active settlement was soon started and, as the British moved in, some of the descendants of the original Dutch (who by this time had become the Afrikaners) began to move out onto the veld. Antipathy between these two European nationalities continued until it culminated in the Boer War in 1899. After that was over four provinces —Cape Province, Natal, the Orange Free State, and the Transvaal— emerged. The Union of South Africa was established in 1910, obtained dominion status in 1931, and in 1961 withdrew from the British Commonwealth and became the Republic of South Africa.

The Hottentots, who, along with the Bushmen, were the first inhabitants of South Africa, are non-Negroid. They were the Africans whom the Dutch settlers first encountered and have now largely disappeared. In 1960 the number of Hottentots was estimated at 55,000 and the true Bushmen at 25,000. The "Cape Colored" people, often referred to as "Colored," are the descendants of the admixture of the early settlers with the native Africans as well as with slaves who were later introduced from Madagascar and the East Indies. The Bantu tribes encountered as the settlers moved northward account for the majority of the native African population today. The white population is predominantly of Dutch or British ancestry and in 1960 accounted for nearly 20 per cent of the total population of South Africa; the Bantu comprise nearly 70 per cent of the population, and the Colored and the Asians (largely Indian) make up the remainder, the Colored outnumbering the Asians three to one.

In spite of South Africa's close and continuous contact with the Western world, because of apartheid she is probably the least understood of any African country. The late Hendrik Frensch Verwoerd, the foremost exponent of apartheid and long the country's prime minister, defined the policy as "political independence (of white and black) coupled with economic interdependence (of both)." The basic idea is to separate South Africa into a commonwealth of politically independent black states and independent white states with close and mutually beneficial economic ties. One such black state, the Transkei, is already so functioning. In the light of the confusion that has followed more immediate forms of independence in Africa, this orderly, long-range policy seems worth a trial. South Africa evidences a stability and security not known in many other countries on the continent.

The Bantu Administration and Development Department of the national government, with ministerial rank, administers all activities involving the care and development of the native Africans throughout South Africa. It has a grass-roots approach to the innumerable problems involved in providing reasonable living conditions for people who knew only a stone-age culture a generation or two ago. The native cattle of South Africa are run much as they have always been by the Bantu people in the large areas that have been established for their exclusive habitation. In this respect, the Bantu have been treated much better than natives are usually treated when a colonizing people moves in. The area that is now South Africa was not heavily populated when the white man appeared on the scene. African tribes were still on the move—and in some areas who got there first, the trekker or the Bantu, is an open question. The tribes have been assigned good land, comparable with that which the white South African has for his own use. The tribesmen have not been pushed out to marginal lands as were the American Indians. The Bantu are given all the help they will accept in husbanding their cattle—dipping vats properly located, free prophylactic treatment for disease, marketing facilities, and demonstration and instruction in elementary management methods. A gradual improvement in their handling of cattle is being accomplished. The dipping vats for tick control are now widely used. The number of cattle being marketed is steadily increasing as wants for the articles of civilization increase.

CATTLE BREEDS

As in many other parts of the continent, breeds or types of native cattle in South Africa are known by the locality in which they are found or by the tribe that owns them. It would appear that there have been four major migrations of cattle-owning peoples into southern Africa—the Hottentots down the west coast, the Nguni down the eastern side, the Basutos (whose cattle have been developed into the Drakensberger breed) down the center, and the Bechuana down the west side, between the routes traveled by the Hottentots and the Basutos. The Bechuana cattle have now been bred out by admixture with the Africander and exotic breeds.[1]

[1] Bisschop, personal communication.

Africander bull.

Africander cow in foreground.

Africander.—The most important commercial breed in South Africa is the indigenous Africander. This is a large breed; a mature cow weighs 1,400 pounds and a bull weighing as much as 2,000 pounds. The color commonly seen is solid red, a dark shade currently being preferred, with only minor white markings on the underline permitted. Light red to golden yellow, yellow, and even grey are known in the breed, however.

Horns characteristically extend laterally from the head, nearly straight but with a forward curve and a slight upturn toward the end; they have an oval cross-section and usually are twisted. There is a well-developed neck hump, larger in the male. The rump has a marked slope and the hindquarters are rather light considering the size of the animal. The dewlap is quite large, but the sheath is not as marked or pendulous as in many Zebu types. The cows are poor milkers although they produce sufficient quantity to raise good calves. With good management and nutrition, calf crops of 80 per cent or more are obtained but average about 65 per cent under the usual ranch conditions with a four-month or longer breeding season. The backbone of the South African cattle industry, this breed is found almost exclusively in the hands of European growers.

Two diverse opinions about the origin of the Africander are discussed in the introductory chapter on Africa. A third theory holds that the Alentejana breed of Portugal was introduced by the early Portuguese who reached the Cape and that these animals, intermixing with the cattle of the Hottentots, produced the Africander. These early Portuguese, however, were not colonizers here, being largely refugees from ships wrecked rounding the Cape of Good Hope. It is highly improbable that they introduced sufficient numbers of their native cattle to have had any effect on those of the Hottentots, or that they even could have done so in the small ships at their command for the cruise that required months to reach the Cape. Furthermore, the Alentejana cattle seen in Portugal today have no characteristics that suggest the Africander except the red color, and any number of unrelated breeds are red. The question could also be raised whether the Alentejana in its present form even existed in Portugal nearly 500 years ago.

It is well established that the Hottentots were occupying the southern tip of Africa and had cattle in large numbers when the Portuguese first arrived there. It is known that later, when the Dutch started colonization in earnest, they made use of the Hottentot cattle for draft purposes. No mention is made of their having brought cattle with them, and being agriculturalists they would have been more inclined and in better position to do so than would the Portuguese. The natural course would have been for the Dutch to have used and developed the Hottentot cattle for their own purposes, and there is much evidence that this is what they did. By selection for draft ability, and much later for fleshing properties, Dutch farmers produced the forerunner of the present Africander, which can certainly be considered an indigenous breed, highly improved by selection.

Nguni.—This indigenous Sanga breed, seen in northern Natal on the east coast of South Africa, is the same as the Landim of adjoining Mozambique. The horns are upswept, vary considerably in length, and are shorter and thicker on the bull. The lyre shape is characteristic; when viewed from the side the horn has a slight curve forward, then backward, and again forward toward the tips. The hump is rather small on the male, often not discernible on the female.

To a casual observer the color of Nguni cattle might be termed widely

Nguni bulls. Photographed at Bartlow Combine.

Nguni cow. Photographed at Bartlow Combine.

varied. The numerous patterns seen are often the result of selection by chiefs for many generations. In a personal communication from Dr. J. H. R. Bisschop they are described as follows:

"The color pattern of all Nguni cattle is a complex but very definite one—a pattern which is found in indigenous cattle throughout the world. It includes black, red, brown, dun and yellow as whole colors. Albinos do occur but are rare. The basal pattern is white with color markings (of any of the colors mentioned) on the muzzle, around the eyes, on the ears, and as spots on the lower front legs. From this basal pattern, color builds up, first as small color specks to bigger color spots on the front limbs and neck and backwards onto the sides of the body; next to full color panels over the ribs, but still with color spots on the neck and limbs; and finally to full color panels including the neck, sides of body, and the upper parts of the limbs or even the whole of the limbs, but always with a white top and belly line, and a white face with color specks. The basal color of the muzzle, eyes and ears persists. The Bantu have specific names for all the whole colors and for the many gradations of the color pattern, which is known as the 'Nkoni' pattern."

The breed has been under selection improvement at Bartlow Combine, in Natal, by the Bantu Administration for twenty years. Improved cows average 1,000 pounds in weight, with individuals reaching 1,400 pounds. Bulls average 1,450 pounds, with aged individuals weighing as much as 1,650 pounds. There is a polled tendency, and in a herd of any size a number of polled individuals are always seen. The cows are good milkers for an African breed, individuals yielding as much as 4,500 pounds a lactation. The Nguni has also been developed by selective breeding at the Mpisi Animal Breeding Station of the Swaziland Administration.

Bapedi.—The improved Bapedi, as developed from the indigenous native breed at the Stellenbosch Farm of the Bantu Administration in the Transvaal, is a medium-sized animal. It is an offshoot of the Nguni, and, therefore, a Sanga descendant. Cows average 1,000 pounds, with individuals up to 1,100 pounds. Many Bapedi cattle are a mottled grey; there are also blacks with white faces and blacks and reds with a white topline and bottomline. Solid reds and nearly solid whites also exist. Although there is this variation in color, conformation is quite uniform and does not indicate dilution by any exotic blood. Formerly, hides were widely used for warriors' shields. Color patterns were selected so that the shields of the *impis* (regiments) of a tribe could be distinguished

Bapedi bull.

Bapedi cows.

from each other, not to distinguish between friend and foe in battle as is sometimes heard.[2]

The manner in which the Stellenbosch herd was originally selected and subsequently maintained—buying the best individuals that the African owners could be induced to sell and then maintaining these bloodlines intact—has resulted in as pure an example of the breed as could be obtained. The bone structure is rather light, and the body is well muscled with a fairly parallel topline and bottomline. The bull has a small but definite hump; the female frequently shows none at all. Very thin, wide, upswept horns of round cross section are characteristic. Like many other indigenous African cattle, the Bapedi breed exhibits a strong tendency toward being polled. It fleshes well if on good pasture. Natural selection has developed it into a hardy breed, highly tolerant of most African cattle diseases.

Drakensberger.—This breed has been developed by selective breeding from the cattle of the Basuto people. It is considered by the Drakensberger Breed Society to be a "dual purpose breed (milk and beef) with good draft abilities" and has a good beef conformation. The improved type is solid black in color except that a small underline of white is permitted. The horns are of medium size, outward thrust and then turned upward, and are white with black tips. The hump is small and located in front of the withers. The breed does well in the sour-veld regions of South Africa; it is the only indigenous breed which has been developed in those areas.

Basuto cow from which the Drakensberger breed was developed. Courtesy J. H. R. Bisschop

Bavenda.—In the northeast corner of the Transvaal, just east of Kruger National Game Park, the Bavenda tribe of the Bantu has a small breed

[2] Bisschop, personal communication.

Bavenda bull.

of cattle known, after it, as the Bavenda. The breed, also an offshoot of the Nguni, is a Sanga type. Black predominates in coloring, black with white markings is next, and red and red with white markings follow. The hump in the male is usually no larger than the seasonal swelling in the neck of a nonhumped bull during the mating season and is barely distinguishable in the female and the ox.

Since long before the white man came to Africa, the Bavenda breed has been husbanded in an area unfavorable for cattle, a nearly tropical climate where the rainfall ranges from less than 20 inches annually to 80 inches in the mountainous regions. The animals are herded on common ground and corralled at night. Natural selection has developed a

Bavenda cow in foreground.

hardy and disease-tolerant type. It is slow maturing, cows usually calving at four years of age. The badly overgrazed condition of the lands available for pasture and the lack of grazing control result in poor nutrition, which certainly has its effect in causing late calving. These conditions are probably the cause of the Bavenda being a smaller and less produc-

Young men of the Bavenda tribe on their cow mounts.

tive breed than the Nguni and the Bapedi. Cows are poor milkers but are exceptionally fertile. Mortality of young calves is very high, however. Bulls run with the cows throughout the year; some are castrated, not as a breeding control measure, but because the oxen are wanted for draft. Both cows and oxen are so utilized in farming.

The horns of the Bavenda are of good size, upswept and usually turned backward at the tips. There is a strong polled tendency. In a random count of more than 3,000 head of native cattle, 16 per cent were naturally polled.

A herder is sometimes mounted on a cow or an ox with a light rope through the center membrane of its nose for control. Young men ride cows as a pastime.

Bonsmara.—Numerous efforts have been made over all of Africa to upgrade or improve native cattle by crossing them with exotic breeds, usually without producing any outstanding results. Many of these attempts were abandoned because the second and third crosses were invariably inferior to the original native animal. An exception to this is the Bonsmara, which can now be called a fixed breed. In 1947 the Mara Experiment Station of the Department of Agricultural Technical Services, in the northern Transvaal, started developing both a Hereford and a Shorthorn cross on the Africander, using bulls of each breed on separate cow groups. These two crosses—the five-eighths Africander—

Bonsmara bull. Photographed at Mara Experiment Station.

Bonsmara cow. Photographed at Mara Experiment Station.

three-eighths Hereford and the five-eighths Africander—three-eighths Shorthorn—were at first maintained as two separate lines. Both crosses were called Bonsmara for the director, Jan C. Bonsma, and the Mara station, where the program was initiated. Later the Hereford and the Shorthorn crosses were indiscriminately mixed. The composition of the Bonsmara is now considered to be three-sixteenths Hereford, three-sixteenths Shorthorn, and five-eighths Africander.

Heifers from the fourth generation of the Bonsmara breed were at breeding age in 1965 and show a definite improvement in fleshing characteristics over the original Africander. The sloping rump has been flattened to some extent and the hump reduced in size, external evidence of it being practically gone in the female.

Some commercial growers have developed a stable cross in their own breeding programs, usually by using Hereford bulls. The size of the Africander has been maintained and earlier maturity has been developed. Steers of a three-eighths Hereford–five-eighths Africander cross reach market weights of 1,050 pounds three weeks earlier than do pure Africanders on parallel feeding programs.

A Bonsmara cattle breeders' society has been formed to foster the further development of the breed, particularly by using Bonsmara bulls for upgrading purposes.

Exotic Breeds.—Many of the beef and milk breeds of Europe and also the American Brahman and Santa Gertrudis of the United States have been introduced to South Africa over the years. There have been numerous crossing experiments, but the only noteworthy outcome of these has been the establishment of the Bonsmara herd.

The national dairy herd consists entirely of the European breeds, with the Friesian predominating. These animals have largely been maintained pure, although there has been some upgrading to the point at which the resulting stock is practically a pure representation of the breed of bulls used.

The most important foreign breed is the Hereford, although there is a fair representation of the Shorthorn, Sussex, Red Poll, and Simmental. These breeds have generally been maintained without admixture with native cattle except for experimental breeding. Many breeders maintain purebred stocks of these breeds.

Although these European breeds have been satisfactorily maintained, South Africa must be considered a marginal environment for them, at least until many generations have elapsed, because of the heat stress, parasites, and endemic disease there. More care is required in prophylactic treatment, dipping, and disease-control measures generally.

MANAGEMENT PRACTICES

The Bantu runs his cattle as the main object in life. The use he makes of them and his methods of handling vary with the distance of his kraal from the centers of population. He utilizes the milk for his family; if he belongs to one of the tribes that grow crops, he uses cows and oxen for draft in farming. All cattle are kraaled at night for protection, and the calves are cut off so that the morning's milk can be taken. They are kept in until quite late in the morning. Opinions differ about the reason for this: it may be because of sheer laziness regarding the milking or because of the belief that dew on the grass is harmful. This practice of nightly corralling leads to serious overgrazing in the areas surrounding the villages. Most native cattle are undernourished in varying degree. When the animals are on pasture, herdboys are in constant attendance. Animals that die of natural cause are eaten. On festive occasions, now mainly for weddings, an animal is slaughtered for consumption. These customs show some gradual change as the creeping effects of a money economy advance. Thus, a celebration may now mean the killing of a goat instead of a bull, since the monetary value of cattle is realized. Money is sometimes even used for lobola, rather than giving the customary 10 or 12 head of cattle.

The grazing lands of the Transvaal are classified as "sweet" and "sour" veld, both types appearing in some areas. In general the sweet- and mixed-veld areas are in the west, where there is an annual precipitation between 10 and 18 inches. The native grasses retain a good nutritive value when mature, and cattle winter well. The sour-veld area is in the east, where the rainfall is more than 25 inches, and the native grasses when mature are low in nutritive value.

In the sweet-veld regions of the country, across the boundary lines of the settlement areas where the Bantu runs his herds, the European grower follows modern practices in cattle raising. His land is fenced and cross-fenced for grazing control and for the separation of his breeding herd and the animals being grown out. It is common practice over much of the Transvaal to have two breeding seasons—ninety days from December through February and seventy-five days from August until October 15. In well-managed operations cows are run in age groups, using 3 per cent bulls. Heifers are bred as two-year-olds. Pastures are rotated to obtain optimum usage, and dry growth is left standing for the winter months. Hay is seldom put up.

The minimum economical unit in these areas of the Transvaal is 10,000 acres, which will run 200 to 250 mother cows and permit the marketing of three- and four-year-old steers off grass. Pastoral pursuits are now in the hands of the third or fourth generation of the original settlers, and many farmers are now in the 2,000-acre bracket. This situation has led to cash-crop cultivation of land that should have been left in grass in the 15- to 20-inch rainfall area. A drought of several years' duration was not over in 1966. Such circumstances have hurt the cattle industry seriously. In general it would appear that a movement to larger land units and the conversion of marginal cultivated land to pasture will be the future trend in cattle raising in this part of South Africa.

In these sweet-veld areas there is as yet very little grain feeding to finish cattle. A few growers finish Africander and Bonsmara steers on corn and obtain shipping weights of 1,000 pounds in twenty months. This practice is growing in what is known as the maize triangle. Now most cattle for slaughter are shipped off grass—in the exceptional operation at two and one-half to three years of age, but more generally at three to four years.

The Africander is the principal breed of the European grower although there are good herds of British beef breeds, in which the Hereford predominates, in limited areas of better rainfall at the higher elevations. The climate there is equable enough for these exotic breeds to do well under good management, which requires frequent dipping for ticks.

Artificial insemination has not gained much of a foothold in the ranching areas of South Africa. Efforts of the Department of Agriculture to introduce the practice have been practically abandoned. Some large breeders are working with it in their herds, and it is employed in only 3.5 per cent of the nation's dairy herd. The general practice in running cattle is not well adapted to the use of artificial breeding. For the Bantu, who still have 40 per cent of the cattle in the country, it certainly is not applicable. Even if their cultural attitude could be overcome, the procedure would be impractical with their small herds in isolated locations. Neither is it very practical for many of the European growers. The small operator runs a few hundred head on large pastures under conditions where detection and breeding would be difficult and costly. The large breeder usually has his cattle too widely scattered and is accustomed to two breeding seasons or to one long one. To go to artificial insemination would involve some drastic changes in his operation.

The Bushveld Bull Breeders Society has recently established a thoroughly modern, well-operated bull-testing station, at Potgietersrust. The first bull sale was held in 1965, the performance record of each bull sold being given. On the first test group, which included approximately 60 bulls at two to two and one-half years of age, belonging to thirteen breeders, gains varied from one and one-half pounds to slightly over three pounds a day. Feed conversion, pounds of feed consumed for each pound of gain the animal made, varied from six and one-half to twelve. This was the second bull-testing station in the country, the first being near Pretoria, at the Research Institute for Animal Husbandry and Dairying, which can accommodate 450 bulls annually.

Some of the work done on experiment farms, both those of the Ministry of Agricultural Technical Services and the ones under the Department of Bantu Administration and Development, is exceptional. At Bartlow Combine of the Bantu Administration, the progeny-testing program has produced some remarkable results in a herd of 1,000 Nguni cows. Calves are weighed periodically until they are weaned at eight months, then at twelve and twenty-four months. Breeding stock is selected on a weighted basis utilizing all of these data. Milk production of every cow is obtained by weighing her calf before and after sucking, twice a day, every other day. The increased milk yields of some individuals have now reached the point at which the cow must be stripped after sucking and the weight of the milk so obtained included in her production. The method of obtaining the quantity of milk produced by weighing the calf is very efficiently handled. One hundred fifty cows and the same number of calves are worked in a unit, the cows being identified by brand, the calves by ear tag. The cows enter from the outside into

Arrangement for obtaining milk production by weighing calves after sucking. Photographed at Bartlow Combine.

a double row of S-shaped pens. After they are all in place, the calves are run in from an inside aisle and find and enter the other side of the S their mother is in. The before and after weights of all the calves are obtained in forty-five minutes.

Cows are graded on their milk production so obtained. Individuals have been developed that now produce 4,500 pounds of milk during a 240-day lactation. The herd average is 3,000 pounds. Milk yield is one of four criteria on which brood cows are selected and graded; the others are the rate of growth of the cow until it reaches two years of age; the rate of growth of her calf as determined by a compilation of the weights at eight, twelve, and twenty-four months; and the individual's general conformation. Final selection of the animals to be retained in the breeding herd and those to be sold to the Zulu cattle owners are made at two years of age. About 75 per cent of the cattle reaching this age are culled. Six lines of bulls are now being used.

The average Bartlow cow now weighs 850 pounds, and the average bull 1,450 pounds, compared to weights in 1950 of 750 pounds for the cow and 1,100 pounds for the bull. Not all of this increase can be attributed to genetic improvement, for better management practices, which have meant better nutrition, have also had their effect. Improvement more directly attributable to selection is seen in the increase in weaning weights from 1958 to 1964: the average weaning weight of eight-month-old Nguni heifer calves increased from 323 to 350 pounds; at the same age, Nguni bulls increased from 346 to 391 pounds.

MARKETING

Marketing of cattle as well as other livestock for slaughter is a remarkably orderly procedure. The Livestock and Meat Control Board sets a floor price for each of the five grades of dressed cattle carcasses. These prices are established one year in advance for the nine large urban areas, where 60 per cent of the cattle marketed in the country are slaughtered in abattoirs operated by the municipalities. This factually fixes cattle prices for an agent at one of the abattoirs, who handles the business from there on. The dressed carcasses of each owner are identified by tag. The morning's kill is auctioned as soon as carcasses are on the rail, being sold to the highest bidder as the butcher buyers follow the auctioneer down the line. The afternoon's kill goes to the cooler and is similarly sold the following morning. If bids above the floor price set by the Board are not received, that organization automatically becomes the buyer. The grower is paid exactly what his animals sold for plus an allowance on a weight basis for hides and offals and less a levy, which in 1965 was one-fourth of a cent a pound. The board buys very little beef (in 1965 only 1.7 per cent) at the floor prices.

Grading is done by employees of the Board. There are five recognized grades: super, which includes about 10 per cent of the kill in Johannesburg; prime, 20 per cent; and grades 1, 2, and 3 combined, 70 per cent. Very few young animals and practically none that have been fed grain except the throwoff from the livestock shows are killed. The super grade would be roughly equivalent to USDA low-choice; the prime grade, middle-good. Prices are low, although they have been steadily increasing in recent years. In 1964 the price on the cold-dressed-weight basis for prime grade was about 20 cents a pound and for No. 1 grade was 16 cents a pound. These amounts are equivalent to 12 cents a pound for prime and 9 cents for No. 1 grade on a liveweight basis.

ABATTOIRS

Modern plants have been built in Cape Town and Port Elizabeth, but the facilities in the seven other controlled areas can be considered only fair. The Johannesburg abattoir is killing as many as 1,800 head of cattle daily in facilities built thirty years ago for 500 head. In spite of this situation, sanitation and inspection are good. Even though the city has expanded so that it now surrounds the abattoir on all sides, good paved yards under shed roofs are provided for holding stock, which comes in by both rail and truck, as it is received for slaughter. Plans are under way to build a modern plant outside the city.

The new abattoir recently completed at Cape Town includes a "clock auction," patterned after the flower bulb auctions in Holland. The buyers sit at desks so they can see both the carcasses as they are brought in on the rail and a large clock whose hand points to a circular price scale.

Each desk is equipped with a push button which electrically stops the clock and records both the price registered at the instant the button is pushed and the desk from which the signal came. This finalizes the sale, which is automatically recorded, and action proceeds without any cessation to the next carcass. The clock is started at a price somewhat above that which will be offered and then moves slowly downward. Any buyer can buy at the price the clock indicates at the moment by pushing the button. The action is fast and orderly.

CATTLE DISEASES

Disease control is systematically organized, and rigid prophylactic measures are the general practice among the European growers. Rinderpest and pleuropneumonia were eradicated at the turn of the century. Foot-and-mouth outbreaks occur sporadically and are successfully combated by slaughter. A special institute is planned for the preparation of vaccines against the local types of the disease. South Africa has been freed of the tsetse fly.

The rinderpest outbreak in 1896 decimated both the cattle and the game populations, completely eliminating both over large areas. Because the tsetse fly can live only on animal blood, it was eliminated when the cattle and game died and has not been a serious menace since. Comparatively small reinfestations of the fly subsequently occurred in some areas of Natal, totaling about 7,000 square miles. These areas were cleaned up by aerial spraying with DDT at a total cost of $1,000 a square mile, excluding the cost of airplanes, which were supplied by the army. The operation was completed in three years. The area involved was acacia veld, and only the brush regions were treated. The game present in sizable numbers was not removed, contradicting the theory held throughout much of Africa that all the game must be killed to attain tsetse control.

One of the most serious cattle diseases is the tick-borne heartwater. The standard treatment practiced by the better grower is to inoculate calves at up to three weeks of age with an injection of infected sheep blood so that they actually get the disease and, with their calfhood resistance, recover without further treatment. Animals so treated carry a high degree of tolerance for life but are not actually immune. Periodic dipping of all animals is necessary for effective control. Older cattle, especially of exotic breeds, are inoculated if they have not been subjected to the calfhood treatment, then corralled and their temperature taken morning and night. Any marked increase in temperature is the signal for heavy antibiotic treatment, which effects a cure in 98 per cent of the cattle so treated; they are then left with a high degree of tolerance to the disease. Complete elimination of the carrier tick is not desired, and the general practice of dipping the animals is an attempt to keep the tick population to a desirable minimum so that the resistance in the animal that has resulted from inoculation will continue. Dipping for tick control is common, varying with the season, location, and type of tick. Once a month may be considered adequate during the summer in the Transvaal, while every five days is the practice in the more heavily infested Natal area.

Prophylactic treatment with cultured vaccine is not compulsory for blackleg and is required for anthrax only in proclaimed areas. Replacement heifers are vaccinated with Strain 19 for brucellosis when they are less than eight months old.

GOVERNMENT AND CATTLE

The government plays a major role in all phases of cattle operations. The Department of Agricultural Technical Services is concerned with those phases which primarily concern the European grower. The Veterinary Services under this department are charged with the control of cattle diseases and are responsible for the Onderstepoort Veterinary Research Laboratory, a modern institution devoted to everything from the preparation of vaccines to basic research. The experiment stations, also under this department, are well managed and do outstanding work.

The Department of Bantu Administration and Development, which is responsible for all activities involving the native African, exercises control over everything that involves his cattle, including the administration of animal-breeding stations devoted to the improvement of indigenous cattle. Some of these stations also have Africander herds. The administration endeavors to orient the native African cattle owner to a money economy without forcing the issue. Like most Africans in their natural habitat, the South African Bantu must have cattle: they are his business, his wealth, and a form of security. None of the varying degrees of civilization to which he has been subjected have changed this custom except in the case of the minority who live in town. If the Bantu's cattle and his management of them can be improved, he will have a better chance of adapting himself to the changes that come as civilization continues to encroach on his way of life. This is the approach which the Bantu Administration seems to be taking. The administration's experimental farms in the reserve areas are devoted to improving indigenous breeds so that better bulls can be supplied the African cattle owners and also to serving as training and demonstration centers. Significant results follow simple improvements in management practices such as allowing only selected bulls to run with the cows for breeding, using better grazing methods, and dipping regularly for parasite control.

OUTLOOK FOR CATTLE

Cattle raising is further advanced in South Africa than in other countries on the continent, with the possible exception of Rhodesia. This progress is especially evident in the European operation, to a lesser extent in that of the African. There is still room for expansion, however. The total cattle population is 12,000,000 head, as compared with 15,000,000 head in Texas, Oklahoma, and New Mexico, a slightly smaller area. Sixty per cent of the cattle population is in the hands of European growers; the Bantu have the remaining forty per cent. The annual offtake of 1,530,000 head comes largely from the European herds, while that from the Bantu herds is relatively small. Both the European and the Bantu have better cattle than are found in most other parts of Africa. The country has been settled for 100 years longer than the rest of Africa south of the Sahara Desert; and this longer establishment, along with the stability in government, accounts for the healthy cattle industry.

South-West Africa

Land area (sq. mi.):	318,000
Population (1966):	610,000
Density (per sq. mi.):	2
Agricultural (59%) (1966):	360,000
Cattle population (1966):	2,550,000
Offtake (10.8%) (1966):	275,000
Year visited:	1967

THE MANDATE OF SOUTH-WEST AFRICA is a real frontier country and one of the most peaceful and pleasant spots on the continent. Angola is on the north, Botswana on the east, South Africa on the south, and the Atlantic Ocean on the west. In the northeast corner lies the Caprivi Strip, 40 miles wide and about 300 miles long, which extends to the Zambezi River and a border with Zambia. Except for the narrow coastal desert on the west, all of South-West Africa is a major plateau, 3,000 to 5,000 feet above sea level, with a temperate climate.

Wherever the European has settled, the amenities of civilization are in evidence. The inhabitants run the full scale of mankind. Some of the 18,000 Bushmen, who live in the east, still live off the game they kill with poisoned arrows and on the roots and plants they gather, have no fixed abode, and wear only loincloths. Tribes toward the north exhibit varying degrees of African culture, ranging from the basically nomadic Ovahimba and Ovatjimba tribes to the Ovambo, a settled agricultural and cattle-raising people and one of the most advanced tribes, in the light of Western standards, in all of Africa. The nonwhite population is 514,000; the European is 96,000, of which 65 per cent are South Africans, 30 per cent are of German descent, and 5 per cent are English.

Although the Portuguese landed on the coast of South-West Africa seven years before Columbus reached the Western Hemisphere, the white man made no substantial inroads in the country until late in the nineteenth century, when the Germans moved in, subdued the continually warring tribes, and started colonization in what became German West Africa. During World War I the area was occupied by the army of the Union of South Africa, and at the end of the war the League of

Nations mandated the area to that country. Then followed a rapid influx of white South Africans. They and the descendants of the original German settlers have within the past forty-five years developed stock raising in the country until it is considered to be approaching the maximum carrying capacity.

The land area nearly equals that of the combined states of New Mexico, Arizona, and Nevada. The human population of these three states is five times that of South-West Africa, but the cattle number is nearly the same. The arid climate is the limiting factor in agriculture, and the raising of sheep and cattle is the major agricultural pursuit. A desert strip runs 900 miles along the Atlantic coast, extending inland as much as eighty miles. The desert strip is the product of climatic conditions similar to those on the western coast of South America—a cold offshore ocean current which eliminates the possibility of westerly winds depositing moisture on the hot land. The rainfall over the rest of the country varies from 2 inches in the extreme south to 20 inches in the north, all brought by northeasterly winds from Central Africa. The drier southern half is sheep country; the northern half is utilized mainly for cattle. The only continuously flowing rivers are the Kunene and the Okovango, along the Angola border in the north, and the Orange River, which defines the border with South Africa in the south. All other rivers in South-West Africa are nothing but sand beds during the long dry season. There is very little cropping except for the areas irrigated mostly with underground water. Making the most of what nature provides, a highly specialized Karakul sheep industry has been developed by the pastoralists in the south, with skins averaging $8.00 on the London market. In 1966 the top price for a stud ram was $15,000.

Below the Angola border an area which stretches for 600 miles all the way across the country and extends for an average of 80 miles southward was designated by the League of Nations mandate to be held in perpetuity as tribal lands for the native Africans then inhabiting it. This region was the traditional homeland of these peoples and reserving it for their use excluded the white man from settling there. This requirement has been rigidly adhered to and even enlarged in its administration by South Africa. Below these native reserves in a 14- to 18-inch rainfall belt that extends to the center of the country is the cattle-raising area.

Although many European breeds have been introduced and do well under good management in the warm, dry climate, the Africander is the base of the cattle industry. Cattle of the Europeans are raised largely for beef, although a sizable dairy industry supplying local demand utilizes mainly the Swiss Brown and the Simmental breeds. The cattle of the

Kaokoveld cattle in a badly overgrazed area near a watering place.

Large ox of the Kaokoveld, probably about eight years old, 900 pounds.

Africans, who have only half as many as the Europeans, are of little commercial importance because of the cultural reluctance to sell them.

CATTLE BREEDS

Some of the purest types of indigenous cattle to be found in Africa are seen in the native reserves in the north. These Sanga descendants were brought in as African tribes began migrating to the area several centuries ago. The two types seen are similar in all characteristics except

Typical Kaokoveld cow in poor condition at the end of the dry season. Weight, 700 pounds.

size and undoubtedly had the same origin. They are referred to here by their local names.

Kaokoveld.—These cattle are seen in the mountainous Kaokoveld reserve in the northwestern part of the mandate. They are of medium size, a mature cow in fair condition weighing as much as 800 pounds, a bull 1,000 pounds. They are multicolored, a characteristic typical of the Sanga in much of Africa. The animals are red, black, or mottled white and red or white and black, with a white topline and bottomline. Horns are usually large, up to three feet in length, and are of two distinct types: they are either upswept and wide apart in a continuous curve or they are distinctly lyre-shaped. Although the Kaokoveld is definitely a humped type, the hump is characteristically small—hardly discernible in the female and in the bull not much larger than the normal sexual swelling in a bull of a nonhumped breed. These cattle are run mainly by Herero and seminomadic Ovahimba tribes, who follow the grass with the seasons.

Ovambo.—Adjoining the Kaokoveld on the east, the Ovamboland re-

Aged Ovambo bull, 500 pounds.

serve is occupied by Ovambo tribesmen, who run well-developed farming operations. The Ovambo cattle are identical with those in the Kaokoveld in all respects except size. They are extremely small animals, a mature cow frequently not weighing more than 350 pounds and a bull not more than 500 pounds.

The reason for the small size of the Ovambo cattle is a matter of much speculation. There can be no question but that they are genetically the same as those animals in the neighboring Kaokoveld. The cattle seen in Ovamboland on the Kaokoveld border are practically of the same stature as in the Kaokoveld itself but decrease rapidly in size to the east. Offspring of Kaokoveld cattle brought into Ovamboland revert to the small size of the cattle native to that region in a few generations.

In the Okavango territory, which borders Ovamboland on the east, the size of the indigenous cattle again increases. There is some opinion that a mineral deficiency may be the cause of the smallness of the Ovambo cattle, but no experimental work has been done to demonstrate this. The soils of South-West Africa are seriously deficient in phosphorus, and vaccination against botulism is common on European farms.

Mature Ovambo cow, 350 pounds.

Africander.—This is the most important breed in the operations of the European farmer. Some of these cattle were brought into South-West Africa during the German occupation, but the important importations came when the settlers from South Africa began to move in after World War I. The breed now outnumbers the European breeds originally intro-

A "captain's" (headman's) bull in Ovamboland, near the Kaokoveld border. Weight, 1,050 pounds.

Africander stud bull, 1,900 pounds.

Purebred Africander cow, seventeen years old, 1,700 pounds.

Polled Africander cow. Photographed at Omatjenne Experiment Station.

duced by the Germans. The pure Africander of South-West Africa is identical with the breed as seen over the rest of Southern Africa. There have been some noteworthy accomplishments to improve the breed for South-West African conditions, however.

A polled strain has been developed at the Omatjenne Experiment Station, in the middle of the cattle belt. A polled bull which was born on a private farm and introduced to the station twenty-five years ago was the start of the selection program for the polled type, and all the polled cattle now trace to this sire. The degree of inbreeding necessary to maintain the polled characteristic has resulted in a somewhat smaller animal than the conventional Africander. Selection was also made for a light-tan color to alleviate to some extent the effects of the strong sunlight, which is unrelieved for months during the dry season. In 1966, 93 per cent of the calves dropped were polled and 73 per cent had the preferred light color.

In addition to equaling the original type in conformation, the Polled Africander has the added advantages of being a light color and having no horns to cause damage when shipped long distances to market.

Holmonger.—This name has been given to a new breed, which has been under development at Omatjenne since 1949. Composition has been fixed at one-half Africander and one-half Swiss Brown and has resulted

Holmonger bull. Photographed at Omatjenne Experiment Station.

Holmonger cow. Photographed at Omatjenne Experiment Station.

in an excellent beef type with a smaller hump and a better rump than the Africander. It is also said to be a faster maturing animal.

While the European breeds do well in South-West Africa, they are more susceptible to heat stress in the hot summers and less tolerant of external parasites and endemic diseases than indigenous cattle. Comparative rate-of-gain tests over a three-year period under carefully controlled conditions were run on nine European breeds and the Africander. The Simmental and the Swiss Brown ranked the highest and their results were practically identical. The Swiss Brown was chosen as one of the components of the new Holmonger breed and the Africander, because of its adaptation to the climate, as the other. A yellow strain of Africander, which had been selectively bred at the station, was used because of the benefit a light color provides under the hot sun. A breeding program to accomplish a half-and-half mixture of these two breeds was established, and systematic selection was made for beef conformation and a light color of the progeny.

New Breed.—This is the only name that so far has been given to another line developed at Omatjenne. It has been fixed at one-half polled Africander, one-fourth Simmental, and one-fourth Hereford. Third-

New Breed bull. Photographed at Omatjenne Experiment Station.

New Breed cow. Photographed at Omatjenne Experiment Station.

generation calves of the New Breed were on the ground in 1967. Also purely a beef type, the animal was bred to produce an early-maturing, polled animal with a smaller hump and a good beef conformation. All of these objectives are in evidence in the animals that have been selected to carry on the program.

Exotic Breeds.—Cattle of several European breeds are maintained both as stud herds and in commercial operations. The Hereford is much favored, the breed being maintained by importations from South Africa. A small farmer, running as few as 200 brood cows, will pay as much as $1,500 for a good Hereford bull.

Simmental cattle also do well and are in demand. They date back in many instances to the stock brought in by the German settlers at the turn of the century.

Among the other European breeds that have gained a foothold are the Swiss Brown, the Pinzgauer, and the Sussex. These cattle are usually maintained in a relatively pure state by European growers and conform to the same breed standards as they did in their native land. The Shorthorn and the Aberdeen Angus as beef breeds and the Red Poll as a dual-purpose breed were formerly popular but have lately lost ground. The Friesian is only used for milk production under conditions of stabling and artificial feeding.

MANAGEMENT PRACTICES

Native cattle of the Africans on the northern reserves are grazed, for the most part, in the manner common to tribal cattle over most of Africa. Whether the movement of the tribe is continuous, with only periodic stops until the grass in an area is consumed, or seasonal, as with the agricultural Ovambo, cattle are customarily under the constant eye of a herdsman and are corralled at night and moved daily from grass to water and back. In remote areas where cattle at times are more or less isolated, however, herds are sometimes left unattended during the day and bedded down in a desirable location at night.

An unusual pride exists among these tribes regarding the possession of large oxen. Headmen, called "captains," usually select the better bulls for castration at one year of age. This is still accomplished by pounding the testicles between two rocks, although the use of the knife for this operation is being adopted gradually. These oxen are then grown out to an old age and are given extra care by special feeding during droughts and the dry season. Undoubtedly this negative form of selection has had the effect of reducing the size of these cattle.

Some use is made of milk, but calves are rarely weaned and are not separated from the cows at night when corralled. The volume of milk produced is therefore small, and many cows are not milked. Cattle are slaughtered for food only on ceremonial occasions; then the poorer ones are killed unless the event is a major one, such as the death of a chief, when old oxen are chosen. The majority of native cattle die of old age, although more of them are gradually being marketed. Under the constant care of the herdsmen, they become very docile. Instead of being driven, they are often moved by an attendant walking ahead and the herd moving methodically behind him.

European growers in the cattle belt below the African reserves utilize modern husbandry practices. Their lands are fenced and pastures cross-fenced. Cattle are in the open the year round, on good grass after the rainy season and on mature dry growth the rest of the year. Pastures are rotated and cattle can be maintained in good condition the entire year if not overstocked. Most of the grazing land is of the open savanna-acacia type, on which the acacia rapidly encroaches with overgrazing. The carrying capacity commonly referred to is 1 animal unit to twenty-five acres. During a foot-and-mouth epidemic in 1960-61, very few cattle were marketed and overgrazing was serious. Areas of considerable extent are seen in which the acacia has practically taken over.

Cattle are normally marketed off the grass, preferably at the beginning of the dry season when steers at three and one-half years of age weigh more than 1,000 pounds and grade 1 or 2 in the South-West Africa system. There is no grain feeding.

Most cattle operations are on a family-size scale of 10,000 to 15,000 acres, with approximately 200 brood cows. There are larger, well-managed units running up to several thousand head, but they are a small minority. Labor requirements are minimal since all cattle are on fenced pastures the year round and require little attention. On the average farm, the labor force ordinarily consists of the owner, a son or two, and a few native Africans.

MARKETING

Nearly two-thirds of the European cattle marketed, about 175,000 head annually, are moved by rail either to Cape Town or to Johannesburg, both in South Africa, a trip requiring five to eight days. The resulting loss in weight and grade is not as severe as might be expected, running about a half grade loss and less than 2 per cent loss in weight. Sales are made through commission houses at the abattoir at the terminal point and are on a cold-dressed-weight (CDW) basis. The remaining European slaughter cattle, estimated in 1966 at 100,000 head annually, are also sold on the CDW basis through the abattoirs in Windhoek, Okahandja, and Otavi or for local consumption in the smaller towns.

A grading system along the lines of that employed in South Africa is used. The three classifications—A, B, and C—of the top grade, super, apply only to young animals in good condition. The next grade is prime, followed by grades 1 to 5. The bulk of the cattle marketed are steers at three and one-half years of age, weighing 1,100 pounds or more and grading 1 or 2, which is in the range of USDA grades low-good to good. In early 1967 prices on a CDW basis ranged from 35 cents a pound for super A to 23 cents for No. 5. A 1,000-pound steer grading 1 or 2 returned to the owner the equivalent of 16 cents a pound liveweight or $160 a head, a price at which the cattlemen were well pleased. Shipments to South Africa after shipping costs are absorbed ordinarily return a little more than this if a shipper is fortunate in his weight and grade loss; if he is not, the return can be materially less.

The total offtake estimated for 1966 was 275,000 head from the 1,750,000 head in the European herds, or nearly 16 per cent.

African cattle owners hold about 800,000 head, of which 665,000 are in the northern reserves. Ordinarily the offtake from these is negligible, but in recent years Portuguese traders across the border in Angola have developed a lucrative business with the African cattle owners in Ovamboland and the Kaokoveld, which are at present isolated from the rest of South-West Africa by a foot-and-mouth quarantine. As the European money economy infiltrates these areas, the African develops wants he never knew before—bright cloth for a woman's dress, more sugar, a bicycle—and to obtain these desires he will part with a few of his prized cattle. Some of the transactions are said to involve the exchange of cattle for cases of beer, wine, or whisky from the cattle trader—commodities which must then be bartered to other traders for the wanted merchandise. A large ox which would have a value of $140 in Windhoek brings $30 in liquor, and even this amount is devalued when traded back to another vendor for cloth or sugar.

Africans occasionally retaliate against this procedure in amusing, if not entirely ethical ways. Cattle of the Kaokoveld are sometimes trained to follow a piping tune played on a reed by a herdboy. When a headman feels that he has been cheated in his cattle trade, he sends his herdboy at night to play his piping tune where the Portuguese trader is holding the stock. The oxen which belonged to the headman then follow the herdboy back across the border. While this contraband trade is small, it is said to be increasing. As many as 4,000 to 5,000 head of cattle have been driven across the border in a month.

Angola's control of foot-and-mouth disease consists only of quarantine of infected areas on outbreaks, and the South-West African authorities have had to deal with sporadic outbreaks by means of vaccination to protect the area south of the reserves from infestation. If foot-and-mouth disease and, more particularly, contagious bovine pleuropneumonia could be cleaned up in the reserves, a sizable number of native cattle would be made available for marketing. Active measures are being employed to achieve this end.

ABATTOIRS

The principal abattoir in the country, at Windhoek, is a thoroughly modern plant with the continuous conveyor system for processing carcasses. The peak capacity is 1,000 head a day on a two-shift basis. The annual kill runs about 100,000 head, 90 per cent of which go to an adjacent canning plant and the remainder for local consumption. The labor employed is indentured Africans from the northern reserves. They work on a contract basis for one year, after which many return to their tribes, although some elect to remain a second year. The proficiency soon attained when a new crew replaces the old one en masse not only shows that the African readily adapts to this type of semiskilled labor but also is a tribute to the efficiency of the management.

CATTLE DISEASES

The worst scourge of African cattle, the tsetse fly, cannot exist in South-West Africa because of the relatively cold winters; thus the country is free of trypanosomiasis. Botulism is serious in many areas because the volcanic soils there are low in phosphorus and the resulting systemic deficiency in the cattle causes them to mouth any available dead animal debris. Losses can be countered by vaccination every year, which is a common practice in well-managed herds, as well as by the feeding of phosphatic supplements. Tuberculosis, introduced by imported dairy cows, is found only in coastal areas where the cattle population is small. Brucellosis has been satisfactorily controlled by vaccinating heifers with Strain 19 at weaning and again as two-year-olds. Blackleg, pleuropneumonia (in the northern native reserves), and anthrax are all encountered but can be kept under control by vaccination. Ticks are prevalent over most of the northern area, and good operators dip or spray every month during the rainy season and every three months in the dry season. Native cattle are treated free of charge by the veterinarian department when called on, but this is seldom done except when there are serious outbreaks.

In 1960–61 there was a serious outbreak of foot-and-mouth disease. The kudu, which ranges throughout the better rainfall areas of the cattle belt, was considered the principal carrier. As an area was cleaned up, eight-foot high kudu-proof fences were constructed all the way across the country to prevent reinfestation by these animals. All of the country south of the northern reserves, against which a strict quarantine is maintained, is now said to be free of foot-and-mouth disease.

GOVERNMENT AND CATTLE

There are six agricultural experimental farms in South-West Africa, all working with cattle. Omatjenne, the largest, 200 miles north of Windhoek, has good facilities, is excellently managed, and is thoroughly modern in the approach it takes to agricultural problems. Some of the outstanding results obtained at Omatjenne are discussed under "Cattle Breeds."

Treatment and control of cattle diseases and parasites are handled effectively and with no more confusion than bureaucratic control ordinarily entails. Slaughter inspection is up to the highest standards.

The Bantu Administration of the South African government is responsible for all activities on the native reserves. Their efforts to advance the African in his cattle raising begin with such fundamentals as providing stockwater. Veterinary services are free, and gradually the African cattle owner is increasing his use of them. Although the native husbandry of cattle still follows traditional lines, the African has more cattle and they are in better condition than would be the case if it were not for the Bantu Administration.

OUTLOOK FOR CATTLE

Local consensus maintains that the area adapted to cattle raising is about fully stocked with the 1,750,000 head belonging to European growers. This is a stocking rate of twenty-five acres an animal unit. For the present condition of the range and under today's economy as it affects cattle raising, this appraisal is reasonable. Large areas of dense acacia with very low forage production could be cleaned by spraying at a cost of $6.00 an acre or by grubbing at a somewhat higher cost, and grasses could then be re-established. Such a practice is said to be uneconomical, and this could well be the case under conditions in 1967, with steers at 16 cents a pound. Certainly there is the potential here for an increase in herds when the economy warrants. Likewise, the storage of plant growth in a more nutritious stage for feeding during the dry season would have its effect in increasing carrying capacity. The practice of pasturing the mature dry grass, which is much lower in nutrients than if it had been cut as hay, is probably advisable in the current economy; but here is another opportunity for increasing production when the price of cattle permits it.

The largest potential for increased cattle production is in the herds of the Africans on the northern reservations. These herds, together with the native cattle on small reservations in the south, are estimated at 800,000 head, nearly half the number of European-owned cattle, although of much poorer quality.

Until the contraband trade with the Portuguese, which has been mentioned, developed recently, the offtake from the native cattle was practically nil. The offtake to the Portuguese traders now probably approaches 4 per cent, a high number for African cattle. Such an outlet must be viewed as temporary, and when a legitimate channel for the African to dispose of his cattle at a fair price is opened, the volume will probably decrease initially. Ordinarily the African cattle owner sells only as many animals as are needed to supply an immediate want, and at a fair price this quantity will be only a fraction of the number he now drives to Angola. Eventually, as the European's money economy enters more into the African's daily life, he may market more of his cattle, but this goal is for a day in the future.

Cameroon

Land area (sq. mi.):	183,600
Population (1966):	5,200,000
Density (per sq. mi.):	28
Agricultural and nomadic (85%) (1966):	4,420,000
Per capita income (1961):	$91
Cattle population (1964):	2,175,000
Offtake (3.4%) (1964):	75,000
Year visited:	1966

THE FEDERAL REPUBLIC OF CAMEROON occupies an irregular wedge-shaped area between Nigeria on the west and the Central African Republic and Chad on the east, ending in a point in the north on Lake Chad. The southern border, 800 miles south of the point on Lake Chad, is with Río Muni (a Spanish province), Gabon, and the Congo (Brazzaville). A 220-mile coastline on the Gulf of Guinea cuts across the southwest corner. A low area, 450 miles along the southern border and 150 miles wide, is rain forest, with as much as 200 inches of rain annually. This area rises to a plateau 2,000 feet in elevation, where the rainfall gradually decreases to 15 inches at Lake Chad.

Following the usual pattern of being first in Africa, the Portuguese landed in Cameroon, at the end of the fifteenth century. Beyond some trading, largely in slaves, the European powers paid little attention to the area for the next three centuries. By 1800 the coast was under British influence and the French were beginning to become interested; but, beginning in 1884, Germany pressed its claims and held Cameroon as a protectorate until World War I, when the British and French took possession. At the end of the war the country was divided under League of Nations mandate, nine-tenths going to France and one-tenth, a narrow strip on the western border, to Britain. When the wave of independence that followed World War II swept over Africa, after considerable terrorism the northern part of the British area went to Nigeria; the southern part, together with all the former French territory, became the Federal Republic of Cameroon and a sovereign state in 1960. French influence is still strong, however. The population is almost entirely African; the

Bantu peoples, located in the southern region, are in the majority. The Fulani and other Moslem tribes are in the north. There are practically no Europeans in the country other than diplomatic representatives and a small number of French engaged in commerce.

In area Cameroon is slightly larger than Arizona and Florida combined and has a population two-thirds as large. Eighty-five per cent of the population is engaged in agriculture, which is largely a subsistence type. The largest export is cacao. The cultivated areas are in the south; the savanna country on the plateau begins as the rainfall decreases and extends to Lake Chad. The cattle of the country, along with somewhat larger numbers of sheep and goats, are run here by nomadic and semi-nomadic tribes.

CATTLE BREEDS

Red Fulani. This breed accounts for the majority of the cattle. It is the same animal as the one described under that name in the discussion on Nigeria.

Red Fulani bull. Photographed at the Duala trader holding area.

Red Fulani cow. Photographed in northern Cameroon.

Nondescript Northern Cattle.—In the drier northern savanna many rather small, nondescript cattle are seen which appear to be the result of indiscriminate crossing of the Red Fulani and the Shuwa cattle of northern Nigeria. Other than being humped and having medium-sized, upswept horns (not as large as those on the Red Fulani), they display no uniform characteristics. Although most are black, and black animals with white faces or white undermarkings are common, color varies widely. Cows weigh 600 to 700 pounds, bulls 700 to 900 pounds.

MANAGEMENT PRACTICES

Cattle are raised almost entirely on the savanna in the northern third of the country. They are in the hands of nomadic native tribesmen. When grazing, the animals are always in the charge of a herdsman. They are gathered at night, bedded down in a suitable spot, and milked for household consumption and for barter when in the vicinity of a village. Frequently cattle of several owners are maintained in one herd. The areas reached by cattle are limited by water and are badly overgrazed.

Millet is the only crop raised in the north; the stalks are saved and fed in the dry season. Seed pods of the acacia are used as feed, the herdsman knocking the pods from trees as he walks ahead of his charges.

Oxen are ridden for human transportation, a custom common only in

Cattle being fed millet stalks. Photographed in northern Cameroon.

Riding oxen. Photographed in northern Cameroon.

a few places in Africa. The animal is controlled with a rope which is passed through the central membrane of its nose and handled as reins. This practice is most common in the northern part of the savanna area.

MARKETING

Cattle are trailed into Garoua, the central market point for the northern cattle country. Here they are purchased by traders, who have them trucked to the south, usually to Douala, the main seaport, 500 miles away. The cattle are held on the outskirts of town by the traders until sold to butchers for slaughter. This procedure occasions an even further decrease in condition, for they are maintained on a continuously overgrazed area close to town and trailed back and forth to water and pasture from each trader's individual holding pens. The result of this lack of maintenance feed while waiting as much as a week for slaughter, after the 200-mile drive and the trucking for several hundred miles, is that the animal goes to the abattoir in extremely poor condition. The weight loss from the owner's herd to the abattoir must be at least 20 per cent. All sales are by the head. The trader pays the owner about $100 for a 900-pound steer or bull in the growing area. Trucking cost is approximately $40 a head to Douala. An animal weighing a little more than 700 pounds is sold to the butcher for $180, or about 26 cents a pound liveweight. Round steak sells in the shops supplying the high-income trade at $1.25 a pound, tenderloin at $1.60.

ABATTOIRS

Animals for slaughter are trailed from the traders' area in Douala to the municipal abattoir several miles away, across the river at Bonaber. This primitive facility is housed in a small concrete building with a small, barbed-wire fenced holding yard. The average daily kill is around 100 head, and only cattle are slaughtered. The average carcass weight is 350 pounds. Waste disposal is by feathered scavengers.

OUTLOOK FOR CATTLE

It is difficult to determine to what extent the savanna areas of Cameroon would support more cattle under good management. As utilized today, the region is overstocked because of inadequate watering facilities and uncontrolled grazing. If these conditions were corrected, cattle numbers could be materially increased.

As is common in most of Africa, cattle are held by their owners for prestige and as a way of life, and numbers are all important. Selling a reasonable offtake at younger ages would permit a sizable increase in slaughter animals. With only elementary improvement in management methods, this could be done.

Neither of these changes in the method of handling cattle appears likely to occur in the near future. The first involves heavy capital expenditures, which are not visible today; and the second means a radical modification of the nomadic heritage of the tribal owners, who are quite satisfied with their way of life. Their inborn instinct to keep their cattle and increase their herds if possible, selling only what is necessary for immediate needs, is in direct contradiction to the Western concept of a cattle industry.

Chad

Land area (sq. mi.):	490,700
Population (1966):	3,310,000
Density (per sq. mi.):	7
Agricultural (98%) (1964):	3,240,000
Per capita income (1964):	$56
Cattle population (1964):	4,000,000
Offtake (5%) (1966):	220,000
Year visited:	1966

THE REPUBLIC OF CHAD, one of four colonies of former French Equatorial Africa, is a landlocked country located in Central Africa within the same general latitudes as southern Mexico and Central America. In area it is equal to that of Texas, New Mexico, and Arizona combined. Chad lies within the vast basin which drains to Lake Chad on the western border. The lake is a shallow body of water which at flood stage covers an area equal to Lake Erie but shrinks to less than half that at the end of the dry season. Libya is on the north, Sudan on the east, the Central African Republic on the south, and Cameroon, Nigeria, and Niger on the west. The population is practically all African and predominantly Moslem; most of the 5,000 Europeans in the country are French.

More than half of Chad's land surface lies north of a line drawn through the northern end of the lake and is desert and semiarid. The annual rainfall varies from 2 inches in the north to an average of 18 at Lake Chad. From here precipitation increases to 50 inches in the extreme south. Most of the population is concentrated in the southern part of the country.

In the rush of the European powers to Africa, the area that is now Chad was overlooked until British and German explorers entered in the mid-nineteenth century and were soon followed by the French. The influence of France prevailed and has been dominant ever since. France wanted the area—not for itself but to consolidate French West, Equatorial, and North Africa. Since Chad was declared independent within the French community in 1960, the government has been unusually stable, although in 1962 constitutional changes elevated the president to the status of a near dictator. The French, however, as in most of their former West African colonies, largely run the country and there is a comfortable feeling of stability from day to day.

Agricultural resources of Chad are meagerly utilized. There are only eighteen inhabitants to the square mile in the southern third of the country, where the rainfall is sufficient to support crops. The economy is almost entirely agricultural, with 98 per cent of the population engaged in some type of farming or livestock raising. The principal cash crop is cotton; the main foods are millet, sorghum, and cassava. Nobody goes hungry in Chad, which enjoys a more ample diet than much of the rest of Africa and is practically self-supporting in food. Little beef is eaten, however, and most of the small offtake from the large cattle population is exported. Fish, which are plentiful in the two main rivers and in Lake Chad, are an important and favorite part of the diet.

The facts that there are no railroads and that for six months of the year, during the rainy season and for a month or two thereafter, all roads are impassable do not particularly concern the rural population. There were no hard-surfaced roads in 1966. Transportation from the outside

Mature Kuri cow. Photographed on the eastern shore of Lake Chad.

Kreda bull. Photographed east of Lake Chad.

Kreda cow. Photographed east of Lake Chad.

is either by air or by truck for 400 miles from the railhead in Nigeria and across the northern tip of Cameroon.

CATTLE BREEDS

The cattle of Chad are indigenous breeds which have been run there by their nomadic owners for centuries. Practically no outside cattle have been brought into the country in recent times, the only noteworthy exception being a few of the N'Dama breed, which the French introduced from West Africa in limited numbers for experimental breeding.

Kuri.—This unusual breed is indigenous to the southern and eastern shores of Lake Chad and its numerous islands. Large, bulbous horns are its distinguishing feature. These nonhumped cattle are considered to be descendants of the ancient Egyptian Longhorn. The Kuri is named after one of the tribes which have run them for countless generations and who were particularly rigid in their selection for the peculiarly formed horns. The breed is frequently referred to in the literature as being large, but Kuri cattle of today are usually small, an average cow weighing less than 600 pounds and the bull weighing perhaps 150 pounds more. The upthrust, quite straight, and often bulbous horns are two feet or longer, and the bulbous part a short distance up from the base of the skull may be as much as eight inches in diameter. The hair coloring is usually off-white, sometimes with black or red markings around the head. Black

N'Dama bull. Photographed at Ferme d'Elevage, Fianga. (See page 718.)

ears are characteristic. Conformation is poor—shallow, narrow, and leggy, usually with a sway-back.

The breed is gradually disappearing; very few Kuri cattle are now seen in the Fort-Lamy cattle market. Even in their traditional habitat around Lake Chad, most herds now contain other types of cattle, many of which are humped.

Kreda.—Nomadic tribes in the semiarid region lying east of the southern end of Lake Chad run these cattle, which are also referred to as Red Zebu. They range south as far as the cultivated areas.

Usually a solid dark red, the breed has horns that are upspread, wide, and often lyre-shaped toward the end. The hump is prominent, but not extreme, in both sexes. The umbilical fold on the cow is unusually large

N'Dama cow. Photographed at Ferme d'Elevage, Fianga. (See page 718.)

and pendulous. Mature bulls on good nutrition weigh as much as 1,200 pounds but in the hands of their nomadic owners rarely reach this figure. The Kreda appears to be the same breed as the Red Fulani of Nigeria and northern Cameroon.

N'Dama.—This breed was introduced by the French at the experiment station near Fianga, and a herd is still maintained there. African owners grow very few of them. The breed is described under "Cattle Breeds" in the chapter on Nigeria. (See pages 716 and 717.)

Unidentified Types.—East and north of Lake Chad a small type of cattle, usually black in color, is maintained by African owners. This livestock is humped and almost a dwarf type; mature cows weigh 500 pounds, bulls 600 pounds.

A small type of black cattle. Photographed east of Lake Chad.

Two hundred miles south of the lake, bordering the cultivated areas, a distinctive, uniformly grey animal is often seen. It is a humped type with medium-sized, upswept horns. Considerably larger than the black cattle, these bulls weigh up to 750 pounds.

Many of the African-owned cattle in the grazing areas are nondescript. Commonly seen, they are almost any color except solid white; a mottled black-and-white pattern is frequent. The only common char-

Grey bull of the type seen south of Lake Chad.

Typical nondescript cattle. Photographed 150 miles northeast of Lake Chad.

acteristics are a medium hump and the small size, the average weight of mature animals ranging between 500 and 800 pounds.

MANAGEMENT PRACTICES

The typical African nomadic practice of raising cattle prevails in the broad savanna grazing belt, with a 10- to 20-inch rainfall that stretches 600 miles across the country from the Cameroon border on the west to Sudan on the east. With rainfall concentrated in the summer months, mature dry feed is all that is available for nine months of the year. There are some government-constructed watering places serviced by deep wells, but most stock must depend on natural sources, which limits the usable grazing land.

Government watering station at Massaquet.

Cattle are seldom slaughtered for food but milk is taken, usually by small boys, who get a pint or two from each cow. The milk is used primarily to make ghee, an oily butter. After the cattle are gathered for the night, the calves are often staked out individually but are first allowed sufficient milk to maintain them in fair condition. In some areas cows are separated from their calves while grazing. The animals of several owners are grazed together on a communal basis in herds of as many as several hundred head. Individual owners usually maintain 40 or 50 head, but an influential chief may run as many as several hundred. They are sold by first owners to traders when occasion demands. No authentic data on cattle numbers are available, but the total population is probably more than 4,000,000 head. The annual offtake is estimated at 220,000, or 5 per cent. Some bulls are now castrated as calves.

A European firm, Compagnie Pastorale Africaine, purchases native cattle at five to six years of age, which are then run for two years on fenced pastures for grazing control and are supplied with adequate stockwater. Even with mature dry growth being the only feed available for nine months, gains of 150 pounds a year are obtained. Under the nomadic system of the African, the annual gain is possibly 50 pounds. A breeding herd, largely of Kreda cattle, is also maintained. It is run on the best pastures so that ample grass is always available. A year-round breeding season is used. Calves are weaned at eight to nine months of age and then grown out in the same manner as the purchased stock. Under this practice a steer reaches a 900-pound slaughter weight at four years, as compared with 850 to 900 pounds for the seven- to eight-year-old African-grown animal that has been finished on good grass for two years.

The total commercial milk production of Chad is 10,000 pounds a day from native cattle, primarily of the Kreda breed.

MARKETING

The largest market is at Fort-Lamy. It is a fenced area at the edge of town and is devoted to the sale of cattle, sheep, and camels. Traders who have purchased cattle from the nomadic owners in the grazing belt sell in the market to the local butchers and to exporters who ship chilled meat by air to Paris. Both groups have their animals processed in the municipal abattoir, then receive the dressed carcasses. Transactions are all by the head and are completed only after much bargaining. Prices are based on a visual estimate of the number of kilos to which an animal will dress out, without much regard to quality. A live animal in the

Fort-Lamy cattle market.

country, depending on the distance from Fort-Lamy, brings the equivalent of 3 to 3.5 cents a pound, or $20.00 a head for a 650-pound bull or steer. This figure could be considered an average in 1966, although some animals weigh as much as 1,000 pounds. The cattle are usually in very gaunt condition when sold. In the Fort-Lamy market the trader gets approximately $28.00 to $30.00 a head after a shrink of 10 to 15 per cent on the trail. These prices are equivalent to 4.5 to 5 cents a pound liveweight.

Removing ticks from a Kuri cow.

Sales of animals which do not reach Fort-Lamy directly are much more complicated. Trader sells to trader as the cattle progress toward market. Some cattle are to a degree communally owned, and settlement is thus an involved process. One hundred thousand head are annually exported on the hoof to Nigeria, a large part of which are smuggled. About 40,000 head are trailed to the Central African Republic.

ABATTOIRS

The municipal abattoir at Fort-Lamy is a well-constructed concrete building with modern equipment and ample holding yards. The plant was designed for gravity-type operation, the cattle to be driven up a ramp to the killing area on the second floor. Instead of utilizing this facility, the workers drag the cattle to a paved area below the killing floor, throw, and kill them in the usual Moslem manner. Then they are hoisted by hand with a rope over a pulley to the second floor, where processing proceeds according to the original design. The difficulty here is that the ramp is so wide that, when part way up, an animal can turn and start back down; for this reason, killing on the second floor was abandoned.

In 1967 the plant was killing 100 head of cattle, 50 sheep, and 5 hogs a day. The number of cattle handled could easily be tripled if animals were available. There are good refrigeration facilities, and more than half the production is chilled and held for four or five days for export by air. Meat for the local trade is delivered the same day it is killed. Sanitation is considerably better than usual in Africa.

At Fort Archambault, near the southern border, there is a comparable, although smaller, abattoir. Both facilities are government operated, all processing being done on a custom basis.

CATTLE DISEASES

The effect of the endemic cattle diseases common to Africa is lessened in Chad by the wide distribution of cattle in the grazing areas. Investigations have indicated a 25 per cent incidence of brucellosis, which must cause a high loss that is not realized by the cattle owners. Vaccination with Strain 19 was made compulsory recently, and it is claimed that 4,000,000 doses were administered in 1965. From time to time there are occurrences of pleuropneumonia and outbreaks of rinderpest. Foot-and-mouth disease is not considered serious, for most infected animals recover after a year and a year's loss in weight gain is of no particular moment to the nomadic owner.

Prophylactic measures adequate for good control of disease in the vicinity of Lake Chad are employed at the Compagnie Pastorale Africaine farm. These practices include vaccination against pleuropneumonia and rinderpest and calfhood vaccination of heifers for brucellosis. In addition all cattle are sprayed for tick control every week in the wet season and once a month in the dry season. Such measures are hardly practical with the nomadic herds.

Ticks are more of a problem on the shores of Lake Chad, where the Kuri cattle are run, than in the savanna areas. African owners sometimes throw and tie an animal so that the ticks can be removed by hand.

GOVERNMENT AND CATTLE

In colonial days the French established a laboratory at Fort-Lamy for the production of vaccines for French Equatorial Africa. This facility continues to supply the same area—Niger, the Central African Republic, and Chad—and now is sending vaccines to northern Cameroon and northern Nigeria as well. Disease-control measures are under the jurisdiction of government veterinarians but are extremely difficult to administer because of the widely scattered and continually moving cattle population.

An experiment station was established in 1950 by the French at Fianga, in the cultivated area near the Cameroon border. A herd of excellent N'Dama cattle is still maintained here and is being crossed with other indigenous African breeds in an endeavor to produce an improved and well-acclimated beef type. A cross of the West African Shorthorn bull on the N'Dama cow has produced a larger animal with a better beef conformation than either of the parent breeds. How much of this improvement is the result of hybrid vigor has not been determined.

OUTLOOK FOR CATTLE

Transportation is the most immediate handicap to development of a cattle industry in Chad. The only outlets now are by air, or on the hoof for distances up to 1,000 miles to the markets in Nigeria or in the Central African Republic. Nomadic cattle owners in Chad have the inborn African desire to keep their cattle, but can be induced to part with them. In fact the indicated offtake from their herds is now 5 per cent, compared with 2 per cent or less from many nomadic cattle owners in Africa. This offtake could probably be increased if more convenient marketing facilities such as auctions in the grazing areas were provided.

The vast grazing areas which are unsuited for cropping would support much larger cattle numbers if they were provided with adequate stockwater and if grazing were controlled. Compagnie Pastorale Africaine has demonstrated in a purely commercial operation what can be accomplished by good management. Their stocking rate is about nine acres a head; the average on the savanna with the same rainfall is certainly in excess of twenty-five acres a head. Some day the world demand for beef will probably force the utilization of this potential.

Congo (Kinshasa)

Land area (sq. mi.):	905,600
Population (1966):	16,000,000
Density (per sq. mi.):	18
Agricultural (84%) (1947):	13,400,000
Per capita income (1961):	$70
Cattle population (1960):	1,050,000
Offtake (7%) (1960):	70,000
Year visited:	1967

THE DEMOCRATIC REPUBLIC OF THE CONGO, formerly the Belgian Congo, lies almost wholly within the vast Congo River Basin. All of the country, the capital of which is Kinshasa (formerly Léopoldville), lies wholly within the tropics, one-third of it to the north of the equator and two-thirds to the south. Most of the area is low plateau of only 1,000 feet elevation, widening out from the Congo estuary on the Atlantic Ocean and extending to the Great Rift Valley in East Africa. The whole country enjoys ample rainfall, 65 inches and more in the rain-forest belt across the north and 40 inches in the wide savanna region to the south. The dry seasons are relatively short, usually lasting fewer than 120 days.

On the west the country is almost cut off from the sea by the Republic of Congo (Brazzaville), the former French colony of the Middle Congo; Cabinda, the small Portuguese enclave; and Angola, leaving a coastline of only twenty-five miles where the Congo River empties into the Atlantic Ocean. The northern border is with the Central African Republic and Sudan, the southern with Angola and Zambia. In the east are Uganda, Tanzania, Rwanda, and Burundi. The third largest country in Africa in area, exceeded only by Sudan and Algeria (both with large expanses of desert), the Congo is nearly one-third the size of the United States.

Although the Portuguese explored the mouth of the Congo River in 1480, the area was practically untouched by Europeans until the Belgian King Leopold II hired Sir Henry Morton Stanley to explore it in 1878, shortly after David Livingstone's expedition there. Leopold, realizing that his country needed to expand, tried to convince his government that development of the Congo should be undertaken. Failing in this attempt, he undertook the task himself and was recognized by the European powers as the absolute monarch of the area in 1885. For all practical purposes the vast area was simply his personal property from then until 1907, when he ceded it to the Belgian government. In 1908 it became the colony known as the Belgian Congo.

In the Katanga region, in the southeast, there were extensive mining developments in copper, cobalt, and industrial diamonds, but very little attention was paid to agriculture until the 1920's. Considerable progress was then made both by European settlers and in improvement of the traditional agricultural methods of the Africans. The country had become self-supporting in food before becoming independent in 1960; since then, however, retrogression of the economy, including agriculture, has been rapid. Only a few Belgian farmers now remain. In 1959 there were more than 100,000 Europeans in the country, 90 per cent of whom were Belgians. No information is now available on the number who have remained, but large numbers are known to have left in the aftermath of independence.

Within a week after independence was recognized, the army mutinied. During the days of their development of the country, the Belgians had maintained law and order. Congolese troops were trained under white officers, and no African rose above the rank of sergeant. With Belgian control gone and supreme power in such unprepared hands, the chaos which ensued was inevitable. The internally bickering forces of the United Nations managed to effect a measure of armed peace during the next four years. Katanga Province declared independence but later was reintegrated. The reins of power were then again turned back to the Congolese politicians, with disastrous results. In 1967 the army was back in power and managing to maintain a semblance of law and order, but there is still no real stability in government. From a solid currency and a healthy balance of payments in 1959, the economy has suffered a one to ten devaluation of the currency and depends on outside aid to feed 30 per cent of the people.

The Bantu-speaking peoples of the Congo, a large element of the population, are not traditional cattle raisers as they are in much of Africa. They were primarily hunters and gatherers until the introduction of manioc and corn from the West. Then, because it was an easier way of life, they became subsistence farmers, employing shifting-hoe cultivation: they cleared a patch of ground; planted it for two or three years, thereby fairly well exhausting the initial mineral content of the soil; and then moved to a new patch. This procedure was continuously repeated. When the Belgians began to expand their interest in agriculture to the Africans, they developed the *paysannat*, an improved form of this traditional method of cultivation. The heads of a family group were taught to clear and cultivate long, narrow fields for their crops, then move to an adjacent parallel strip, and so continue until soil conditions permitted returning to the first field. The fallow period varied from a few to as many as twenty years. This system led to a marked increase in production. By 1958 there were 200,000 of these *paysannats*. The practice of co-operatives using community tractors was also developed under the Belgian agriculturalists; and in 1959, before independence, there were 10,000 commercial African farmers. Although demonstrating the agricultural possibilities of the country, all this was hardly more than a pilot enterprise, considering the size of the Congo.

The latest estimate of the cattle population, made in 1960, was 1,050,000 head. Perhaps one-fifth of these were in the hands of European owners. The total offtake in 1960 was estimated at 70,000 head (7 per cent) and is now probably less than this approximation. In 1967 the major abattoir at Kinshasa was killing 7,000 head a year. No current estimate of the number of cattle in the country or of the number slaughtered is available but, with the exodus of European farmers since 1960, both have certainly decreased.

CATTLE BREEDS

Cattle have been little employed for draft, for there were few in the country except those of the Belgian settlers and those held by the pastoral tribes in the mountainous area on the eastern border. In other parts of the country, even today, only a few Africans own cattle. Except for indigenous cattle held by tribesmen in the east along the Great Rift Valley, the existing herds have come from those animals introduced by the Belgian settlers in the past, primarily for the production of beef.

In the Kinshasa area N'Dama cattle, the small, hardy indigenous West African breed, were brought from French Guinea. (The N'Dama is described under that name in the chapter on Nigeria.) In the southeast Sanga types were introduced from East Africa, and in the extreme south some Africander cattle were imported from Northern Rhodesia (Zambia). All of these improved types were in the hands of European settlers. Two other African breeds, known locally as the Guinean and the Dahomey, were also introduced.

Dehorned N'Dama bull. Photographed at JVL Farm.

Dehorned N'Dama cow in foreground. Photographed at JVL Farm.

MANAGEMENT PRACTICES

Of the fourteen remaining plantations said to be raising sizable numbers of cattle, the Jules Van Lancker, or JVL, farm is an outstanding example of what can be accomplished in cattle raising in the tropics and exemplifies the great capability of the Congo for development of a cattle industry. Located at Kolo, 135 miles south of Kinshasa, the farm was founded in 1923 by Jules Van Lancker. It is now a fully integrated agricultural enterprise producing palm oil, sugar, coffee, and beef cattle, which are processed in the farm's own abattoir and sold to the carriage trade in Kinshasa through the farm's own retail outlet. The largest and probably the best herd of N'Dama cattle in Africa, totalling 25,000 head of which 6,500 are mother cows, is husbanded here.

The foundation cows were brought from French Guinea in 1928. Systematic selection for fleshing qualities was initiated in 1952 and evolved into the outstanding breeding practices which are conscientiously adhered to today. Bulls are produced from ten "selection herds," each of a different family. Every cow in these groups is branded with the number of the herd to which she belongs, the year of her birth, and her own identification number. Her calf is branded the same, plus the quarter of the year in which it was born and its sire's number. Each selection herd contains 30 cows, and is run separately, with 1 bull from a different selection herd. The calves are weighed at eight months. Heifers weighing 275 pounds or more and bulls 285 pounds are separated as breeding stock; the lighter bulls are steered. The selected heifers are placed in the proper age group of the production herd, where they are available for replacements in the selection herd from which they originated. A second cut is made on the bulls at one year of age; a third cut is made on their three-year-old weight and conformation. In this manner the 50 bulls needed annually to maintain the 5 per cent bulls used in the 6,500-cow

N'Dama yearling heifers. Age 18–24 months, weight 550 pounds.

production herd are meticulously selected. This procedure has resulted in an increase in the weight of market steers from 660 pounds at five years of age in 1952 to 775 pounds at four years in 1965.

All cattle are run in separate age groups of 500 head each—mother cows, replacement heifers, and market steers. Heifers are bred when two years old to calve as three's; both cows and bulls go out of the herd when ten years old. The herd is remarkably uniform. The color is generally solid, varying from a light coffee-milk shade to a reddish tan, occasionally with minor white undermarkings. Bulls tend to be darker on the head and shoulders. No prophylactic treatment is employed except for the dipping of every animal once a week the year round.

When the herd was being established, there was a high loss from trypanosomiasis, for the heavy infestation of the tsetse fly was too great for even the tolerant N'Dama. This situation was overcome by taking out much of the brush.

An 85 per cent calf crop is now obtained, most of which is dropped in the spring—September to November—although bulls are left in the herd all year. There are four dipping and branding stations, one for every 1,500 head of mother cows. All cattle, even those in the selection herds, are year-branded and dehorned as calves.

The main pasture grass is the native Hyparrhenia. There is good provision for water in all pastures, which are fenced with barbed wire and concrete posts. Stocking rate is maintained at 1 head to five acres. This ratio is more than ample as far as plant growth is concerned but is kept at this figure to prevent the severe erosion of the hilly terrain which would occur if spots were overgrazed. During the dry season—May to July—only mature dry growth is available for feed, but with the adequate watering facilities the cattle come through in good condition.

The offtake at JVL Farm is 5,500 head annually, or 22 per cent, a high figure even for a European-managed herd in Africa. Slaughter steers are fed three pounds of corn and three pounds of coconut-oil cake for thirty days before slaughter. On the four-year-old grass-fat animals, this practice produces an excellent-quality beef. About half the cattle marketed are slaughtered in the farm abattoir, and the meat is sold in the company store in Kinshasa; the remainder goes through the municipal abattoir.

MARKETING

Butchers purchase cattle in the country for 6 cents a pound, the price being government controlled. Transportation to the abattoir at Kinshasa averages about $1.00 for 100 pounds.

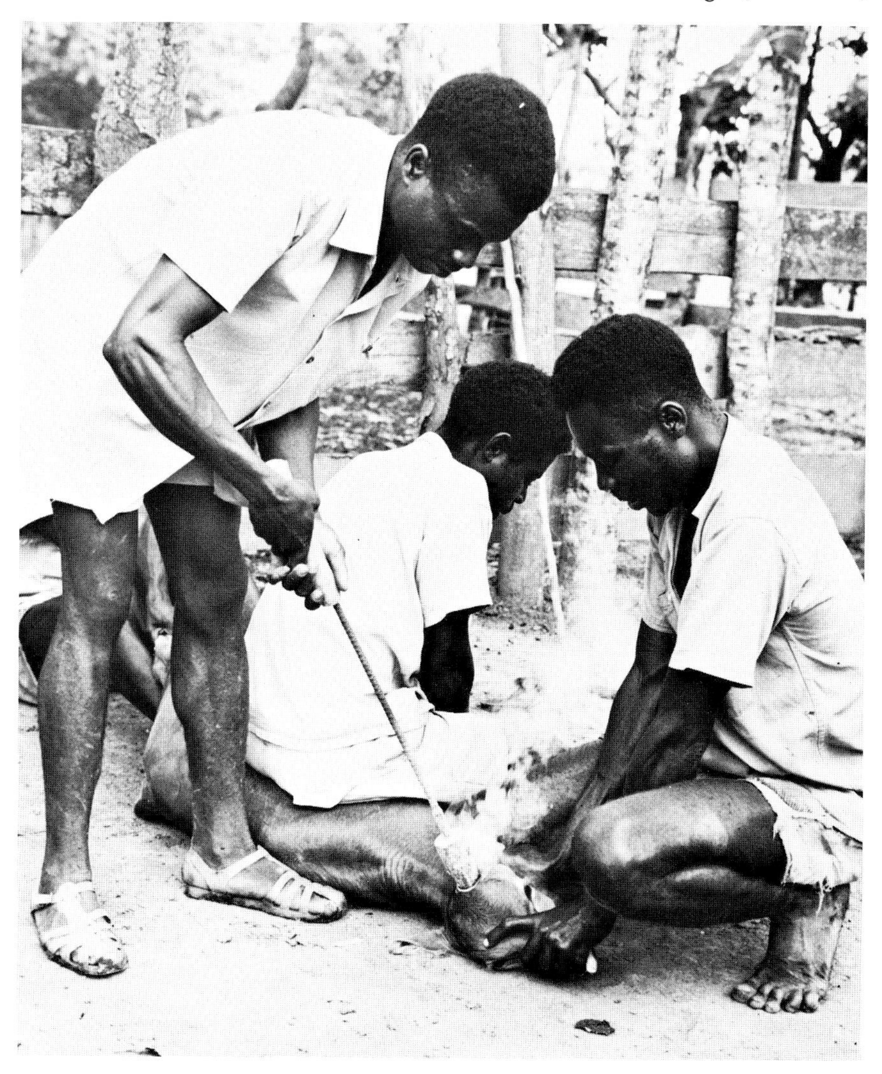

Dehorning with a hot iron. Photographed at JVL Farm.

ABATTOIRS

Le Proposi d'Abattoir Publique de Kinshasa, the municipal abattoir in Kinshasa, is of recent construction and is a good unmechanized facility —tiled, with adequate water and floor drainage. In operation, however, it is reminiscent of the slaughter slabs seen in many other countries of West Africa. A chute through which the cattle could be driven into a stanchion for killing by captive bolt is provided but is not used. Instead, an African crawls up on the retaining wall and is handed a chain that has a loop on the end and is fastened to a long branch. Eventually he maneuvers the loop over an animal's head, and the beast is then dragged by ten Africans to the killing stanchion. The captive-bolt gun has disappeared, and the animal is dispatched by severing the spinal cord with a knife, several jabs often being necessary if the operator is inexperienced. Skinning and dressing are done on the floor without even the use of a cradle. Electric hoists for raising the carcass and mechanic saws for splitting it were in evidence but were inoperative. After being skinned and eviscerated, the carcasses are raised by hand wench for splitting. The heart, liver, and lungs of the animal are inspected, and slashes are made in the flesh for the detection of measles. There are, however, no facilities for disposal of condemned carcasses. Most cattle slaughtered are in very poor condition and would grade a low-standard or below. Two Congolese soldiers carrying submachine guns are kept busy running pilferers off the dressing floor.

An average of 600 head of cattle and an equal number of sheep and goats are killed each month. The minor part that meat plays in the Congolese diet is evidenced by the fact that this is the major slaughtering facility for an area harboring nearly 3,000,000 inhabitants.

GOVERNMENT AND CATTLE

The Belgians established several experiment stations and still supervise the one at Gandajika, in south-central Congo. A herd of 600 head of cattle is maintained, and the carrying capacity on improved pastures is meticulously determined.

Stations at which management was assumed by a Congolese staff after independence have deteriorated sadly. A station at Bateke, fifty miles east of Kinshasa, had a nice herd of N'Dama cattle, and was organized in an endeavor to get more cattle of good quality into native hands. By 1967 most of the herd had been stolen and the facilities for watering stock had become inoperative, but the staff still remained.

A program called *metayage* has been initiated for getting more stock into native hands; it is a method designed to enable the African to acquire cattle legitimately by returning the offspring to the government. The small farmer is given 3 or 4 heifers; after five years, the herd is divided equally—half for the government, half for the farmer.

OUTLOOK FOR CATTLE

The vast areas of the various types of savanna south of the rain forest could support a large and successful cattle industry. Because agriculture is concentrated near the villages, large expanses of these grasslands, with their excellent rainfall pattern, are unused except by wild life. Some breeds of cattle are well acclimated to the tropical climate, as evidenced by the success attained with the N'Dama at Kolo.

The varying degrees of tsetse fly infestation throughout the country would have to be controlled and sizable investments and the know-how of Western pastoralists would have to be acquired to shape the Congo into a profitable cattle country. These conditions will not be forthcoming until a reasonable degree of political stability is attained, however.

Ghana

Land area (sq. mi.):	92,100
Population (1966):	7,740,000
Density (per sq. mi.):	84
Agricultural (50%) (1962):	3,870,000
Per capita income (1964):	$210
Cattle population (1964):	538,000
Year visited:	1966

THE REPUBLIC OF GHANA is the consolidation of the former British colonies of the Gold Coast and Ashanti and some small, adjacent protectorates. Lying 5 to 11 degrees north of the equator, the country is in the center of the southern coastline of the African hump. Borders are with Upper Volta on the north, Togo on the east, and the Ivory Coast on the west. A coastal plain rises from the Gulf of Guinea, the southern boundary, to a low, hilly area and then to savanna in the north. The principal drainage is to the Volta River, which flows through the center of the country in a generally north-south direction. The average rainfall varies from 29 inches on the Accra plains to 42 inches in the central part of the country, but severe droughts are rather frequent. Annual precipitation on the Accra plains has been as low as 13 inches. The area of Ghana is nearly equal to that of Mississippi and Louisiana combined, and the population, which is almost entirely Negro, is one-fifth larger than that of those states.

None of the European powers showed much interest in the Gold Coast other than for purposes of trade—initially for gold from the natives, later for slaves. The Portuguese landed first in 1471 and, without assuming actual control, were the dominating influence until Dutch, British, and Danish traders moved in during the seventeenth century. These foreign merchants maintained themselves without any formal action by their governments until the British began to consolidate their position early in the nineteenth century. This move culminated in establishment of the British Colonies of the Gold Coast in 1874 and of Ashanti in 1901. British policy was to control by indirect rule through the tribal kingdoms, and there was little, if any, settlement by Europeans. This situation led to a considerable degree of self-government even before independence was granted in 1957.

Ghana is a one-crop country—cacao—and that crop accounts for two-thirds of the exports and is the basis of the economy. Ghana was one of the most prosperous of Britain's former West African colonies and had over $500,000,000 in her treasury when she gained independence. Corruption throughout the government and image building by the president-dictator had reduced the country to bankruptcy by 1966. Agriculture is in the hands of African farmers with small holdings. Cattle raising is not an important element, there being only 7 head for each 100 inhabitants.

CATTLE BREEDS

The relatively small number of cattle in Ghana were probably brought into the country during tribal movements in fairly recent times. For the most part, they are nondescript animals, the result of uncontrolled breeding.

West African Shorthorn.—The cattle of the country are locally called West African Shorthorn or Gold Coast Shorthorn. They can hardly even be classified as a type, their only common characteristics being that they are small, nonhumped animals. Conformation, shape of horns, color, and size vary widely. If on good feed mature cows average 450 pounds and do not exceed 600 pounds; bulls reach heavier weights but usually do not exceed 550 pounds, with extremes of 700 pounds. There are black, nearly solid white, and mottled black-and-white animals; but almost any variation in color is seen. Horns range from short stubs to eight inches in length.

The absence of a hump classifies West African Shorthorns as descendants of either the shorthorn or the longhorn humpless or a cross of the two. In the course of several millenniums, migrations of their owners carried both of these ancient types from Egypt across North Africa and down the west coast. Occasionally herds are seen with an infusion of White Fulani blood, but the semblance of a hump definitely places these cattle as crosses. This Fulani cross on the West African Shorthorn is locally called a Sanga, but the animal has no relationship to the Sanga type from which the cattle of East and Central Africa are descended.

A dwarf strain of West African Shorthorn, with a maximum weight of 325 pounds, carries some tolerance to trypanosomiasis and can live at the edge of the rain forest. Wherever found the West African Shorthorn is hardy and well acclimated.

West African Shorthorn cow, 400 pounds. Courtesy Jack Walker

N'Dama.—Some N'Dama cattle, the same as the breed discussed in the chapter on Nigeria, are found in Ghana, mostly at the experiment stations. They are good utilizers of inferior types of forage such as the mature grasses that Ghanaian cattle must subsist on for much of the year. Mature cows average 500 pounds, bulls 600 pounds.

Sokoto.—This is the name used in Ghana for the White Fulani, brought in from Nigeria to be used in upgrading the West African Shorthorn at the Nungua Experiment Station. The animal is described under "White Fulani" in the discussion of Nigeria.

MANAGEMENT PRACTICES

The natural handicaps to cattle raising in Ghana are no greater than those encountered in many tropical or subtropical countries and not as bad as in some. Much of the soil is deficient in phosphorous. Stockwater, particularly in the dry season, from December through April, is a limiting factor on the Accra plains and in most of the savanna in the north. The regions utilized for grazing are generally level or rolling, with fair natural-grass cover interspersed with brush and trees along the water-

Sokoto bull, ten years old, 1,100 pounds. Photographed at Nungwa Experiment Station.

courses. There are practically no fences, cattle are always under the control of herdsmen, and the land is largely "stool" domain. Stool lands are those areas which are controlled by the chiefs of the numerous tribal groups but which can be taken by the government if needed.

The Ghanaian African is not the traditional cattleman found in much of the continent. For subsistence he has depended on a rudimentary agriculture and, to a considerable extent, still does. When the needs of a money economy entered his life, he began to raise cacao trees. Cattle have occupied a lower position in the economy than even their numbers would indicate because little commercial use is made of them. Influential men of the tribes have long been known as traders, and the more wealthy of them have raised cattle and still do in large numbers.

Fulani tribesmen engaged by the Ghanaian owner to care for his stock usually handle the cattle of the country. The Fulani are traditional cattle people from countries to the north. Naturally nomadic, with an aversion toward cultivation, they have wandered into Ghana as hired herdsmen, getting in return for their services the evening's milk and a percentage of the offtake. The common practice is to separate the cows from calves when on pasture. When the cows are brought in at night, they are milked, theoretically leaving some for the calf; in practice, however, this amount is very small. The calves are not taken off until morning, when the cows are returned to pasture; and in the crowded kraals the calves get what milk they can during the night. The cows are poor milkers at best, and these practices brought into the country by the Fulani herdsmen result in an undernourished calf. Particularly during the dry season, animals must be driven long distances to find even poor pasture. Bulls are left uncastrated and eventually sold at six or seven years of age. The calf crop averages about 35 per cent.

West African Shorthorn cattle returning from water.

There is the occasional Ghanaian stockman who manages his own herd, in accordance with better practices. Cows are kraaled at night for security but are not milked, and the calves run with their mothers on pasture. Such a stockman lives near a stockwater dam, has a substantial amount of land under fence, and manages his pastures so that mature growth is available to carry his stock through the dry season. Bulls are run separately from the cow herd until marketed at five or six years of age—one or two years younger than those of the typical Ghanaian owner. Only the best bulls are left with the cow herd. On the Accra plains a stocking rate of six to ten acres a head carries the small cattle satisfactorily in a normal year. Under this kind of management the West African Shorthorn reaches weights of more than 600 pounds at four and one-half to five years of age.

Such operations are exceptional but are gradually increasing because of the high price cattle are bringing. Successful businessmen are beginning to enter agriculture, including cattle raising; and when this happens, management along the above lines is followed.

Some state farms patterned after those in the Soviet Union have been established. Varying in size from a few hundred to several thousand acres, they have not been very successful because of ineffective management. One of these state farms in the north attempted cattle breeding on a large scale but with very poor results. Brucellosis infected the herd, and calving rates were less than 35 per cent.

Commercial dairying does not exist in Ghana. The only milk produced is that which the Fulani herdsmen receive as compensation for their services, and this is consumed for household use or miscellaneous sale in the villages. Various attempts to introduce the European milk breeds have been made, including the use of dairy-type bulls for crossing on native cattle; but without exception, the exotic animals have deteriorated in the tropical environment for lack of special care and the necessary prophylactic measures. Milk production of the native cattle for a six-month period, as determined in an experiment station of the former Ministry of Agriculture, averaged five pounds a day for the West African Shorthorn, six pounds for the N'Dama, and nearly seven pounds for the Sokoto. These averages covered a 182-day lactation period and

Typical Mali steer at Accra cattle market.

were for cattle on pasture only. A limited supplemental ration of four pounds of grain a day practically doubled the milk yield.

MARKETING

Most of the cattle slaughtered originate in Mali, Niger, and Nigeria and are trailed across Upper Volta and into northern Ghana. From there they are trucked to cattle markets at Accra and other population centers. Of the 120,000 head of cattle slaughtered annually, 100,000 are from these sources. At the market traders sell imported cattle and any animals they obtained in the country to the butchers. Sales are by the head at extremely high prices: a young, emaciated animal weighing 440 pounds that would not dress out 50 per cent sold for $154 in 1966, equivalent to 35 cents a pound liveweight. Older bulls brought the equivalent of 30 cents a pound.

ABATTOIRS

Several years ago the use of small slaughter slabs dispersed throughout Accra was outlawed. Now the principal abattoir in that city is located on the seashore and handles cattle, hogs, sheep, and goats. The new facility is an open building with concrete floors, the only sanitary provision being gutters for the drainage of blood and refuse to the sea. Killing is done in the approved Moslem manner, described under "Abattoirs" in the section on Sudan, with the notable exception that after being thrown and tied, the animal is stunned by the captive bolt before its throat is cut. The animal is dressed and skinned on the floor, and the carcass is quartered and hung on hooks. Delivery to retailers starts at 6:00 A.M. and everything is disposed of the day it is killed. Conditions are unsanitary. A pretense is made at inspection but is quite perfunctory. Live animals are inspected by veterinarians of the Ministry of Animal Husbandry and Health; but post-mortem inspection is conducted by technicians of the Public Health Service, a division of authority which practically eliminates any effective control over meat to be condemned. All processing is handled by employees of the butcher, who purchases his animals in the market for slaughter. Practically the entire carcass, with the exception of the contents of the digestive tract, is sold at retail and is consumed. The meat and the bones are cut into chunks and sell for 90 cents a pound. The high-income trade is supplied with either frozen imported beef or selected cuts processed to meet their demands.

A government-financed abattoir was completed in 1965 in Bolgatanga, a trucking point in the extreme north for cattle trailed in from Upper Volta. The objective was to do the processing close to the source of supply of the imported stock. Very few cattle were being killed there in 1966. Only a limited number of refrigerated trucks are available for transporting meat the 500 miles to Accra. The main drawback, however, is that a sufficient supply of cattle is not available. The cattle traders can do better taking their animals down to the coastal area. Another modern government abattoir is being built at the Tema industrial complex, near Accra. The plan is for this plant to operate in conjunction with a new cattle market at Tema and eventually to slaughter all meat for the Accra area.

CATTLE DISEASES

Because of the sparse cattle population, an average of 6 animals to the square mile, endemic diseases do not take the toll that might be expected. Rinderpest, which has now been stamped out in much of Africa, is the most serious. This disease, pleuropneumonia, and anthrax are handled by prophylactic treatment in the areas involved when outbreaks occur. Brucellosis is widespread but there is no program for eliminating it. Foot-and-mouth disease is said not to be serious.

Veterinary service for vaccination against the common diseases is free, as is castration of bull calves, but neither service is widely used. An intercountry control program, composed of several West African countries including Ghana, has been initiated for the control and eventual elimination of infectious diseases. Eight quarantine stations have been established along the frontiers and effect some degree of control on cattle being brought into the country.

GOVERNMENT AND CATTLE

In Ghana the usual propaganda of the new African states concerning agricultural development has included cattle raising, but very little has been accomplished. In 1965 a Ministry of Animal Husbandry and Health was separated from the Ministry of Agriculture to assume control for accelerating livestock production. This action merely added more heads and vehicles to draw on badly depleted government funds.

At the Nungwa experiment station, established by the British in colonial days, an artificial insemination program was initiated in 1965 under the direction of a United States AID advisor. The objective is to upgrade the West African Shorthorn by the use of bulls of exotic breeds. Frozen semen in nitrogen containers is being imported from the United States. The breeds being utilized are American Brahman, Santa Gertrudis, Hereford, and Angus; some Holstein-Friesian are used for milk production. In 1966 this work had not progressed to the point at which an evaluation of crosses from these different breeds could be made.

OUTLOOK FOR CATTLE

Ghana contains the physical elements necessary for a cattle industry which could easily supply the country's beef requirements. The country could even grow sizable surpluses for export, instead of importing 100,000 head annually. Fair to good grasslands could be made much more productive than they are at present by fencing for grazing control and by providing adequate watering facilities. All parts of the country have sufficient rainfall to supply stockwater if properly located dams and wells are provided. Much could be accomplished under present conditions by selecting the best bulls for breeding; castrating the others, or at least running them in separate herds; and breeding for shorter periods to permit calves to be dropped at the end of the dry season.

As grown today in Ghana, the West African Shorthorn is a poor beef animal but, because it comprises most of the cattle population, would have to be the starting point of any upgrading program. Even this small animal, however, responds to adequate nutrition and could bridge the gap to better livestock until it could be improved.

In some west coast countries the well-acclimated N'Dama has been developed into a good beef animal. As a milk animal, which the country sorely needs, the Sokoto, or White Fulani, the milk producing capacity of which has been demonstrated in Nigeria, is already acclimatized and, if properly selected, would prove more productive than the European dairy breeds.

The obstacles to such attainments as these seem insurmountable today. The cultural fundamentals of the African are not attuned to the type of management necessary. That the Ghanaian tribes are not traditional cattle-raising people might not be a handicap, however, for if they begin to grow cattle, they will not have to unlearn the poor management practices of the nomadic tribes who run the large African cattle herds in other countries. As mentioned before, there are a few Ghanaian stockmen who do follow good management practices.

Stability in government would be essential to provide the atmosphere of law and order which would have to precede the heavy financial investment required to make significant advances in the cattle industry. A start was made in this direction in 1965 when the communistic dictator was removed, but it is too early to appraise the true course of this change. All things considered, the development of a cattle industry in Ghana is a long-range program for the future.

Liberia

Land area (sq. mi.):	43,000
Population (1966):	1,065,000
Density (per sq. mi.):	25
Agricultural (75%) (1964):	800,000
Cattle population (1964):	28,000
Year visited:	1966

THE REPUBLIC OF LIBERIA lies at the southwestern corner of the African hump, 4 to 8 degrees north of the equator. The country is bounded by the Atlantic Ocean on the south, Sierra Leone on the west, Guinea on the north, and the Ivory Coast on the east. Most of the surface is low plateau, largely rain forest in its original state, and rises abruptly from the coast. The climate is tropical with more than 100 inches of rain in the east, decreasing to 75 inches in the west. There is normally a six-month dry season, ending in May. The area is nine-tenths that of Mississippi and the population is slightly more than half that state's.

The origin of Liberia goes back to the freed slaves returned to the continent from America and from slave ships intercepted on the high seas. The descendants of those first citizens, now possibly 45,000 of a total population of a little more than 1,000,000 are the elite, recognized as "honorables," and constitute the backbone of the government.

Independent since 1847, Liberia has had self-government longer than any other African-ruled state, with the exception of Ethiopia (if the five years of Italian occupancy of Ethiopia are overlooked). While life in general goes along in quite an elementary fashion, stability in the basic human relationships is maintained. There have been no coups and no expropriation of foreign enterprise. Progress may have been slow, but it has been much surer than in the newly independent states.

CATTLE BREEDS

Liberia probably has not more than 28,000 head of cattle, less than 4 for each 100 human inhabitants, the lowest ratio in all of Africa. Two indigenous breeds of African cattle, neither native to the area, are seen.

Muturu.—The tough little Muturu has worked its way into the rain-forest coastal area from the Ivory Coast. It is of little commercial importance and only a few head reach the slaughter slabs in Monrovia. This breed, described in the chapter on Nigeria, is remarkably tolerant of the trypanosomiasis of the tsetse fly and can subsist on the coarsest forage. It is predominantly black or patched black and white, with small horns and a rather compact body of poor conformation. Females in good condition weigh between 350 and 400 pounds, and bulls weigh not more than 500 pounds.

The few thousand head that inhabit the southern coastal region seem to be something of village or household pets.

Muturu cow of a native owner brought into Suakoko Central Agricultural Experiment Station for breeding.

N'Dama.—This breed apparently entered the country in recent times across the northern borders with Sierra Leone and Guinea. It is trypanosomiasis tolerant, as any bovine species has to be to exist in Liberia, all of which is infested in varying degree with the tsetse fly. More than 20,000 head of this breed are said to be scattered over the northern part of the country.

N'Dama bull with his horns blunted. Photographed at Suakoko Experiment Station.

N'Dama cow herd. Photographed at Suakoko Experiment Station.

The N'Dama is discussed in the chapter on Nigeria.

No milk cattle are maintained in Liberia, and no serious effort has been made either to develop local cattle for milk or, with one minor exception, to introduce any of the dairy breeds for this purpose. Some years ago efforts were made at the Suakoko Central Agricultural Experiment Station to introduce exotic blood for upgrading native cattle. Hereford, Angus, America Brahman, and Brown Swiss cattle were imported in small numbers, but practically all of them died of trypanosomiasis within a few months. One Brown Swiss cross and three head of the American Brahman–N'Dama cross were all that remained in 1966.

American Brahman first cross on a N'Dama cow. Photographed at Suakoko Experiment Station.

MANAGEMENT PRACTICES

Village dwellers own the cattle of the country, mostly N'Dama and a few of the Muturu breed. They are not cattle people, and the few head they hold are incidental to their main farming pursuits. Their small individual holdings are pastured on abandoned cultivated plots which have come back in grass and are reverting to jungle. As in much of the good rainfall area of Africa, the local system of cultivation consists of clearing a small area and cropping it for a few years before abandoning it and moving to a fresh plot. There is also roadside grazing.

Two small herds of cattle are run under modern management methods. The Suakoko Experiment Station, after abandoning the idea of introducing exotic blood, concentrated on improving the N'Dama by selection within the breed. This work was proceeding in 1966 with a herd of 70 head run under fence on a limited area of improved pasture. Both cows and calves were weighed monthly in a selection program designed to develop a larger animal. The research department of the Firestone Rubber Company has another small herd of N'Dama cows, which are pastured in areas unsuited to the planting of rubber trees. These cattle are maintained in excellent condition and show what the N'Dama will do on reasonable feed.

MARKETING

The limited supply of slaughter cattle is purchased by a butcher, either from the owner or from a trader. The butcher does his own processing and retails the meat. A few of the small Muturu cattle find their way to Liberia from the south; the rest of the slaughter animals are N'Damas from the northeast and imports either driven across Guinea from Mali or brought by small ships from Senegal. Cattle are usually trailed on the hoof to the vicinity of the slaughter slabs by the Mandingos, African tribesmen who know no particular boundaries and specialize in cattle handling. Occasionally a small animal will arrive by bus, standing in the aisle between the two rows of seats occupied by other passengers. A thin ox weighing 700 pounds sold for $220 in 1966, or 30 cents a pound liveweight.

ABATTOIRS

There are two slaughter slabs in Monrovia, both of which lack all sanitary provisions, and a somewhat better one at the Firestone Plantation trading center. The annual kill is approximately 5,000 head, the majority of which are imported. The butcher takes the fresh meat to his shop as soon as it has been dressed. All meat consumed by the higher income brackets is imported frozen from Denmark, Holland, or the United States. Steak sells for $2.25 a pound, hamburger for $1.00 a pound.

As an aid project the Yugoslavs built a modern abattoir in Monrovia. Completed early in 1966, it was designed to kill 80 head of cattle, 200 sheep, 50 hogs, and 3,000 chickens a day. The sad feature of this project is that no provision had been made for a supply of slaughter animals or for the operation of the plant; the Yugoslavs only intended to provide the facility, not to maintain or run it. The capacity of the slaughterhouse was designed for five times the annual kill of cattle in the country.

CATTLE DISEASES

Because of the limited number of cattle, disease is not a major problem. Suakoko Experiment Station found brucellosis in its herd and now vaccinates heifer calves at seven to eight months of age and sprays for external parasites. These are probably the only systematic prophylactic measures employed for cattle in Liberia except for those used on the Firestone Plantation herd. The cattle in the country are widely segregated, and there is very little movement of them from one location to another. In a few instances on record in which small numbers of cattle were moved appreciable distances, losses were heavy, probably because of rinderpest, pleuropneumonia, or anthrax, although no accurate identification of the cause of death was made. Generally, however, the tough and well-adapted Muturu and N'Dama cattle seem able to care for themselves.

GOVERNMENT AND CATTLE

The Central Agricultural Experiment Station, at Suakoko, is the larger of the two facilities of this nature maintained by the Department of Agriculture. It was established fifteen years ago by a team sponsored by the United States Department of Agriculture and for a number of years has been under the direction of American-trained native Liberians. The general indifference to cattle raising in the country could be the cause of a flagging interest in such endeavors.

OUTLOOK FOR CATTLE

Formidable obstacles stand in the way of establishing sizable cattle-raising operations in Liberia. First, there is the general languor of the people and the fact that cattle raising is not in their tradition, and second, there are no natural grasslands. Establishing and maintaining productive pastures would require a continuing fight with the jungle. Disease problems would follow any large increase in the number of cattle grown. For some time to come, any expansion in stock raising will probably remain in the planning and report stage, where it has been for a number of years.

Nigeria

Land area (sq. mi.):	356,700
Population (1965):	57,500,000
Density (per sq. mi.):	161
Agricultural (42%) (1963):	24,100,000
Per capita income (1963):	$98
Cattle population (1960):	9,500,000
Year visited:	1965

THE NIGERIAN FEDERAL REPUBLIC lies on the southern coast of the African hump, wholly within the tropics, and covers an area nearly the size of Mississippi, Louisiana, and Texas combined. By some estimates the population is placed at three times that of those states. The country is bounded by Niger and Dahomey on the west, Niger on the north, the Gulf of Guinea on the south, and Chad and Cameroon on the east. The coastal plain rises to a hilly belt opening in the north onto a savanna plateau, which finally becomes desert. Rain forest extends as far as 150 miles inland, with more than 60 inches of rainfall, which gradually decreases to 20 inches on the northern savanna.

What is now Nigeria was a major slave-trade center during the first part of the nineteenth century. In 1861 Britain made Lagos the capital of the colony and by 1901 had brought all Nigeria under control and ended the slave trade. The colony and protectorate of Nigeria, which included all of the present republic except the British North Cameroons, dates from 1914. After a plebiscite in 1959, the Northern Cameroons joined Nigeria; the British Southern Cameroons elected to join the Federal Republic of Cameroon.

Except for the Fulani tribes in the north, who basically are a non-Negroid people, the population is almost entirely Negro; there are fewer than 30,000 non-Africans in the country. The tribal society was highly organized and the policy of the British was to use the native chiefs and kings for local administration in establishing and maintaining control. This plan was not easily accomplished because strong antagonisms existed then, as they do today. Such differences are the basis of the political conflicts that have kept Nigeria in turmoil ever since independence was attained in 1960. Four regions, each of which to a large extent is self-governing, along with the Federal Territory of Lagos form the Nigerian Federal Republic. They are all subject to wide cultural, religious, and tribal differences and are in a perpetual state of unrest. The unresolvable differences between the Moslem element, which is the majority, and the Christian element of the population greatly aggravate these tribal conflicts. In 1965 the length of time the present federation could continue appeared to be an open question. There were two military coups in 1966. In January an uprising dominated by members of the Ibo tribe deposed the government, and six months later the Ibos were in turn deposed. The Eastern region, the Ibo stronghold, seceded from the Federation in 1967 as Biafra. After two and one-half years of brutal warfare, with many losses among the civilian population from starvation, Biafra was crushed to the ground and surrendered in January, 1970.

The economy is largely agricultural. The typical Nigerian farmer cultivates from one to five acres, although in western Nigeria some cacao farms are as large as thirty acres. The principal exports are peanuts, cacao, palm oil, and palm nuts. Cattle, which are raised in large numbers in the northern savanna country by the Fulani tribe, are not an important item in the economy.

CATTLE BREEDS

White Fulani (Sokoto).—The outstanding breed of Nigeria, the White Fulani, or Sokoto, could be developed as a dual-purpose, milk-meat animal. The cows are milked by the nomadic owners, and oxen are used to some extent for draft.

Though the breed is generally considered a Sanga derivative, its horns are not as large as those in most Sanga types. Usually they are of medium length and thick, rising from the head upward and forward. The hump is fairly large, quite noticeable in the female and often lopped over in the male. The dewlap is prominent in both sexes, as is the sheath in the male and the umbilical fold in the female. The hair color is a nearly solid off-white, often with small, scattered red or black markings around the head. Cows weigh from 900 to 1,000 pounds, bulls up to 1,400 pounds.

Red Fulani.—This more typical Sanga derivation has long, outswept horns, quite heavy and up to three feet or more in length. The hump is large; the dewlap, sheath, and umbilical fold are prominent. Although rangy with a poor beef conformation, the breed is the principal

White Fulani bull.

White Fulani cow.

slaughter animal in Nigeria. Milk yield is small and not much used. The Red Fulani is somewhat larger than the White Fulani, cows in good condition weighing 1,000 pounds or more and bulls weighing as much as 1,500 pounds.

Red Fulani bulls being trailed to the Lagos market.

Muturu.—This dwarf breed is found in the rain-forest belt, which follows the coastline across Nigeria for distances of as much as 100 miles inland. The breed is locally called the West African Shorthorn. It is a descendant of the short-horned, humpless cattle that were brought into Egypt before the dawn of recorded history and was carried by the early migrations along the Mediterranean coast and then down the Atlantic coast to West Africa. Natural selection and a low nutrition level apparently have resulted in the dwarf type seen today. Mature cows weigh 350 to 400 pounds, bulls a maximum of 500 pounds. The height at the withers is not much over three feet. The horns are small and outswept. Black and black and white are the predominant colors, although reds and tans, both with black or white markings, are also seen. The animal is compact and well built and, although not utilized for beef to any extent, has an unusually high dressing percentage when slaughtered. There is no hump and hardly any dewlap.

The Muturu is a very hardy animal and is probably the most tolerant of all African cattle to the trypanosomiasis of the tsetse fly, which infests the whole rain-forest area. Although trypanosomes, the protozoans carried by the tsetse fly, are found in the blood streams of these cattle, they exhibit no symptoms of the disease and suffer no loss in condition.

Practically no commercial use is made of the Muturu, although its total population is considerable. Many native villagers own a few head of

Muturu cattle and use them primarily as pets or as scavengers around the huts. The cows are rarely milked and the breed is seldom utilized for draft. Their ownership is another example of the African's inborn desire to have cattle, even though they are held by agriculturalists and not cattle people.

Muturu cow.

Keteku.—Supposedly the result of indiscriminate crossing of the White Fulani with the Muturu, the Keteku is larger than the latter, averaging perhaps 100 pounds more. Less distinctive in type than the Muturu, the Keteku is more varied in color, which is predominantly off-white with small black markings. The Keteku is raised by the villagers in the transition zone between the rain forest and the highlands in the north. Although the breed is definitely humped, the hump is small and frequently not noticeable in the female. The animal also carries a moderate tolerance to trypanosomiasis, although not to the same degree as the Muturu. Very little use is made of these cattle except incidentally for milk for household use.

N'Dama.—This breed is indigenous to the coastal rain-forest area west of Nigeria. It was introduced in Nigeria because of its tolerance to trypanosomiasis and because it is more productive than the Muturu or the Keteku. Authorities differ about whether the N'Dama is a descendant of the humpless longhorn cattle of ancient Egypt or of the Brachyceros. Admixture with Zebus and with Sanga types probably did not

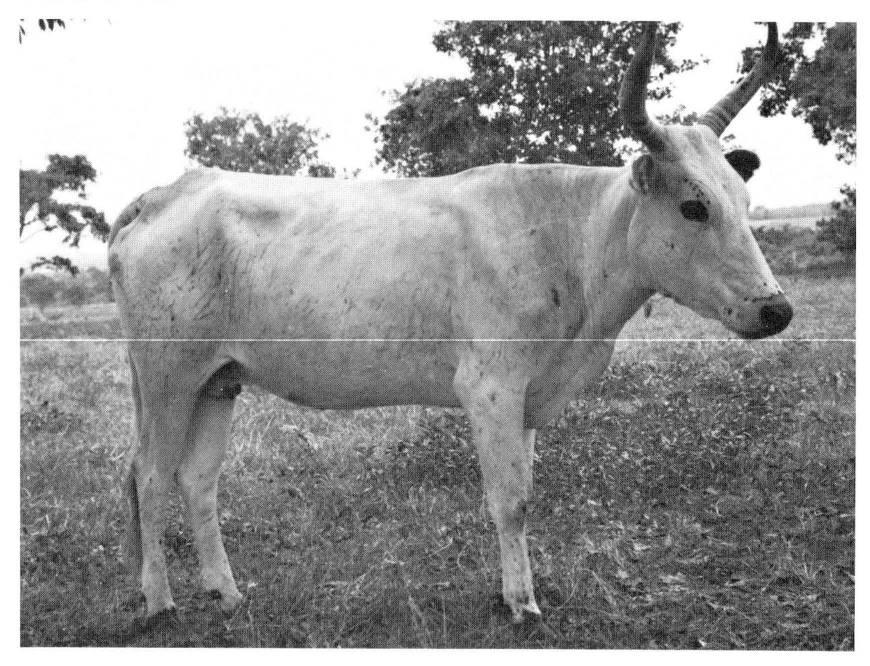

Keteku cow.

occur to any appreciable extent, perhaps because these animals are susceptible to trypanosomiasis and therefore did not inhabit the same areas as the longhorn cattle.

The N'Dama is nonhumped, quite compact in conformation, well-muscled, and has a rather fine bone structure. Uniformly tan, reminiscent of the Jersey, the breed sometimes has white spots on the underside and around the neck. It has good fleshing qualities; on fair nutrition mature cows weigh 600 pounds or more, and bulls weigh as much as 900 pounds. The breed does well on rough forage and the low-protein grasses of the tropics. The horns are usually of medium size, spread outward and upward. Although the breed carries a strong polled tendency in other localities, the animals in Nigeria seldom exhibit this trait.

Holstein-Friesian.—With minor exceptions there are no representatives of European cattle in Nigeria. A few years ago a small herd of Holstein-Friesian cows was introduced in an experiment station of the Department of Agriculture; but, undoubtedly because of climatic stress, it has not done well, even with better-than-average care. The cows were producing

N'Dama bull. Photographed at Fashola Experiment Station.

N'Dama cow. Photographed at Fashola Experiment Station.

an average of 3,000 pounds a lactation in 1965. Another small Holstein-Friesian herd was imported in 1964 from the United States for an AID project near Lagos. It was being maintained with exceptional care under confined conditions; yet, after being in the country for one year, the herd was yielding an average of only 7,000 pounds of milk for a 315-day lactation. In 1967 the Ministry of Agriculture and Natural Resources of Western Nigeria imported a group of about 100 head of Holstein-Friesian, Jersey, and Brown Swiss cows from the United States for experimental work at Ikanne.

MANAGEMENT PRACTICES

Most of the cattle of the country are run by the nomadic Fulani tribe on the savanna plateau to the north of the cultivated areas. This relatively high region has a rainfall of 20 inches or more, which falls mainly from June to September, with no precipitation from October through May. Utilization of the potential grass growth is poor because of overgrazing, lack of watering facilities, and no planned provision for reserve areas to supply feed during the dry season. Whatever condition an animal achieves in the growing season from June to October is maintained fairly well until December by grazing on the matured grass or stubble fields where available, but is lost by the time the rains start again. The seasonal movement is south at the beginning of the rains and north when the grass is gone. There is some grazing on crop residues where the savanna borders the cultivated areas. The trail routes are determined by the availability of water, and the resulting concentration leads to extreme overgrazing.

Poor nutrition under such management results in cows calving only every two or three years, and calfhood mortality is about 50 per cent. Bulls are not castrated and normally run with the cows the year round, although there are exceptional tribesmen who hand-breed their best bulls to the best cows. The principal interest of the native cattle owner is in cattle numbers—not quality. Ownership varies from 20 to 50 head for a tribesman, while a chief's herd may run well up in the hundreds. Cattle are sold reluctantly and only when there is a need for money. The annual offtake is probably less than 2 per cent—practically all of which is bulls, since it is illegal to kill females capable of reproduction. Milk is taken regularly for family use or for barter with the villages passed in the constant trailing. There is no slaughter for food except on ceremonial occasions. The lobola system prevails in these nomadic tribes.

The cattle in this grazing region probably number about 7,000,000 head, 80 per cent of the cattle in the country. They are all of the White or Red Fulani breeds, with the latter predominating. Their susceptibility to trypanosomiasis confines them to the savanna, where the tsetse fly cannot exist.

Efforts made to settle the Fulani people have had some minor success, but the large majority trail their cattle continuously, searching for grass as they have done for past ages.

In the southern, higher rainfall area, Muturu, Keteku, and some N'Dama cattle are in the hands of the small farmers. An individual owner has a few head of cattle at most. Very little use is made of these cattle except occasionally for draft. The cows are neither milked nor eaten unless they die of natural causes; they are seldom sold, for the owner gets his cash requirements from his crops. Often the livestock is seen in very fair condition because of the individual care they receive.

MARKETING

The normal flow is for the traders to buy cattle from the nomadic tribesmen in the north and hire herdsmen to trail the stock several hundred miles to the Ibadan or Lagos markets. Bulls, usually Red Fulanis, are practically the only animals sold for slaughter and vary from four to ten years of age, averaging seven or eight. This trek to market is ample evidence of the hardiness of these cattle in all but one respect: on nearing their destination, they pass through the tsetse-infested belt that extends across Nigeria above the coast and become infected with trypanosomiasis. By this time, however, they are so close to market that they can be slaughtered in a matter of days before the disease has seriously affected their condition. Some play out along the trail and are sold for village slaughter, but most arrive at the markets in better condition than would be expected considering the distance traveled. Some Kuri cattle destined for slaughter are trailed from Chad, and there are other imports of live cattle from time to time to augment the beef supply of the country.

Cattle markets are open areas, unfenced, on the outskirts of the large towns. All sales are by the head. Bargaining between the trader and the butcher starts late in the afternoon. Prices in the spring of 1965 were equivalent to 14 cents a pound on weights varying from 450 to 1,000 pounds and averaging 600 pounds, or $84.00 an animal. Immediately after the sale is made, the herdsmen whom the butcher employs start for the slaughter slab. The animal is driven through town, one man leading with a rope around the bull's head, another following with a rope tied above the hoof on a rear leg. On arrival at the slab, the animal is tethered until he is slaughtered the next morning.

ABATTOIRS

Southern Nigeria contains no modern abattoirs. Animals are killed and dressed on the municipal slaughter slabs, each of which has a concrete floor, the surface of which is raised slightly above ground level. Shallow trenches are provided to remove liquids. The central slab outside Lagos has a shed roof and electric lights, but most slabs have neither and do not operate until daylight. Killing is in the approved Moslem manner, described under "Abattoirs" in the section about Sudan. The city collects $1.68 for each animal processed—$.28 for inspection and $1.40 for use of the facilities.

Each butcher employs three or four butcherboys, frequently relatives. They assist in getting the animal into position on the slab and in holding it for killing, skinning, and dressing. They are not paid in cash but, in accordance with established custom, are allowed to keep the tail, trimmings, and some intestinal parts, together with the slices of flesh they filch. These gleanings are sold by the boys to small retailers. Slaughter slabs are usually too small for the number of animals handled and, with five people working around each carcass, confusion is rife; yet everything proceeds to the point at which the meat is ready for removal. In Lagos delivery to the butcher's shop is made by municipal trucks and costs 21 cents a carcass. At the smaller slabs in the country, meat is carried away on the heads of the butcherboys. Veterinary inspectors are on duty at the slabs, but condemnations are few. Sanitation varies but is usually better than would be expected, considering the nature of the facilities. Except at the stores which cater to the higher-income bracket and have their own coolers, meat is sold the day the animal is killed. No distinction in the price of the different parts of the carcass is made except for the tenderloin. The carcass is cut into small chunks and sells for approximately 28 cents a pound, the fillets for 60 cents. Most meat would grade no higher than USDA standard. The average kill at the Lagos slab is 170 head a day for a city of 500,000 persons.

Among the modern abattoirs in the north are the one at Kaduna, built by United States AID, and the one at Bauchi, built by the federal government. The four other abattoirs in the region have modern equipment but are not fully utilized. These six facilities, however, process only a small part of the total number of animals killed.

CATTLE DISEASES

Trypanosomiasis is the most serious of the cattle diseases in Nigeria, but it is largely confined to the rain forest and the transition belts along the coast and along the rivers in the north. The trypanosomiasis-tolerant Muturu and Keteku breeds, and to a limited extent the N'Dama, are the only cattle in most of these areas. No efforts for control of the tsetse have been made in the south. In the north some measure of control has been effected in isolated pockets and in an 800 square mile area in the Hadejia River valley.

Rinderpest is widespread but control measures have considerably reduced its incidence. There is some vaccination for blackleg and anthrax. Foot-and-mouth disease is quite general but is mild. Its effect of causing an animal to lose condition for a year is not too serious a matter to growers who hold their cattle only for prestige. Parasitic diseases are common, but there is very little spraying or dipping. A government laboratory at Vom, in central Nigeria, produces several million doses of vaccine a year, but this amount is entirely inadequate for disease control of the more than 9,000,000 cattle.

GOVERNMENT AND CATTLE

Efforts are made at disease control by the Veterinary Services but are not extensive enough to cover the large areas involved. Some of the experiment stations' work is noteworthy, however. Inspection of meat for human consumption is perfunctory.

For twenty years the Fashola Stock Farm of the Ministry of Agriculture and Natural Resources, north of Ibadan, has maintained a strain of N'Dama cattle originally imported from Sierra Leone. Good progeny records had been kept on the herd, which numbered 1,000 head in 1965. Selection for earlier maturity had only recently started but the herd was a remarkably uniform group of cattle, weighing at least 150 pounds more than the average of the breed because of the improved nutrition.

Development of the White Fulani breed for milking ability was undertaken at the University of Ibadan in 1950. The average production is now 3,000 pounds a lactation, with individuals yielding as much as 5,000 pounds. Moore Estates, a nearby livestock center of the ministry, has a small herd of Holstein-Friesians, the seedstock of which was imported from the United States. Management at the two establishments appeared to be equal. Although the average milk production of these cows was the same as that of the White Fulanis, the Holstein-Friesians are definitely inferior in condition. This comparatively low production for Holstein cows is typical of what often happens when an exotic northern breed is brought into a tropical environment. Although the yield of these third- and fourth-generation White Fulanis may not appear very impressive, it is a sizable increase for a breed the best of which ordinarily produce 1,000 pounds a lactation.

OUTLOOK FOR CATTLE

The northern savanna region, the natural grazing area of Nigeria, is badly overstocked under present watering and management conditions. Even if these conditions were modernized and cattle were maintained in good breeding condition, it is doubtful whether numbers could be increased materially. The offtake and weights of already existing herds, however, could be greatly improved.

The productiveness of the present cattle population could be multiplied several fold if the increase were sold at a reasonable age—three and one-half to four years instead of an average of seven or more—and the calving and mortality rates were improved. The present offtake of possibly 2 per cent is low only because most animals are allowed to die from natural causes. Adequate watering facilities for the dry season and control of grazing would give much better utilization of plant growth. Such changes in management methods would infringe on the way of life which the Fulani tribes have known for centuries and will be a long time in attainment.

There is no dairying, and with the growth of the urban population and a gradual improvement in the economic status of the working class, there is a need for milk. Establishing dairy herds would require clearing areas in the high rainfall area and introducing an acclimatized breed, possibly the improved White Fulani, for milk production.

All things considered, it does not appear that the future holds much promise for improvement in any branch of Nigeria's cattle industry for some time to come.

Senegal

Land area (sq. mi.):	76,100
Population (1966):	3,500,000
Density (per sq. mi.):	46
Agricultural (85%) (1965):	2,970,000
Per capita income (1963):	$172
Cattle population (1967):	2,200,000
Offtake (5.5%) (1966):	120,000
Year visited:	1966

THE REPUBLIC OF SENEGAL is bisected by the 15th northern parallel and lies in the transition zone between the Sahara Desert and the tropics of Africa. In area it is equal to Louisiana and half of Arkansas. The rainfall increases from 14 inches on the Senegal River, the northern boundary with Mauritania, to 60 inches at the southern boundary with Guinea and Portuguese Guinea. The eastern boundary is with Mali and there is a 400-mile coastline on the Atlantic Ocean on the west. The narrow enclave of the small independent country of Gambia extends 300 miles into Senegal from the coast. Most of Senegal is savanna or rolling hills, with varying degrees of acacia encroachment. In the extreme south the tsetse fly exists in a narrow forest belt.

In 1444 the Portuguese established the first European foothold in Senegal. Two hundred years later the French began to take over and have been the dominating influence ever since, although the British had control for a short period in the middle of the eighteenth century. Dakar, now the capital, was the seat from which the French administered all of French West Africa in later years, and the people of Senegal have had longer and closer contact with Western civilization than have the other territories that once comprised French West Africa. Since Senegal's independence was proclaimed in 1960, Léopold Sédar Senghor has been president. He had to depose his prime minister who attempted a coup d'état in 1962 and modify the constitution to eliminate the office, but these alterations were made quite peaceably. The population is almost entirely African. Only 25,000 Europeans, mostly French, live in the country, and there is an equal number of Lebanese and Syrians.

Although Senegal manufactures more home-grown products than most other African states, her economy depends mainly on one crop, groundnuts (peanuts). The better land with sufficient rainfall is utilized primarily for this crop, which accounts for 75 per cent of all exports. Much of the remaining land, particularly where the moisture is insufficient for cropping, is utilized by semimigratory tribesmen for cattle raising and to a lesser extent for sheep.

CATTLE BREEDS

The cattle seen in Senegal are indigenous breeds common to various parts of West Africa. There has been practically no introduction of exotic blood except for some recent importations of Zebu breeds for experiment station breeding work.

Gobra.—The dominant breed in the northern savanna is locally called Gobra. This is the same animal as the White Fulani seen in Nigeria and other parts of West Africa and was evidently brought to Senegal by Fulani tribes, a sizable element of the population. The Gobra, of Sanga derivation, is rather small, ranges from light grey to practically white in

Improved Gobra bull. Photographed at the Centre de Recherches Zootechniques de Dahra-Djoloff.

Improved Gobra cow herd. Photographed at the Centre de Recherches Zootechniques de Dahra-Djoloff.

color, and has large, upswept horns, often with a lyrelike twist toward the end. In fair condition the native cow weighs 650 pounds, the bull 850 pounds.

N'Dama bull of the north. Photographed at the Centre de Recherches Agronomiques Tropicaux de Bambey.

N'Dama cow of the north. Photographed at the Centre de Recherches Agronomiques Tropicaux de Bambey.

A variation of the Gobra, the Peul (French for Fulani), is seen in limited numbers. It is similar to the Gobra in all respects except color, which is a light tan, often with white or red markings.

Maure.—Conformation and general appearance of this animal are similar to those of the Gobra. The Maure is, however, red or red mixed with white. The breed is a better milk producer than the Gobra and is seen in the northern part of the country but in fewer numbers than the Gobra.

N'Dama.—In the tsetse fly infested rain-forest area of the south, only the sturdy little N'Dama cattle, with their highly developed tolerance to trypanosomiasis, are maintained. Better developed specimens of the breed are found in limited numbers in the northern grazing region, but in the denser forest area, nutritional deficiency for generations has resulted in a small animal, often polled or with small, loose horns. The N'Dama accounts for perhaps one-third of the total cattle population but is of little importance commercially.

Djakoré cow, the result of unplanned crossing of the N'Dama and Gobra breeds. Photographed at the Centre de Recherches Agronomiques de Bambey.

Djakoré ox team. Photographed at the Centre de Recherches Agronomiques Tropicaux de Bambey.

Djakoré.—In the area between the savanna and the rain forest, the Gobra and the N'Dama have interbred for generations under uncontrolled conditions. This practice has resulted in a distinctive animal, locally known as the Djakoré. It carries the tan color of the N'Dama, solid and quite uniform but usually lighter. The hump is much smaller than that of the Gobra and frequently is not discernible in the female. The horns are intermediate in size between those of the parent breeds. This animal carries a fair measure of the trypanosomiasis tolerance of the N'Dama. The Djakoré does not account for more than one-tenth of the total cattle population. It is an excellent draft animal and is used for that purpose; it is not generally sold for slaughter.

Gobra-Djakoré Cross.—The Centre de Recherches Agronomiques, an experimental station established at Bambey by the French in colonial days, is undertaking the development of an improved draft type by crossing Gobra bulls on Djakoré cows. It intends to eventually establish a new breed of this cross, which as yet has not been named. Although draft animals are not widely used in Senegal, there undoubtedly is a place for them in agricultural development, since most cultivation is still done by hand. The objective is to produce a better draft animal

First cross of a Gobra bull on a Djakoré cow. Photographed at the Centre de Recherches Agronomiques de Bambey.

while retaining the trypanosomiasis tolerance which the N'Dama blood has given the Djakoré.

Exotic Breeds.—The only exotic breeds which have been brought in, at least during recent years, are three Zebu types now grown at the Centre de Recherches: Red Sindhis and Sahiwals from Pakistan and Guzerats from Brazil have been imported since 1963. Small purebred herds of these breeds are maintained for the production of bulls that are employed in experiments aimed at developing a better milk animal from the Gobra cow. There are no representatives of the European milk breeds in Senegal.

MANAGEMENT PRACTICES

The savanna area in the northern half of the country is mainly utilized by tribal cattle owners. (There are no European cattle owners in Senegal.) During the rainy season, which varies in length from three to four months during July to October, the good grasslands of the Valée du Ferlo are utilized by tribesmen owning 40 or 50 head of cattle or more and chiefs, who may maintain 1,000 or more. As the rains cease, the cattle movement is in two directions—north to the Senegal River valley and south to the cultivated areas where, after harvest, the peanut vines are consumed as forage. Cattle owners usually have habitations in villages in both the dry- and wet-season areas in which they operate. Only at the experiment stations are modern practices in animal husbandry used. The African tradition of keeping cattle just to have them is showing some evidence of change because of an increasing tendency in Senegal toward having a money economy. The owners have more need for cash; therefore they sell more animals to satisfy it.

Milk is taken by the African growers for their own use and to some extent is traded for other needs. Cattle are not ordinarily slaughtered for food by the owners.

Stockwater is always the limiting factor in cattle raising. During the eight- to nine-month dry season, the general practice is to water cattle every two days. They are driven long distances, perhaps ten miles, either to government watering stations supplied by deep wells or to small tribal wells utilizing a rope and a bucket. Usually a good volume of water is found at approximately 600 feet; at the government wells, concrete troughs in the form of a cross, spreading 100 feet in each direction, are supplied by deep well pumps with ample water for several hundred head at a time. Throughout the day owners drive their herds to these points, and the cattle are held waiting their turn to drink. It is surprising that the cattle are easily controlled when this method of watering is used. More than 1,000 head, which have had no water for forty-eight hours, will be congregated at these watering points in various-sized groups according to ownership. Each group is controlled by two or three men or boys, usually on foot but sometimes mounted on riding oxen. The same procedure is followed for the smaller numbers watered at the hand-operated wells in the more remote locations.

Gobra cattle at a government watering station in northern Senegal.

It is evident that the pasture areas near the watering places are poorly utilized. For a considerable distance around them, every vestige of plant growth except a few trees has been removed; and, as the dry season advances, the perimeter of the area which can support stock gradually recedes. At the extreme limits of the grazing areas, good grass is never touched. Much of the savanna is classified as *Forêt Classe*, which means that cultivation is prohibited. Without such a limitation, population pressure could soon bring ill-adapted land under cultivation for dry-land grains, with eventual dust bowl results, and even though the land is left in grass, severe erosion will eventually take its toll. The high concentration of cattle in the grazing areas of Senegal accentuates this problem.

Although cattle are seldom used for human transportation in Africa, in the cattle country of Senegal this practice is sometimes seen.

MARKETING

Cattle marketing is in an elemental stage. A small dealer in the growing area purchases a few head from individual owners, frequently men under stressed circumstances involving an immediate need for cash. The going price in 1966 was $30.00 to $50.00 a head for a steer five years old or more, weighing 450 to 750 pounds. Most male cattle are eventually

Riding ox.
Photographed in northern Senegal.

castrated, but this is not usually done until the animal is one or two years of age. A mallet is used to pound the testicles on a wood block. It is illegal to sell females for slaughter without a veterinarian's certification that the cows are nonproductive or to sell any animal less than four years old for slaughter.

The first trader consolidates his purchases and sells to a second trader, who moves the livestock toward Dakar or another of the larger towns where commercial slaughtering is done. It is not unusual for cattle for slaughter to thus pass through the hands of several traders before reaching the market near the abattoir in Dakar. Here the ten butchers who dominate the meat trade pay $60.00 to $90.00, depending on size, for an animal that returned only half this price to the original owner and has probably shrunk 20 per cent since the original sale. Live cattle prices would, therefore, be the equivalent of 6 to 6.5 cents a pound to the grower and 11 to 13 cents to the last trader, who sells to the butcher for slaughter. All sales are by the head, with quality having only a minor influence.

ABATTOIRS

The municipal abattoir in Dakar, by far the best in the country, was built by the French in 1956. The only modern elements are the well-constructed building, with an ample display of white tile, and good holding yards, all of which could easily accommodate several times the 150 to 200 head of cattle that comprise the average day's kill. Slaughter is handled in the typical Moslem fashion, described under "Abattoirs" in the discussion of Sudan. There are overhead rails and hoists; but the carcass is split with a hand meat saw, a process which requires fifteen minutes an animal. Refrigeration for some imported frozen meat and also for a twenty-four-hour cooling period for most of the output is available, although some carcasses are sold warm to retailers who so demand. Offals, hides, heads, intestines, and feet are handled in a deplorable manner in a paved area adjacent to the killing floor. Veterinarians make reasonably competent inspection. The average liveweight of animals slaughtered is 625 pounds, yielding a dressed carcass of 330 pounds. The fee for slaughtering is 1.5 cents a pound. No by-products are recovered.

The butcher who buys the animals for slaughter and has them processed in the municipal abattoir sells the dressed carcass to a retailer at 25 to 30 cents a pound. Selected halves sold to dealers catering to the European trade bring a higher price, about 35 cents a pound.

The total number of cattle slaughtered annually in Senegal is estimated at 170,000, of which 40,000 are driven in from Mauritania.

CATTLE DISEASES

Because of the conditions under which cattle are handled in Senegal, disease control does not appear to be a serious problem, in spite of the near-tropical climate. The fact that all the cattle are indigenous and well acclimated also is a major factor. Rinderpest and pleuropneumonia are the most serious diseases, and the cattle in the research centers and those of a few more advanced growers are vaccinated against these diseases once a year. Senegal, Mauritania, Mali, Guinea, Gambia, Ivory Coast, Sierra Leone, and Chad joined together in 1966 in a three-year campaign against rinderpest.

In northern Senegal there are not many parasites. Toward the south there is some tsetse fly infestation. The trypanosomiasis-resistant N'Dama is the principal breed here and takes fairly good care of itself. Brucellosis is said not to be much of a problem, but with the conditions under which cattle are raised it is doubtful whether losses due to the disease are known. Foot-and-mouth disease does not cause much trouble. The numerous veterinary stations established under the Department of Animal Industries in the cattle areas furnish free prophylactic treatment, but it is not widely utilized.

GOVERNMENT AND CATTLE

The government experiment stations, which have been mentioned, work along practical lines. The Centre de Recherches de Dahra-Djoloff has shown that under controlled conditions in the 20-inch rainfall area, eleven acres of the savanna will adequately support a 900-pound cow and her calf. If grazing is not controlled and sufficient mature growth left for the dry season, twenty acres are required and the cow will be in poor condition.

Through several generations of selection within the breed, the Gobra has been developed to the stage at which the average weight of the station's cow is 900 pounds and the bull 1,100 pounds, compared to 650 pounds for the cow and 850 pounds for the bull of the African cattle owner. Part of this increase results from better nutrition, but the genetic improvement also has been substantial.

Efforts to develop a milk-type animal involve, first, selection within the Gobra breed and, second, crossing it with exotic Zebu types which will acclimatize more readily than the European breeds. There are no commercial dairy cattle in the country, and a good milk cow is something badly needed.

At the Bambey research station the N'Dama breed has been under selection for size for more than twenty years. A greatly improved type has resulted. There is also a program here to develop a better draft animal from the Djakoré breed—a grass-roots approach to a real need of the country.

The desire of the government to aid the cattle industry in its day-to-day operation is seen in the many veterinary stations which have been established, the reserving of grasslands from cultivation, the attempt to increase the breeding herds by limiting the slaughter of females, and the provision of watering places.

OUTLOOK FOR CATTLE

The indicated offtake of slaughter animals is 120,000 head, only 5.5 per cent of the estimated 2,200,000 cattle population. It is a little above average for African-owned cattle. This cattle population is only two-thirds that of Louisiana and the half of Arkansas with which Senegal has been compared in area. With good management—installation of fences for grazing control and adequate provision for stockwater—cattle numbers could be materially increased. Two elements are lacking for such an accomplishment. First, with the economy primarily dependent on peanuts, there is not the financial ability to support it; and second, the African cattle owners are well satisfied with their seminomadic life. Although they sell a somewhat larger proportion of their cattle than some other African growers, they do so rather reluctantly. Given the means to increase their herds, it is questionable whether they would sell many more until they move closer to a money economy.

PART FIVE: Asia

Asia

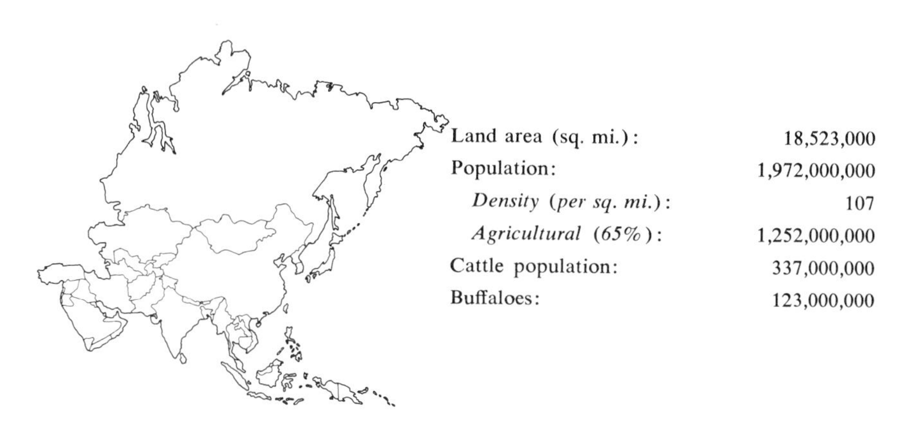

Land area (sq. mi.):	18,523,000
Population:	1,972,000,000
Density (per sq. mi.):	107
Agricultural (65%):	1,252,000,000
Cattle population:	337,000,000
Buffaloes:	123,000,000

THE great land mass of the Eastern Hemisphere that lies between the Mediterranean Sea and the waters of the Indian Ocean on the south and the Arctic Ocean on the north has been arbitrarily divided at the Ural Mountains into Asia and Europe. Asia is the area to the east of the Urals and the Ural River. After the Ural River enters the Caspian Sea, the generally accepted boundary between the two continents follows the Caspian Sea and the Caucasus Mountains, which separate Russia in Europe from Turkey in Asia. Then the waters of the Aegean, Mediterranean, and Red seas finally conclude the western boundary. Also included in Asia are the islands lying in the arc off the eastern and southeastern coasts—Indonesia, the Philippines, Taiwan, and Japan.

A wide belt of desert-to-arid land extends eastward from the northern part of the Arabian peninsula to Afghanistan, then north to the Soviet Union, and then east again across China. This region is sparsely inhabited, for the most part by nomadic pastoral peoples of many countries who travel seasonally with their camels, sheep, and goats in a constant search for water and grass. For countless centuries there has been little change in their way of life.

To the south of this arid belt lie the Himalayan Mountains, which separate it from the more fertile and better-watered lands of India. Continuing east of the Himalayas, rough, mountainous terrain divides Southeast Asia from China to the north.

North of the arid belt and paralleling it through the Soviet Union, Mongolia, and China, there is a fair-to-good agricultural region. Above this fertile area, coniferous forests extend to the Arctic circle, and farther north the tundra stretches to the Arctic Ocean.

The civilizations and cultures of Asia and Egypt dominated the world for the first millenniums of recorded history. As Greece and then Rome flourished, there were wars with the empires of western Asia, but eastern Asia largely remained isolated from the rest of the world until the explorations of the Portuguese, Spanish, and British during the sixteenth and seventeenth centuries. These explorations paved the way for the colonizations by the European powers on the periphery of southern and southeastern Asia and the major islands to the east. Britain, France, the Netherlands, and Spain were the principal nations to establish such representation. The aftermath of both world wars then set the pattern for these colonies to become independent. As a result of the contact with the European countries in the colonial days and the increased trade of the West with Asia, Western civilization came to have a strong influence on commerce and the life of the people. In all the countries that border the southern coasts of Asia, including most of the large islands to the east, this Western influence is in evidence in varying degree.

Agricultural and pastoral practices range from those man first employed, after he had domesticated animals and scratched seed in the ground with a stick, to instances of modern irrigation works, large-scale farming with heavy equipment, and modern animal husbandry. While these more modern practices are not widespread today, they are slowly increasing; but if the multitudes resulting from the population explosion are to be fed, the use of up-to-date methods will have to progress much more rapidly in the immediate future.

Bovine animals, representatives of the tribe Bovini, appear either to have originated independently in Asia, Europe, and Africa or to have found their way to these continents from some unknown point of origin. Except for Africa, where the buffalo has not been domesticated, Asia is the only continent on which representatives are still found in their wild state. In Southeast Asia there are still wild gaur, banteng, and, possibly, remnants of the kouprey; on the Tibetan plateau there are wild yak; in northern India and in Africa there are wild buffalo. With these exceptions, all the other once-wild species have disappeared and are now only represented by the domesticated animals of man.

Asia's major contributions to the useful cattle of the world are the Zebu type and the buffalo. The Zebu was domesticated in either India or Southeast Asia long before the dawn of history. In the United States, Zebu cattle are commonly called "Brahman," a disparaging term when applied to cattle, for it is the name of the highest, strictly vegetarian caste in the Hindu hierarchy. A Zebu breed which was developed in the United States from imported Indian cattle is officially known as the American Brahman. The Zebu of India and Pakistan is characterized by a prominent shoulder hump; loose skin; a pendulous dewlap, sheath, and umbilical fold; and a sloping rump. Zebu types of cattle were carried to Africa long before the Christian Era. Within the past century, several of the Zebu breeds have been imported to both North and South America and have had a pronounced influence on the cattle populations of these continents.

The buffalo appears to have been domesticated at a somewhat later date than the animals commonly referred to as cattle. There is some conflict of opinion concerning whether this domestication first occurred in India or in southern China. The use of the buffalo as a draft animal has always been closely associated with the cultivation of rice and may have been the reason that this animal spread to major rice-growing areas of the east—Indonesia, the Philippines, and Southeast Asia. The buffalo also moved west but not to the same extent. It is still an important draft animal in some countries of the Middle East and on the Balkan Peninsula, although it is far outnumbered there by cattle.

The buffalo, whether used for draft or milk production or both, is a more efficient user of rough forage than are domesticated cattle. In areas in which little attention is given to providing adequate grazing, the buffalo maintains itself in good condition on stubble fields or swamp grasses where conventional cattle, even of well-acclimated types, deteriorate rapidly. The buffalo has not taken to domestication as well as cattle; the former is not as docile, and individuals displaying a streak of viciousness are much more common. For this reason, only cows are worked in some countries.

The two types of Asiatic buffalo are frequently referred to locally as the "swamp" buffalo and the "river" buffalo. The term "swamp buffalo" refers to a breed within which there are wide variations, particularly in size, ranging from the small carabao of the Philippine Islands (900 pounds) to the large (2,000 pounds) buffalo of Thailand. "River" designates the milk-type buffalo of India of which there are several breeds, the Murrah and the Nili being the most prominent.[1]

The swamp (herein used to designate the type) buffalo is generally larger than the river animal. The wide, sweeping horns of the swamp buffalo are usually marked with chevrons, extend backward, and have a nearly rectangular cross section. The animal is used primarily for draft and, to maintain its well-being, must have water in which it can

[1] Temple, personal communication.

wallow. The river buffalo also likes water, although the element is not as essential in this animal's husbandry as it is for the swamp buffalo. The river buffalo has much shorter horns, which have a curled tendency.

Swamp buffalo.

River buffaloes of the Murrah breed.

Excellent milk strains have been developed from it, and in some areas large herds are maintained solely for milk production. Both types are used only incidentally for meat. The African buffalo, much like its Asiatic counterpart, by blood grouping belongs to another branch of the tribe Bovini.

The banteng is another branch of the tribe Bovini, not directly related to either cattle or buffaloes. It is still found in the wild state in Southeast Asia and on the island of Java. It is thought to have been first domesticated on Java or Bali, and the Bali cattle as seen today are direct descendants. Except for being somewhat smaller in size, these animals on the island of Bali are the same as the feral banteng.

Two other representatives of the genus Bibos are found in Asia in both the domesticated and the wild state. The gaur, a close relative of the banteng, has been domesticated for many centuries and is an inhabitant of the forest and mountainous areas of northeast India and the Indo-chinese peninsula. The inhabitants of Assam who own tame animals use them to some extent for draft but take little, if any, milk from them.

The yak of central Asia has been used as a pack animal and for milk as far back as there is any knowledge of the people in this area. It is small and shaggy haired and weighs 500 to 800 pounds. The yak becomes quite docile on handling. The wild yak, in appearance the same as the domesticated type, inhabits the more sparsely populated areas.

Discussion of the cattle in Asia has been divided into four loosely defined geographical regions—the Middle East, South and Central Asia, Southeastern Asia, and the Far East. To a limited extent this arrangement permits the grouping of similar types of cattle and the methods of handling them.

Turkey, which has only a small part of her territory and cattle population in Europe, is discussed under Asia. The Soviet Union is covered under Europe, since most of her cattle are in the European part of the country.

Middle East

Afghanistan

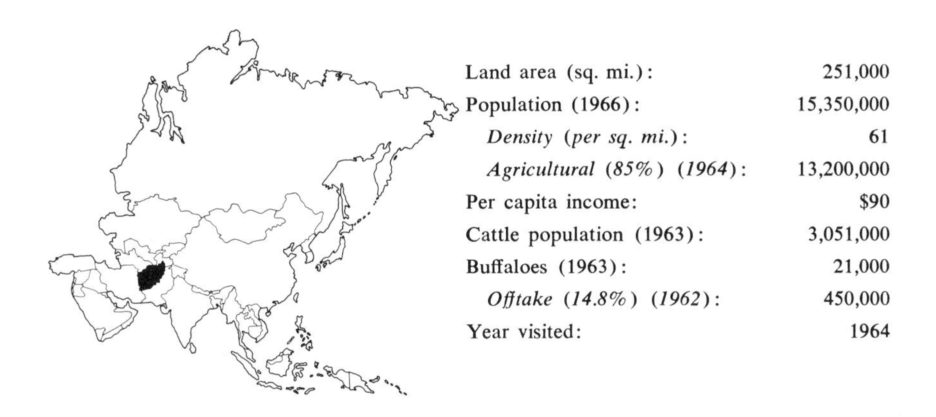

Land area (sq. mi.):	251,000
Population (1966):	15,350,000
Density (per sq. mi.):	61
Agricultural (85%) (1964):	13,200,000
Per capita income:	$90
Cattle population (1963):	3,051,000
Buffaloes (1963):	21,000
Offtake (14.8%) (1962):	450,000
Year visited:	1964

THE KINGDOM OF AFGHANISTAN is a landlocked country in south-central Asia, somewhat larger in area than Arizona and New Mexico combined and lying in the same general latitude. It is enclosed by Pakistan on the east and south, Iran on the west, the Soviet Union on the north, and a fifty-mile frontier with China in the extreme northeast. The Hindu Kush, a westward terminal range of the Himalayas as much as 25,000 feet high, separates the northern part of the country, which has a 10- to 17-inch rainfall, from the southeast, which is arid and receives less than 10 inches of rain yearly.

What is now Afghanistan has had many foreign conquerors, including such legendary figures as Genghis Khan and Tamerlane in the thirteenth and fourteenth centuries, respectively. Although there was contact with Russia in the north and Britain on the east through India, most European colonizing powers were kept out during the eighteenth and nineteenth centuries, probably mainly because of the rugged borders. The government was an absolute monarchy until 1919, when a trend toward democratic forms began. Though a constitution was promulgated in 1931, autocratic government continued for the most part. It remains to be seen if the new constitution of 1965 will effect much change.

The economy is almost entirely agricultural and pastoral. There are no railroads, and the main outlet is by road through the Khyber Pass to Pakistan. The country, with practically no foreign element in the population, is in the elemental stage of emerging from primitive husbandry practices. Away from the cities, barter and orders for grain, instead of money, still are the mediums of exchange. One-seventh of the people are nomads, living off their flocks as they seasonally drive them from the plains to the mountains and back. The largest export item is karakul skins. Although sheep and goats, with camels for transport, are the principal livestock of the caravans, cattle are a sizable factor in the settled sector of the agricultural economy. Most farmers have a few cattle; they are used mainly as draft animals and are milked for household use. There are no dairies and few commercial suppliers of milk.

CATTLE BREEDS

The cattle of Afghanistan as generally seen are a small, nondescript Zebu with no distinguishing characteristic other than the hump. Some larger animals, also Zebu but of no discernible breed, are seen with the caravans of the nomads. Indigenous breeds are recognized, however, and are seen in some localities.

Zebus belonging to a caravan. Photographed at the Kabul cattle market.

Kandahari.—Most animals of this breed are predominantly black, although some are shades of dark brown to tan. These rather small animals (a mature cow weighs 800 pounds, a bull 1,100 pounds) have short, stubby horns. The hump is more pronounced in some individuals than in others.

The true milking potential of the Kandahari has probably never been determined because of the shortage of feed in the dry seasons. On good feed better specimens have yielded 25 pounds a day at the peak of the lactation period, which does not exceed five or six months. In the cultivated areas it is both a work and a milk animal and is native to the lower elevations.

Kandahari cow.

Konari.—The Konari is a dwarf type, a mature cow rarely weighing more than 450 pounds, a bull, 650 pounds. The animal is predominantly dark red and usually has some white markings, often a white or partially white face. The horns are small and upturned, the hump small.

Found principally in mountainous areas, the Konari is a hardy breed and has the reputation of being able to withstand the effects of strong sunlight. Although mainly a draft animal, it is milked for household use. The best cows yield about ten pounds of milk a day on good feed.

Konari cow.

Nondescript.—Most of the cattle of the country are nondescript. Usually small in size, they seem to be a degenerate mixture of the Kandahari and the Konari and possibly of the larger Zebus that are seen with the caravans. The color is varied, although most are dark brown or black. Sufficient uniformity does not exist to permit classifying these cattle as a breed.

Nondescript cow as seen in the countryside. Note the charm over her right eye.

MANAGEMENT PRACTICES

The cattle of the nomads are relatively unimportant in numbers; most of the cattle of the country are owned by small farmers. The animals are grazed on nearly barren wasteland and crop residues, usually with a young herder in attendance. Although used primarily for draft, the cattle to a large extent are milked for household use. A small bag containing a charm is often attached to a rope tied over the forehead.

Under the direction of Soviet Union specialists, the Agriculture Ministry has established an artificial insemination center at Kabul. Twelve bulls, representing three exotic breeds are used—Russian Black and White and Swiss Brown, both imported from the Soviet Union, and Holstein-Friesian from the United States. Adequate facilities have been provided for a modern artificial insemination center, but the management is so poor that the results are negligible. The Beni Hassar Experiment Station, also under the direction of the Agriculture Ministry, tried to use the Kabul facility but returned to natural breeding because of the unsatisfactory service there. The conditions are deplorable. In accordance with Soviet practice, the speculum method of impregnation is employed. The instruments are not even sterilized between cows, but are merely dipped in water from an open ditch. Semen collection and dilution are carelessly handled. Only 20 per cent of the cows bred, many of them several times, become pregnant. The Afghan technicians, Soviet trained, are not available on holidays or on Friday, the religious day.

MARKETING

The livestock market in Kabul presents a picturesque and casual scene. Not only are animals of the farmers brought in, but passing caravans of Kochis (nomads) stop to add cash to their funds. Trading is leisurely, and there is a distinctly social atmosphere. Refreshments are available from passing vendors, who also have hookahs (water pipes) for rent. Livestock is bought by the municipal slaughterhouse across the road from the sales area, which is along the river bank. Many animals that are taken back to the country are traded between owners. Sheep and goats predominate, but buffaloes, a sizable number of cattle, and quite a few camels are also available. Sales are by the head. The exceptional cow in fair condition and weighing 900 pounds or slightly less sells for $90.00; a thin Konari cow that might weigh 425 pounds sells for $26.00. Animals are driven to water in the river and then returned to their location in the market.

The livestock market in Kabul.

ABATTOIRS

The municipal slaughterhouse in Kabul is a modern establishment that was provided by Yugoslavian aid. Although the throat of the animal is cut in the typical Moslem manner described under "Abattoirs" in the chapter on Sudan, from this point on the processing conforms to Western practice. Carcasses are handled on overhead rails to the coolers. Meat is chilled for twenty-four hours by refrigeration and then goes to the stores or is quick-frozen and put in frozen storage. All equipment is of Czechoslovakian manufacture. An average daily run is 1,500 sheep and goats and 100 large animals, mostly cattle and buffaloes but including a few camels. The practice of the live animal being bought by the abattoir, processed, and sold to the trade is unusual in this part of the world.

GOVERNMENT AND CATTLE

Nine government farms in Afghanistan, although they function mainly as experiment stations, bear evidence of having been organized along the lines of Soviet state farms. Five of these farms are in the vicinity of Kabul. All depend on a central machinery station for the usual farm equipment—tractors, combines, and plows. The Soviet Union has largely abandoned the use of a central station for farm machinery; Afghanistan, too, has found such stations to be a serious detriment to the operation of the government farms, for all of them need the same tools at the same time and there are not enough to go around.

At Beni Hassar, near Kabul, breeding experiments have been undertaken to upgrade Kandahari and Konari cows for milk production. Both breeds are being maintained in their original types and are also being crossed with Holstein and Swiss Brown imported bulls.

The first cross of either the Holstein or the Swiss Brown bull on the Kandahari cow produces a very fair-looking animal in which the milk production is doubled. The cattle are kept on the loose-housing plan and milked by hand. The yard is badly overcrowded and poor management is much in evidence; the animals are fed much better than are the cattle of the small owner, however. Although the days are hot in the summer, the climate is temperate enough that the European breeds would be expected to do well; and first crosses of Holstein-Friesian and Swiss Brown bulls on Kandahari cows give evidence of this.

OUTLOOK FOR CATTLE

Afghanistan is emerging from a primitive past and for many centuries has had one outlet to the outside world—that over the Khyber Pass to what was once India (now Pakistan). A second road was opened in 1964 through the Soviet frontier in the north, and a third, to Iran, is under way. The country has enjoyed air transportation only in recent years. Aid, in the form of grants and personnel, is accepted freely from both the Western world and Communist countries; but the Soviet Union, looking down on the country from the 1,200-mile frontier on the north, could well be the dominant factor in the future. The government, in form a constitutional monarchy, has not changed things much for the small farmer. The cattle of the country are used for very little except draft. Government efforts to promote a better milk cow have not accomplished much. Considering the sizable cattle population, it seems that the animals should play a more important part in the economy of the country as it improves. Under present management practices cattle numbers could probably not be greatly increased, although a reasonable grazing control should maintain them in better condition. Improvement in quality would certainly result from upgrading by the use of bulls of more productive breeds.

Iran

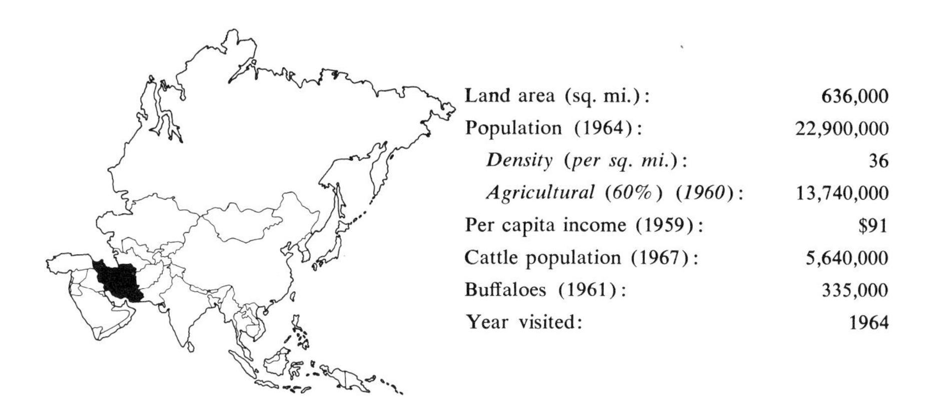

Land area (sq. mi.):	636,000
Population (1964):	22,900,000
Density (per sq. mi.):	36
Agricultural (60%) (1960):	13,740,000
Per capita income (1959):	$91
Cattle population (1967):	5,640,000
Buffaloes (1961):	335,000
Year visited:	1964

THE EMPIRE OF IRAN, formerly Persia, lies in southwestern Asia in the same general latitudes as Texas, New Mexico, Arizona, and Colorado; in area it is slightly larger than these states combined. The northern border, with the Soviet Union, is broken by the Caspian Sea. Afghanistan and Pakistan lie to the east, Turkey and Iraq to the west, and the Persian Gulf and the Gulf of Oman to the south.

The Elburz Mountains, with peaks over 18,000 feet high, separate the narrow fertile plain on the Caspian Sea from the Iranian plateau. The area between the Caspian and the mountains was the historical garden spot of Persia where silk was produced; such crops as rice, cotton, citrus, and tea are now grown there in an intensified agriculture. South of the Elburz range the plateau region is a harsh country, mountainous and hilly on the periphery. Rainfall varies from 4 to 25 inches, decreasing to the east and south. There are large arid areas and in east-central Iran, some desert.

The ancient Persian Empire began to disintegrate during the fifth century before Christ. Invading Greeks, Romans, and Turks in turn had their hand in subduing what remained of the empire. The internal government was often fraught with intrigue and suffered from many incompetent rulers. This situation continued until the days of Portuguese sea supremacy in the East. During the sixteenth century the Portuguese dominated the Persian Gulf in spite of opposition from Turkey. A century later the English took control of this trade route, with commerce, rather than conquest, the principal object. From the seventeenth century through the nineteenth century, the neighboring countries of Turkey, Afghanistan, and Russia, as well as France and Britain, at various times invaded Persia; but the area that is now Iran emerged intact. A beginning was made at constitutional government after the first of the twentieth century, and the country is now classified as a constitutional monarchy. The petroleum resources have ensured a stable economy for the past forty years.

Although the economy of Iran is principally agricultural, there is no pattern to agriculture or stock raising in the country today. Small, subtropical farms along the shores of the Caspian Sea are cultivated with wooden plows; and modern equipment is used in the successful dryland farming on the Iranian plateau, south of the Elburz Mountains in a region of 10-inch rainfall. The most primitive husbandry practices, transmitted from time immemorial, exist beside large-scale modern farms. Even after the Shah initiated land reforms in 1962 to legally limit individual land ownership to 500 acres, in the hinterland there were still feudal holdings whose owners counted their possessions by the number of their villages. The areas of land involved and the number of livestock have never been defined or counted.

Irrigation, which is essential for cultivated crops in most of the country, has been employed for centuries but could be vastly improved; and large nonproductive areas could be brought under water by using modern works. Although a number of large irrigation and hydroelectric dams have been built, many more are needed. Water for irrigation in some localities is still supplied by the age-old kanats, hand-dug tunnels leading from an underground water-bearing strata to the land, many miles away, which is to be irrigated. These tunnels are constructed by three-man teams. Shafts thirty inches in diameter are dug with hand tools to the grade level of the tunnel. The horizontal bore, also usually thirty inches in diameter, is then dug by one man, the dirt being scraped to the shaft, placed in a bucket, and raised to the surface by a crude wooden windlass, operated by his two partners. These kanats run distances of as much as fifty miles, in places at depths up to 100 feet, and take years to construct—and they are still being built.

In the semiarid regions south of the mountains, much of the public domain is grazed bare and has been for centuries by the nomadic herds of camels, sheep, and goats. Over much of this area the rainfall would be sufficient to permit the growth of good dry-land grasses and planting would be made possible if grazing were controlled. The majority of the cattle are owned by village dwellers and are used primarily for draft. There is a growing trend near the cities to establish dairies.

CATTLE BREEDS

Most of the cattle of Iran are indigenous, often of a nondescript character. Exotic breeds of the European dairy type have been brought in recently, since an interest in increasing milk production has developed.

Kurdi.—This indigenous breed of Kurdistan, which is located partly in northwestern Iran, has been the pack animal of the Kurds for countless generations. It is said to be of Brachyceros origin and in Iran is found the farthest east of any of the nonhumped indigenous breeds of cattle.

Quite uniform in appearance, the Kurdi is small, black or nearly all black, and nonhumped. The legs frequently shade to a greyish color and there is often a grey marking on the forehead and on the underline. The horns are short and thin, and polled animals are common. The breed is fine-boned and has a poor conformation by Western standards. Mature cows weigh less than 600 pounds, the small size apparently being the result of generations on very low nutrition. The Kurdi is used for draft in areas where there is cultivation.

Kurdi cattle in western Iran.

Nondescript Cattle.—Over most of Iran the native cattle are usually of a nondescript character, humped and varying considerably in size and color. Nearly solid black animals are seen, but a wide range of mixed colors is common. The horns are of medium size and upturned but often with a forward thrust. Mature animals range in weight from 500 to a maximum of 800 pounds. Village dwellers own most of these cattle.

Nondescript village cattle. Photographed in central Iran.

Exotic Breeds.—A number of sizable herds of European milk breeds have been established in recent years in the vicinity of Tehran and the larger cities. The Swiss Brown, Friesian, Holstein-Friesian (imported from the United States), and Red Danish breeds are most prominent. In the temperate climate all of these northern cattle do reasonably well with good care.

Buffalo.—The swamp buffalo and some Murrahs and Murrah crosses are found in the highly cultivated area along the Caspian Sea and in the Khuzistan valley in the southwest. They are used for cultivation in the rice paddies and are milked to some extent.

MANAGEMENT PRACTICES

In much of the country cattle are subject to a retrograde type of selection: the better males are selected for work and the poorer ones left to do the breeding. When they come in milk, the better cows are often sold to the city dairies. Except for the dairy industry near Tehran and other large cities, cattle to a large extent must shift for themselves on the sparse grazing available around the villages. Their principal function is to serve as draft animals for farm work and transport, although they are milked for household use or for some peddling in the villages. The worn-out animals and unwanted young are used for slaughter if transportation is available to a city livestock market. Bulls, unless kept as work animals, are killed at six months of age. Sheep and goats outnumber cattle 7 to 1, and their meat is universally preferred. In the country sheep milk and goat milk are preferred to that of the cow; but in the cities, either from preference or because its quantity production is more practical, cow's milk is generally used.

The small dairyman near the larger cities often buys cows in the country when they freshen, brings them in and milks them for one lactation, then sells them for slaughter. Under today's conditions this apparently is a sound practice economically but has a negative influence as far as improvement in the cattle of the country is concerned. The small herd owner cannot afford to keep his own bull. Neither can he depend on artificial insemination services, which are available only in the vicinity of Tehran. He is also reluctant to feed a cow during her dry period. For breeding his cows, he usually must rely on the nondescript bulls near his village.

Some straw and cottonseed hulls are used in the winter feeding of cattle, and there is evidence of vitamin A deficiency in many dairy herds. Death loss in young calves in the country is high, undoubtedly because of nutritional deficiency.

A number of modern dairies near Tehran maintain herds of exotic milk breeds, the base stock of which was imported. These establishments are usually owned by individuals of financial means who have become interested in the movement for increased milk production. Herds of Swiss Brown and Red Danish from Europe; Friesian from Israel, Den-

Third-cross heifer of Swiss Brown bull on native cow.

An upgraded dairy herd on which Swiss Brown bulls were used.

mark, and the Netherlands; and Holstein-Friesian from the United States are seen. Some dairies are upgrading native cows by continued breeding to bulls of these milk breeds. Brown Swiss bulls have been used to a considerable extent for crossing on native cows to obtain a larger draft animal, as well as for increased milk productivity. Dairy cattle are almost never pastured. All feed, principally alfalfa, is brought in, and the use of loafing yards is becoming common. Corn silage is put up in pit silos, and a number of dairies now have milk parlors.

Except for the animals in the well-managed dairies near the cities, very few cattle in Iran enjoy an adequate diet. The grazing areas are badly overstocked, and such winter feed as is available is inadequate for normal maintenance. This situation results in poorly developed animals and a low reproduction rate.

An artificial insemination center at Mehrabad, near Tehran, is operated by the Ministry of Agriculture. It is adequately equipped and has several good Swiss Brown bulls. In 1964, 25,000 services were being made annually, although it was not known how many of these were repeats. Conditions were such that the conception rate could not be determined. Lack of adequate communications made it necessary for the inseminator to visit his round of farms daily so that any cows detected during the past twenty-four hours could be bred. There is some reluctance on the part of the small farmer to employ a procedure that is so foreign to his cultural background. If this unwillingness to change can be overcome, artificial insemination could be a most effective tool for the improvement of the small dairy herds.

MARKETING

The Tehran livestock market is in the center of the city. Much of the stock is brought in by rail and truck; only that from nearby comes in on the hoof. Owners hold their cattle, buffaloes, camels, sheep, and goats in groups in a large open area adjacent to the abattoir until sold. There is much traffic by traders who have bought animals from farmers in the country and sell to whomever they can—a "butcherman," another trader, or a farmer. The butcherman is a middleman who buys in the market, has his purchases dressed out, and sells the carcasses to the meat stores. Farmers also bring their animals in and sell direct to the butcherman or the traders.

Many of the cattle offered for sale are females, cast off from small dairies after finishing a lactation. Cattle are seen in much larger numbers than in most Near East markets, probably because of the large European sector in Tehran's population. All sales are by the head. In 1964 a dry cow weighing 750 pounds and in poor condition brought $120. A small native cow that could not give over ten pounds of milk a day, with calf by side, sold for an average of $190.

ABATTOIRS

From the livestock market, animals are driven directly to the municipal abattoir, where they are processed on a custom basis. All animals are killed by cutting the throat; the large ones are then hoisted for skinning and eviscerating. Dressing is completed by 8:00 A.M. The carcasses go directly to waiting trucks, some of which are refrigerated, and are sold the same day. An average day's kill includes 700 cattle, 100 to 150 buffaloes, 10 to 15 camels, and 8,000 sheep. There is also an indeterminate amount of slaughtering by both private and small butchering operations. The wholesale price of meat is fixed by the government.

Construction of a modern abattoir in Tehran was under way in 1964. The site adjoins the old plant, which was built in 1944. Large animals will be slaughtered in a revolving killing box; the animal will be stunned by electric shock and its body dumped on the floor before another animal enters. The throat will be cut in the approved Moslem manner described under "Abattoirs" in the section on Sudan; and carcasses will then be processed on an overhead rail to the cooler, where they will be held for twenty-four hours at 30 degrees F. before being sent to the stores. All equipment will be of Belgian manufacture. Facilities for making sausage and various meat products are being provided. When completed, the abattoir will be one of the most modern in western Asia.

CATTLE DISEASES

Loss occasioned by disease and parasites is serious, particularly in the case of the exotic breeds. The tick-borne theileriasis, the East Coast fever of Africa, is particularly virulent in cattle that are not acclimated to the environment of Iran. Government assistance in prevention and treatment of cattle diseases is available to a considerable extent but is not widely used.

GOVERNMENT AND CATTLE

The Ministry of Agriculture is endeavoring to increase milk production in the country. The Karaj Agricultural College at Karaj has demonstration herds of both Swiss Brown and Danish Friesian cows that are maintained on loose housing. Corn silage and hay are fed during the winter, and green chop is brought to the cows during the growing season. Irrigation is necessary for all forage crops in the 10-inch rainfall area. A number of small farms in the vicinity are milking as many as 25 cows and are closely following the management practices at Karaj. A few of these farms have milk parlor installations.

The Animal Husbandry Research Institute at Heidarabad, near Karaj, also under the direction of the Ministry of Agriculture, is raising Swiss Brown bulls for the ministry's bull studs and to supply villages and individual breeders, but few of the animals are being sold. Good husbandry practices, such as loose-housing arrangements, separation of the young stock by age groups, and a milk parlor, are much in evidence. Milk production of the cow herd averages 8,800 pounds of 3.7 per cent butterfat milk for a 305-day lactation. Swiss Brown bulls are used instead of Friesian to obtain increased milk production because the former produce better draft animals as well.

OUTLOOK FOR CATTLE

Iran has a large potential for livestock development, particularly on the unutilized areas of the Iranian plateau. Cattle should play a major part in any co-ordinated program along such lines. If nomadic grazing could ever be controlled, dry-land grasses could be grown on much of the land that enjoys 10 inches or more of rain annually, all of which is now kept barren by the seasonal movements of the large flocks of sheep and goats. Such control, however, where the nomads have been free to travel as they willed for centuries would be very difficult to effect.

With more irrigation works and drainage where necessary, large areas now unproductive could be brought under cultivation. While sizable developments along this line are already in operation, much more remains to be done. Cattle would be the logical means of draft for the small farmers, who must play a large part in any such schemes. The development of a better draft animal would be a boon to farmers today as well as in the future. The use of tractors has increased materially in recent years and can be expected to continue to do so, but for many years to come cattle and the buffalo will furnish the major part of the power required for cultivation in Iran.

For the present it appears that progress in cattle raising will be along dairying lines. As the economy improves, the requirement for more meat will likely be met by sheep and goats, which the populace prefers to beef; but additional milk for the cities must come from cows. The movement toward increased milk production will undoubtedly continue, mainly through upgrading native cows by using Swiss Brown bulls.

Iraq

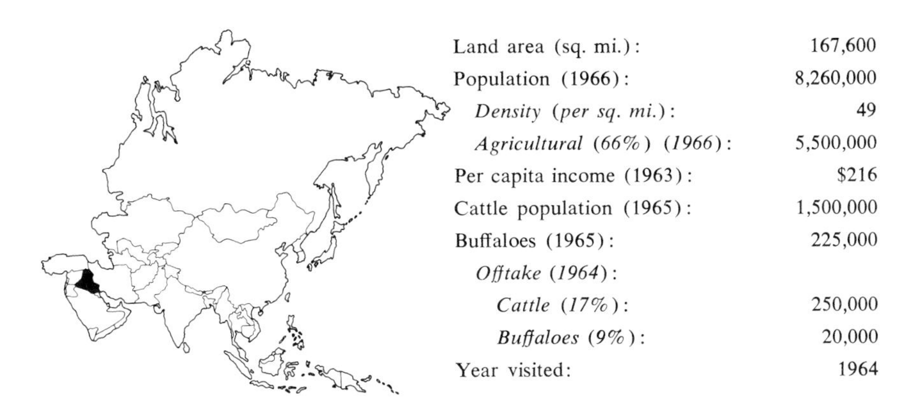

Land area (sq. mi.):	167,600
Population (1966):	8,260,000
Density (per sq. mi.):	49
Agricultural (66%) (1966):	5,500,000
Per capita income (1963):	$216
Cattle population (1965):	1,500,000
Buffaloes (1965):	225,000
Offtake (1964):	
Cattle (17%):	250,000
Buffaloes (9%):	20,000
Year visited:	1964

THE REPUBLIC OF IRAQ, in the same general latitudes as Texas, lies between Turkey on the north and Saudi Arabia on the south; Syria and Jordan are on the west and Iran on the east. In the southeast corner it borders on Kuwait. Iraq, the Mesopotamia of the ancients, is the legendary site of the Garden of Eden. It has been in the path of conquering armies since the beginning of history and has been part of many empires, the last of which was the Ottoman. After World War I it was mandated by the League of Nations to Great Britain and subsequently became an independent kingdom in 1932. After an army coup, one of several in two decades, the country was declared a republic in 1958. Its political life has been troubled ever since. The trend in foreign relations has turned decidedly from the West to the East in the past decade. The majority of the people are Arabic speaking; very few Europeans live in Iraq.

In area Iraq is equal to Arizona and half of Nevada. The Euphrates and Tigris rivers drain the country from the northwest to the southeast. Annual rainfall varies from 40 inches in the mountains in the north on the Turkish border to an average of 6 inches at Baghdad. Nearly half the country, that which lies west of the Euphrates, is desert. Rainfall is sufficient for grain crops in the high country below the mountains in the north, but farther south irrigation from the two main rivers is necessary for farming. Many nomadic tribes, with their sheep, goats, and camels, move from the plains areas in the winter to the high country in the summer. Settled farmers usually maintain cattle for draft and buffaloes for milk.

In 1918 during their period of occupation, the British imported both Jersey and Red Sindhi cattle from India to feed their army. These animals have left their influence, particularly on the cattle in the southern part of the country. At one time some Friesian cattle were imported from Palestine, and the result of their introduction is also seen. Later, Ayrshires were tried; but apparently they were more susceptible to disease and to heat stress, and few survived. In 1953 Friesian bulls were again introduced for upgrading native cattle.

Four cattle breeds are indigenous to Iraq. Many of the cattle seen in the countryside are a nondescript type, both humped (from admixture with Zebu cattle) and nonhumped.

Jenubi.—This small, rather rangy animal, light red to tan in color, is predominant in southern Iraq. A mature female may weigh 800 pounds. As seen today, the Jenubi probably carries some Red Sindhi influence. There is only small evidence of a hump in the female, however, although it is more prominent in the male. The shape and size of the head, the color, and the small horns extending straight out from the poll suggest a possible relationship to the Shami of Syria. The milk production of the better Jenubi cows, only ten to twelve pounds a day, is, however, much less than that of the Shami. The lactation period is short, never exceeding nine months, and is influenced to some extent by availability of feed. The

Jenubi cow.

breed is well acclimated and heat tolerant and carries much resistance to external parasites. On good feed it fattens well, a two-year-old steer weighing 700 pounds.

Sharabi.—This is a larger animal than the Jenubi; a mature cow weighs 1,000 pounds. The breed is found in the northern farming area, along the Tigris River. The Sharabi has a very rangy conformation and small horns, which point upward. The hump is evident but quite subdued, suggesting that the origin was an admixture of humped and nonhumped cattle. Coloring is black and white, a distinctive white streak running down the back and along the belly, often leaving the two sides a solid black. This peculiar marking does not suggest any Friesian blood but could raise a question concerning whether there is some common ancestry with the Syrian Julani. The Sharabi is larger than the Julani and does not have its shaggy coat, but the black-and-white coloring and the small horns pointing upward are Julani characteristics.

Milk production is small, although exceptional individuals yield 15 pounds a day.

Sharabi cow.

Rustaqi.—A medium-sized animal, mature cows weighing from 900 to 1,100 pounds, the Rustaqi is tan to dark red or light grey to white. Used primarily for draft, the breed is found in the central part of the country and is a poor milk producer, yielding six to ten pounds a day.

Kurdi.—The same breed as the Kurdi discussed under "Cattle Breeds" in the section on Iran, this animal is small and predominantly black and is found in the mountainous region in the north. The Kurds use it largely as a pack animal.

Native Cattle.—Throughout the country there are many nondescript cattle which are locally called Native Cattle. They vary widely in color, conformation, and size but are usually humped.

Buffalo.—The buffalo is the river type of milk buffalo and is an important dairy animal near Baghdad. It is not used for draft and is maintained solely for milk production. The natural assumption is that the buffalo was introduced from India, although there have been no importations in modern times. A white patch on the forehead is common and indicates influence of the Nili breed.

Buffalo dairy herd. Photographed outside Baghdad.

MANAGEMENT PRACTICES

For the past decade the agriculture of the country has been in a sorry plight. Land reforms initiated in 1958 at the time of the dissolution of the Iraq-Jordan union resulted in a disastrous decrease in all agricultural production. Once an exporter of rice and wheat, Iraq had to begin to import grains. When the large landholders left, with them went the management essential to a productive agriculture. In the extremity thus created, the government accepted the Soviet offer of aid. Approximately 2,000 Russian-trained agricultural technicians were sent in to establish the Soviet type of farm organization. The Arab was not amenable to the regimentation required by this system, and the Russians could not exercise the rigid control essential to it. The program soon failed without having made any significant accomplishment.

The agricultural policy of the government then turned toward more Western methods, and production has been improving. The agrarian reform program in 1964 provided for putting a family on forty acres of irrigated land and eighty acres of dry land, which is sufficient to provide a reasonable livelihood and to permit the maintainence of a few head of cattle or buffalo in addition to the sheep. The traditional share-crop system is declining. Lack of education is the largest problem; very few of the small farmers can read or write. Ten vocational agricultural schools, averaging about 100 students each, are now functioning and teach the practical rudiments of agriculture and animal husbandry. The impact of this program on an agricultural population of 5,500,000 may be small, but it is a beginning at least.

Under such conditions the cattle of the country have received little attention. Those of the Kurds in the north are pastured on the rough land not suitable for cultivation. The draft animals of the small farmer on the plains and irrigated areas graze on dry land and crop residues. Some milk is taken, but the milk of sheep and goats is preferred. Only unwanted animals—worn-out work cattle or young males—are sold for slaughter.

Near the towns dairying is followed in a more progressive manner, with buffaloes playing the most important role. In the Baghdad area 80 per cent of the milk production is from this source. Cattle are usually driven out to pasture during the day and returned to the owner's small shed for milking at night. Buffaloes are usually maintained in walled enclosures and feed is brought to them. Both cattle and buffaloes are fed a good supplemental ration—barley, milo, or corn with sunflower-seed cake for protein. During the months when pasture is not available, tibbin (chopped wheat or barley straw) is fed as roughage. Although alfalfa originated in Iraq, it is seldom used for either hay or pasture. Because milk yields are measured in "bottles" which vary in size, production data are difficult to obtain. The average cow probably yields approximately eighteen pounds a day, the best buffalo as much as 25 pounds. The producer is paid 8 cents a pound for his milk.

Herd of a small dairyman in Baghdad.

A small artificial insemination center, established in 1958 at the Abu-Ghraib Experiment Station, uses frozen semen from progeny-tested Friesian and Jersey bulls shipped from England and the United States. Seven Friesian bulls are also used. Two other bull studs within forty miles of Baghdad produce fresh semen for local use. A total of 75 cows a month were being bred from the three studs in 1964, the small usage resulting from cultural prejudice. The rectal method of insemination is used and a conception rate of 50 per cent is obtained.

Cattle section of the Baghdad Livestock Market.

MARKETING

The Baghdad livestock market, located in the city, is an open area in which animals are held in groups by their owners. All types of livestock are handled—cattle, buffaloes, camels, sheep, and goats. Farmers sometimes transport their cattle in small trucks but usually drive them in on the hoof. Sales are by negotiation between the owner and a middleman who is buying for resale to a shop owner. Sales are also made to traders who buy for a short feeding period or to other farmers who wish to add to their stock.

Many of the cattle and buffaloes sold are of fair quality, probably grading good or low-good. All sales are by the head. Bulls for slaughter, usually weighing less than 800 pounds, sell for the equivalent of 20 cents a pound. Cows in milk being bought to go in the buyer's herd bring disproportionately high prices: in 1964 a Sharabi cow in poor condition, weighing 800 pounds, would bring $275.

ABATTOIRS

The municipal slaughterhouse adjoins the Baghdad livestock market. The bulk of the city's meat supply comes from here, but there are numerous illegal private slaughterers near the city who kill from 1 to 10 animals a day. Killing is done in a large building by throwing the animal, cutting its throat, then hoisting the body for skinning and evisceration. The butchers who purchase animals in the market have them killed and the carcasses returned for a fee. Government veterinarians make rudimentary inspections. Killing is started at 2:00 P.M. and work is completed by early morning. Meat is taken at once to the butcher's shop, sold, and consumed the same day. An average day's run is 1,850 sheep, 100 cattle, 25 buffaloes, and 15 camels.

GOVERNMENT AND CATTLE

The universal trend of people leaving the country for the cities is taking place in Iraq. Part of the government's efforts to increase milk production is an enterprising program to counter this pattern. The program was initiated at Abu-Ghraib, near Baghdad. The Dairy Administration had completed a dairy colony of 500 units in 1964. Each unit consists of an open shed and a small yard designed to accommodate 15 cows, either cattle or buffaloes, and an adjoining dwelling with living room, bedroom, open kitchen, shower, and toilet. These facilities are a marked improvement over even the better homesteads of the village farmer with small holdings. Running water is available for both people and cattle;

there is also electric power. Feed is provided by the Dairy Administration, which buys the total milk production. Concrete wallows are provided for the milking buffaloes, which travel to and from them daily, spending two to four hours in the water.

Buffalo wallow provided by the Dairy Administration. Courtesy Dr. Sami Kassir

The Abu-Ghraib Agricultural Experiment Station operates the largest dairy in Iraq, the only one commensurate with Western standards. The objectives are to produce breeding stock, improve native cattle by upgrading, and serve as a demonstration center for dairying operations. There is a nucleus of imported Friesian stock which is being crossed with the native Jenubi and Rustaqi breeds. In addition to 125 milking cows, bulls are raised for the bull studs in the area and are also supplied to nearby farmers. Altogether, three hundred head are kept in the herd. Three types of Friesian bulls are used: the Netherlands, the German, and the Danish. The cattle are maintained for ten months of the year on good pastures, the cows in milk being brought into modern milk barns for night and morning milking. Berseem, an annual Egyptian clover, is used extensively for pastures. Milk is processed in a modern dairy plant operated by the Dairy Administration. Milk is also bought for processing from nearby farmers. Most of the output is sterilized and bottled so that it can be kept without refrigeration.

Heat stress has a pronounced effect on the exotic milk breeds, causing a deterioration in condition and a decrease in milk yields. Crossing one of the European dairy bulls on the native cow usually doubles the progeny's milk production. Experimental breeding, using Friesian, Ayrshire, Jersey, and Swiss Brown bulls, has been undertaken to determine what the future program should be. Data for this purpose are to be obtained from the bulls loaned to nearby farmers as well as from the station's own herd.

OUTLOOK FOR CATTLE

The national preference for mutton over beef favors the development of a dual-purpose, milk and draft cattle. Some degree of progress has been made in bettering the dairy animals. The draft cattle of the country remain just as they have come down through the centuries, mostly of nondescript and degenerate types. They are probably deteriorating in some degree because of negative selection, the better bulls naturally being selected for work. Draft animals will play a major part in the agriculture of the country for some time to come, and development of a better type for this purpose would seem warranted.

The political turmoil which prevails over the country casts its cloud over all agricultural activity. The feeling of uncertainty about what the future holds is widespread and reaches down to the small farmer with his few head of cattle. The progress that has been made on his behalf has reached only a minute minority.

Israel

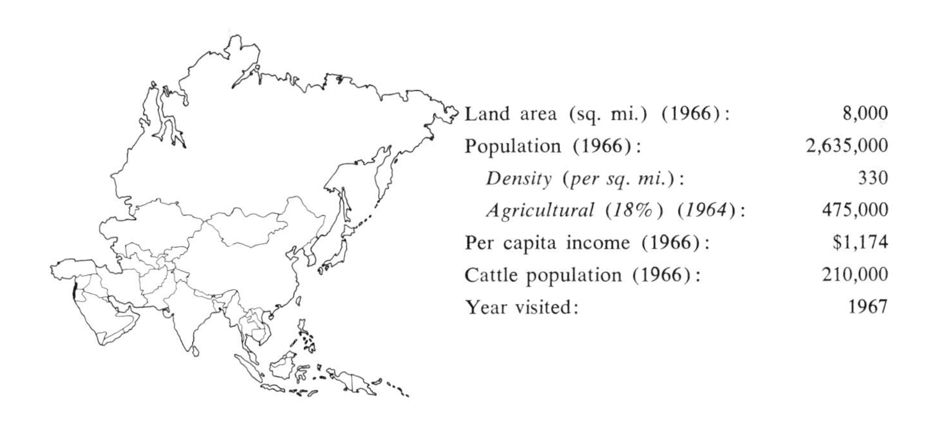

Land area (sq. mi.) (1966):	8,000
Population (1966):	2,635,000
Density (per sq. mi.):	330
Agricultural (18%) (1964):	475,000
Per capita income (1966):	$1,174
Cattle population (1966):	210,000
Year visited:	1967

THE STATE OF ISRAEL is a narrow strip of land at the southeastern end of the Mediterranean Sea and ends at a point on the Gulf of Aqaba. As recognized by the 1948 armistice agreement, Israel is bordered on the north by Lebanon, on the east by Syria and Jordan, and on the west by the United Arab Republic and a coastline on the Mediterranean Sea. In 1967, Israel occupied the United Arab Republic's portion of the Sinai Peninsula to the Suez Canal, and also pushed Jordan back to the east bank of the Jordan River.

The latitude and climate of Israel are similar to those of California south of Los Angeles. The land area is somewhat less than that of Massachusetts. Over half of Israel is in the Negev desert region, which extends northward from the head of the Gulf of Aqaba to the Dead Sea.

At the beginning of the sixteenth century, Egypt lost Palestine, of which today's Israel is part, to the Ottoman Empire. The area remained under Turkish rule for 400 years, until taken by the British during World War I. In the interval between the world wars, it was under mandate to Great Britain. Neither the British nor the League of Nations had any success in establishing governments acceptable to either the extremely nationalist Arab or Jewish elements of the population in the general area. The situation culminated with Israel declaring its independence in 1948 and fighting to armistice recognition of its borders, which are as yet undefined by treaty. The Arabs in large measure fled Israeli territory and the Jews left that of the Arabs, but this action has proved to be no solution to the deep cultural hatreds involved. In the first eighteen years after World War II, 850,000 Jewish emigrants from central and eastern

Europe entered Israel and were followed by Jews from North Africa and from Near East countries. More than 700,000 Arabs left Israel, but about 180,000 remained. The population is now almost 90 per cent Jewish. Ever since the State of Israel was established, there have been nearly continuous border incidents with her Arab neighbors and two open wars, in both of which Israel was victorious. The territory of the United Arab Republic on the west and that of Jordan on the east, which Israel is currently occupying, have long been recognized by the world powers as belonging to the Arab countries. The eventual resolution of this conflict lies in a troubled future.

Israel's economy has rapidly changed from one that was basically agricultural to one that is primarily industrial. The largest export item is now cut diamonds, followed by oranges. Effort is directed at increasing both industry and intensive farming. Some dry-land grain is grown but in general cultivated crops are irrigated.

Historically, since biblical times the Jewish people have been anything but farmers; yet agriculture was the basic requirement in developing the new homeland if the large influx of immigrants who came in after World War II were to be accommodated. The means employed to attain this end were completely foreign to age-old Jewish instincts. Immigrants arriving after the war came with little but the clothes on their backs. They banded themselves into their *kibbutzim* as the most effective method of getting started. A *kibbutz* (plural *kibbutzim*) is a total communal grouping of families on a purely voluntary basis. All human requirements and activities—housing, food, clothing, livestock, farming, buying and purchasing—are under the control of a committee elected by all the adult members. All individual requirements are supplied by the *kibbutz*; the individual owns nothing. If he decides to leave, he takes only his personal belongings. He lives in an apartment or cottage but eats all meals in a common dining room. His children from birth onward are cared for in the communal nurseries and schools except for a few hours when they visit their parents after the day's work is completed. A prosperous *kibbutz* is composed of as many as 150 families and a population of 700. An arrangement more foreign to the established concept of the family as the focal center of Jewish culture cannot be imagined. The accomplishments, however, have been material and exceed those of the somewhat similarly organized socialized farms of the Communist countries.

Founded on the fruits of agriculture, many *kibbutzim* are now wealthy and have branched out into varied nonagricultural enterprises, such as manufacturing plastics, water meters, and furniture—any pursuit or investment that can be found to employ the surplus funds as they accumulate. After the *kibbutz*, the *moshav* farm organization entered the picture. In the *moshav* the farmer owns his land and livestock but belongs to a rigid co-operative, composed of the farmers in a village, for marketing and purchasing. These farms are including an increasingly larger sector of the agricultural economy, possibly as much as two-thirds in 1967, although the *kibbutzim* still have a majority of the cattle in the country. There are very few private farmers.

CATTLE BREEDS

The dominant breed is the Israeli Friesian. Indigenous cattle have practically disappeared. Quite recently exotic beef breeds have been introduced in a limited way in an effort to increase beef production.

Arab Cattle.—This is the local name for the nondescript cattle seen grazing on rough land unsuited for cultivation. Except for some individuals showing the influence of Friesian blood, these animals are rather uniform in appearance. They are black or nearly black and have short, stubby horns. Most animals have no semblance of a hump, although individuals are seen with some hump tendency. A mature cow rarely exceeds 600 pounds in weight. As the owners can manage to do so, they upgrade these cattle by using exotic bulls. There are probably fewer than

Arab cattle in a typical grazing area.

50,000 head in the country now, and they will probably disappear within the next few generations.

Israeli Friesian.—The first importations of Friesian cattle came from the Netherlands in 1932 and were principally employed to upgrade the native cattle. At the same time, the Shami breed is said to have been brought into Israel by immigrants from Syria. Whatever the base of the Israeli Friesian, the fact that it was indigenous to the area contributed much to its ability to adapt to the warm climate. Later, Holstein-Friesian cattle were imported from Canada and, still later, from the United States.

In conformation the Israeli Friesian tends more toward the well-muscled Friesian type of Europe than the rangy Holstein-Friesian of Canada and the United States. Its ability to withstand hot climates and to maintain its milk production is recognized in many subtropical areas in both Africa and Asia.

Israeli Friesian cow.

Exotic Breeds.—In recent years some British beef breeds, American Brahman, and a few Charolais have been introduced with the idea of developing better beef animals by crossing them on Israeli Friesian cows. The Hereford was originally preferred for this endeavor, but the

American Brahman and the Charolais later became more popular. The American Brahman is used only to a limited extent in the dairy herds but is quite generally employed for upgrading native cattle. Young Israeli Friesian bulls feed out into excellent beef animals and will probably not be superseded by any of these exotic beef breeds or by crosses by them on Friesian cows. The American Brahman crossed on native cows, however, effects a marked improvement.

MANAGEMENT PRACTICES

The handling of livestock on a typical *kibbutz* is somewhat like that on a collective farm in the Soviet Union. Men and women are appointed to various tasks under the authority of the appropriate committee. The difference is in their attitude. The *kibbutznik* (member of a *kibbutz*) is where he is voluntarily, and there is no state control of his *kibbutz*. The worker on the Soviet collective farm is there because he has to be, and factually the state is the sole authority.

Cattle are maintained both in barns and under loose housing, depending on the arrangement of individual farms. If land is available for a limited amount of pasturing, small fenced pastures are utilized. The barns and yards are frequently of modern construction. Practically all cows are milked; growing out the bull calves for slaughter is the usual practice. Both milk parlors and milking machines for cows held in stanchions are used. Farms milking 300 cows with 200 bulls on feed are not unusual for a large *kibbutz*.

Dairy barn of a *moshav* farmer.

Operations of the individual member of a *moshav* are much the same as those of the *kibbutz*, the owner in the former handling a smaller number of cattle—usually milking 20 to 40 cows—and having a proportionate number of bulls on feed.

All cattle belonging to the *kibbutzim* and *moshavim* are well fed. If they are not pastured, green chop is hauled in; during the winter, hay and silage are fed. Most concentrates are imported, and the balanced grain ration fed both cows and slaughter bulls consists of whatever grains can be bought the cheapest—usually barley, corn, or sorghum—and various oil cakes for protein. Prepared feeds are universally utilized. Orange skins and pulp, supplemented with cottonseed meal, are fed during the canning season by farms in the vicinity of a canning factory.

Practically all dairy herds are Israeli Friesian and 90 per cent are bred artificially. It is not unusual for either a *kibbutz* or a *moshav* to average 11,000 pounds of milk a lactation, and some of the best cows reach more than 13,000 pounds. Calves are taken from their dams at birth and are pen-fed individually until three months of age, then in groups. Females are fed a good growing ration; bulls for slaughter receive a heavy grain ration to a finished weight of 1,000 to 1,100 pounds at sixteen to eighteen months of age. It is a growing practice to cross American Brahman, Hereford, and Charolais bulls on Friesian cows to the extent possible while still using Friesian bulls to obtain the

Finished Israeli Friesian feeder of a *moshav* co-operator. Weight, 1,100 pounds.

necessary replacement heifers. Straight Israeli Friesian cattle, fed bulls, and discarded cows, however, supply 70 per cent of the slaughter animals.

MARKETING

Israel has no cattle markets as such. To ensure orderly marketing of slaughter animals, the Meat Production and Marketing Board, an independent organization on which the Ministry of Agriculture is represented, prepares lists of salable animals offered by the various growers—*kibbutzim*, *moshavim*, or individual farmers—showing the date on which the animals are to be offered for sale. The kind and quality of each lot is indicated by number. In indicating quality, young bulls weighing more than 815 pounds are No. 9; bulls under this weight are No. 8; "used bulls" (discarded sires) are No. 7. Heifers range downward from No. 6 to No. 4, cows from No. 3 to No. 1, according to quality. These lists are distributed to approved traders who mark their bids on a price-per-kilogram, dressed-weight basis opposite the lots they wish to purchase and submit their figures to the board before the sale day. Animals are allotted to the best bidders; on the average, 75 per cent of the cattle offered for sale on a given day are disposed of in this manner. The remaining 25 per cent is purchased by the meat board. A floor is put under this marketing procedure by a minimum price fixed by the board. The government compensates the owner of any cattle that return less than the fixed price, whether they were sold by bid to a dealer or on a cold-dressed-weight basis to the board itself, by making up the difference.

ABATTOIRS

Slaughtering is in private hands and is usually done at a co-operative. The facilities generally are modern and have good equipment. Killing conforms strictly to kosher requirements and involves an odd use of the English killing box, where it is employed. After an animal enters, the box is revolved 180 degrees until it is upside down. The animal's head is then secured by a rope while a rabbi cuts its throat by drawing his knife quickly across it. A second rabbi inspects the lungs, and any lesion is cause to reject the carcass as not being kosher. Following this procedure, there is good veterinary inspection, but the health of the Israeli cattle is unusually good and there is little condemnation.

The carcass is manhandled out of the box onto a cradle and then to the overhead rail, where it is processed in the usual manner. An average plant works three days a week on live cattle, averaging about 40 head a day, the bulk of which are Friesian fed bulls. Three days a week are spent in packaging frozen meat from the Argentine; only front quarters are handled since the hindquarters carry too much fat for the Israeli market. This packaging procedure is in reality a rekoshering process; the meat is defrosted, washed, cleaned, and repackaged for local trade.

The holding yard of an abattoir has a substantial shed roof, concrete paving all around, and a stanchion to hold each animal.

Carcasses are graded principally for muscling and penalized for both fat covering and marbling. Bulls are graded from a top of 1 through 2+, 2, and 3+ to a bottom of 4; cows are graded A, B, C, and D. A No. 1 grade is equivalent to a USDA average-good, although the fat cover and marbling will be somewhat less and the muscling considerably better. These carcass grades have no definite relationship to the quality designation of live animals by the meat board.

The average dressing percentage of the fed-bull carcass is 56 per cent and never exceeds 59 per cent. In 1967 a bull grading No. 1 brought the grower 45 cents a pound liveweight. The retail price of beef averages $1.45 a pound. As mentioned, the demand is for exceptionally lean meat, and in feeding out the Friesian bulls, care is exercised not to allow them to get too fat. First-cross Hereford bulls are marketed at 850 to 900 pounds because at heavier weights they carry too much fat cover. In their Friesian cow Israelis have an ideal animal for the production of the type of beef their market demands; and efforts to improve this animal by introducing exotic blood would appear certain to eventually increase the cost of production.

GOVERNMENT AND CATTLE

Operation of the farms raising cattle is left in the hands of the *kibbutzim*, *moshavim*, and the few private operators; but the government maintains a firm hand over the industry. The part of the meat board in marketing has been discussed. All feed-stock ingredients are purchased by the government and sold to private concerns which prepare the mixed feeds for both dairy and feeder cattle.

OUTLOOK FOR CATTLE

The carrying capacity of Israel for cattle is almost completely utilized, at least until the time when more land can be brought under irrigation. On some of the rougher ground that cannot be cropped, good dry-land grasses could be established to add to the scarce pastures; but there is not enough of this in a 10-inch or better rainfall area to increase the over-all carrying capacity materially.

Strong emphasis has been placed on increasing beef production. The fallacy that exotic breeds will fill this need has gained a considerable foothold. Apparently it will take some time for the axiom to be accepted that feed is the main requisite for producing beef.

The Israeli is not a meticulous farmer. He has been learning a business during the last 20 years that he had forgotten over the past 2,000, and it takes many generations to make a farmer. There is no denying that he has done a remarkable job with the Friesian cow in both milk and meat production, and his having had ample funds to work with does not detract from the accomplishment. The fact remains, however, that his day-to-day husbandry practices are not commensurate with the quality of his livestock and the facilities that he has at his command. Few barns are kept in good order. Feeders, which are often found in yards where the manure is a foot deep, cost the owner many pounds when they go over the scale. Wastage of feed by careless handling is common. The remedy of such conditions eventually will mean an even better record in both milk and beef production.

Jordan

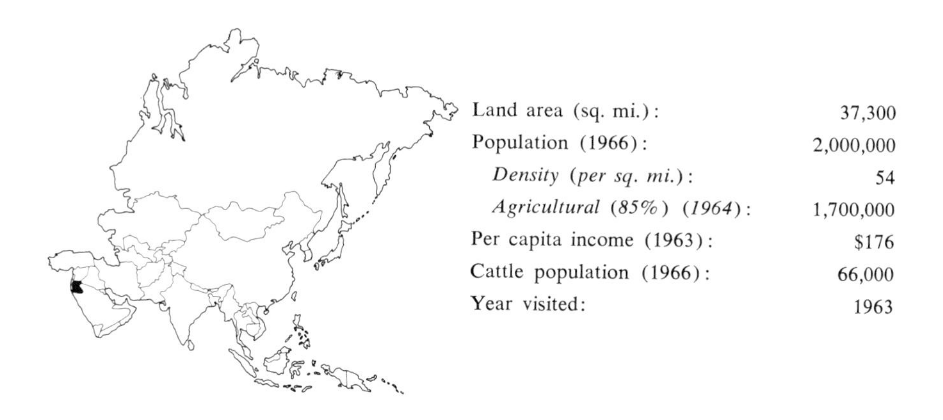

Land area (sq. mi.):	37,300
Population (1966):	2,000,000
Density (per sq. mi.):	54
Agricultural (85%) (1964):	1,700,000
Per capita income (1963):	$176
Cattle population (1966):	66,000
Year visited:	1963

THE HASHEMITE KINGDOM OF JORDAN is a landlocked country except where it touches the Gulf of Aqaba in the southwest corner, giving it access to the Red Sea. Syria lies to the north, Iraq to the east, and Saudi Arabia to the east and the south. On the west, the Israeli border separates Jordan from the Mediterranean Sea. In area the country is one-third larger than the southwestern corner of Texas that lies between the Pecos River and the Mexican border. Jordan is in the same general latitude as that part of Texas. Almost all its inhabitants are Arabs.

Transjordan was a province of the Ottoman Empire until the end of World War I, when it was mandated to Britain by the League of Nations. It became independent in 1946 as the Hashemite Kingdom of Jordan. The creation of the state of Israel was unsuccessfully resisted by Jordanese forces; but in 1949, 2,000 square miles of territory lying west of the Jordan River—including old Jerusalem, Bethlehem, and most of the early Christian shrines—were added to Jordan by treaty with Israel. Half of the Jordanian population now lives in this area, known as West Jordan. West Jordan was occupied by Israel in 1967, and its future is involved in world politics.

Seven-eighths of Jordan is desert, which is traveled by nomads with their flocks of sheep and goats, with camels for transport, in a continuous search for grass and water. The cultivated areas are in the west and the north, where the yearly rainfall is about 25 to 30 inches but decreases rapidly to less than 5 inches in the Jordan valley. The better crop land is irrigated. Farming practices vary from the wooden plow drawn by a camel to a few large-scale dry-land farming operations using modern, heavy equipment. On the marginal lands available, Jordan has accomplished more than any of the other postwar Arab nations; good use has been made of some of the American aid that was poured in after World War II.

In recent years income from tourists traveling to the Jerusalem area has been an important foreign exchange earner. This source of revenue disappeared overnight when Israel occupied West Jordan. Basically the economy is agricultural, wheat and barley being the principal crops. Sheep and goats are the important part of the livestock sector and are utilized for both milk and meat. Cattle play only a minor part in the economy.

The cattle of the country are the same nondescript animals as those discussed under "Israel" as Arab Cattle and are utilized to some extent for farm and cart draft. They also serve as "the peasant's purse," being disposed of only when there is a need for cash. The small herds graze on open areas under the care of a herdsman and are returned to villages at night. Some miscellaneous mixing with the Friesian cattle of the dairy herds has occurred.

Jordanian native cattle returning to villages. Friesian cross on right.

In the larger towns there are a few dairies with Friesian stock, the base of which has been imported from England. These dairies consist of a walled enclosure with an adjoining barn. Roughage is hauled in daily from the country, and a grain ration is fed.

There is an adequate municipal abattoir in old Jerusalem, well maintained and with good inspection. Sheep and goats are the principal

Dairy owner and Friesian herd in the town of Tulkarm.

animals handled, although a few cattle and camels are slaughtered every day. All meat is disposed of to vendors the morning after slaughter.

In 1963 the government was making some attempt to increase the milk production. An experiment station operated in conjunction with Hussein College at Tulkarm had a herd of 12 Friesian cows and 1 bull, all of which had been imported from England, which were to be the nucleus of a dairy demonstration center.

If there is any future for cattle in Jordan, it will be along the dairying line. Even this is problematical. The predominantly Arab population has a strong preference for the milk and meat of sheep and goats. Horses—or a camel if the farmer can manage it—are preferred over cattle for farm draft. If West Jordan is eventually completely lost to Israel, there will be very few cattle left in the country.

Lebanon

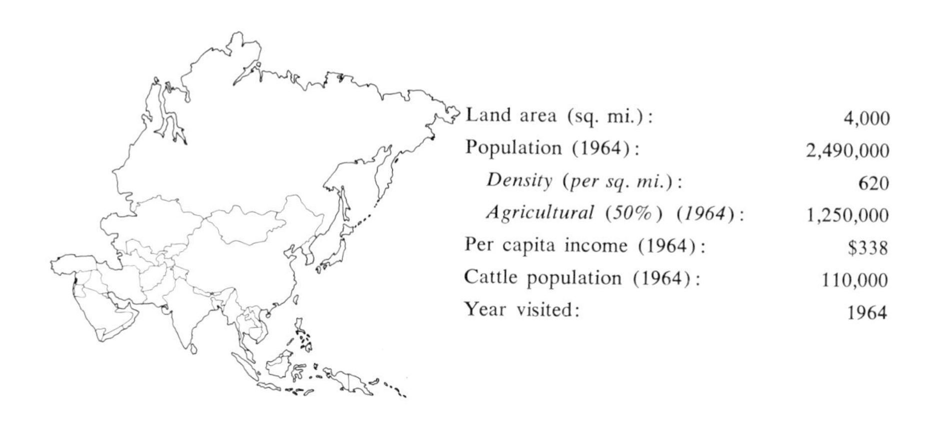

Land area (sq. mi.):	4,000
Population (1964):	2,490,000
Density (per sq. mi.):	620
Agricultural (50%) (1964):	1,250,000
Per capita income (1964):	$338
Cattle population (1964):	110,000
Year visited:	1964

THE REPUBLIC OF LEBANON is located at the eastern end of the Mediterranean Sea and is enclosed by Syria on the north and east and Israel on the south. It lies in the same latitude as Los Angeles and in area is not quite as large as Los Angeles County.

The region was the home of the Phoenicians and was held and lost successively by each ancient Mediterranean nation as it rose and fell. Lebanon was a province of the Ottoman Empire until the end of World War I, when it was mandated to France by the League of Nations. France did not agree to the unilateral declaration of independence in 1943, and it was not until 1946 that factual independence was attained.

Most of the people are Arab, although slightly more than half the population is Christian and has been so since the time of Christ.

From the narrow coastal plain along the Mediterranean coast, the land rises sharply to the high Lebanon Mountains then falls to the Bekáa Valley, which is drained to the south by the Litani River. In spite of the rough terrain, more than one-fourth of the land area is cultivated and one-half the population is engaged in a highly intensified agriculture. There is much irrigation; in many places steep slopes have been terraced and cultivated for centuries. Cattle are widely used for farm draft, but goats and sheep outnumber cattle nearly 6 to 1. The goat population is more than 500,000, or 125 to every square mile.

The indigenous cattle of the country are referred to as Baladi cattle. They are the small, black animals commonly seen in the western edge of the Middle East and described in the section on Israel as "Arab Cattle." Cows as well as oxen are used for farm work by farmers handling only a few acres. What milk can be obtained is taken for household use. Even working cows are milked. The bulls are grown out to weights of about 500 pounds uncastrated and are sold for slaughter.

Baladi cattle are being bred out by crossing them to Friesian bulls. In the larger dairy herds the Friesian has first place, Friesian-Baladi crosses second, and Red Danish third. Half of the dairy cattle are in hands of owners each of whom owns only one cow. There are a few of the Shami cattle of Syria in the country.

Baladi cow.

The Agricultural Research Center of the American University of Beirut, at Hoch-el-Sneid in the Bekáa Valley, maintains a small herd of Shami cows and another of Friesians. They were being employed in comparative tests on production and feeding and for experimental breeding work.

Artificial insemination was introduced in Lebanon in 1956. Possibly 10 per cent of the cows in the country are now being so bred. The

Friesian herd. Photographed at the Agricultural Research Center, American University of Beirut.

speculum method is used, and the conception rate is approximately 20 per cent. Farmers with few animals have very little interest in the program. The national calving percentage by both natural and artificial breeding is estimated at 70 per cent.

Only a gesture is made at the control of cattle diseases. Government veterinarians are supposed to vaccinate for foot-and-mouth disease and anthrax when called on to do so, but the program is not fully appreciated by the farmers.

The Friesian is destined to become the milk cow of Lebanon, and the Baladi will probably remain her draft animal although not a very efficient one. Very little room for raising cattle exists in the small country.

Pakistan

Land area (sq. mi.):	365,000
East Pakistan:	55,000
West Pakistan:	310,000
Population (1964):	100,000,000
East Pakistan:	54,000,000
West Pakistan:	46,000,000
Density (per sq. mi.):	
East Pakistan:	982
West Pakistan:	148
Agricultural population (77%) (1964):	80,000,000
Per capita income (1964):	$76
Cattle population (1964):	33,500,000
Buffaloes (1964):	8,400,000
Offtake (cattle and buffaloes) (4.8%) (1964):	2,000,000
Year visited:	1963

THE REPUBLIC OF PAKISTAN consists of two widely separated tracts on opposite sides of India—West Pakistan, which occupies an area 1,000 miles long with an average width of 300 miles and separates India from Afghanistan and Iran; and East Pakistan, 55,000 square miles in area on the extreme east of India. Kashmir, formerly Jammu and Kashmir, the 82,258-square mile area still in dispute between Pakistan and India, lies to the northeast of West Pakistan. The predominantly Moslem areas of both East and West Pakistan were carved out of British India in 1947 to separate, as far as was practical, the Moslem and Hindu elements into two dominions, Pakistan and India. Nine years later the Islamic Republic of Pakistan was proclaimed and the new country remained a member of the British Commonwealth.

West Pakistan, with a coastline on the Arabian Sea and mountainous frontiers bordering Afghanistan, is an arid region averaging less than 5 inches of rainfall. In area it is equal to Texas and Louisiana and lies in the same range of latitude as Virginia to Florida. Agriculture in the Indus River valley, which is the main drainage from north to south, depends on irrigation.

East Pakistan, not quite as large as Florida, lies north of the Bay of

Bengal and occupies the fertile plains in the delta area of the Ganges River. Rainfall is more than 70 inches annually. Over half the population of Pakistan—ten times the number of persons in Florida—lives here.

When Pakistan attained dominion status in 1947, the economy was almost entirely agricultural. Since then considerable industrial progress has been made, but 75 per cent of the people are still engaged in agriculture. Cattle and buffaloes are utilized mainly for draft. They are held for the most part by village-dwelling farmers who cultivate small farms.

CATTLE BREEDS

Most of the cattle of the country are nondescript Zebus and many border on a degenerate type because of countless generations of malnutrition.

Red Sindhi.—This breed of Pakistan has gained quite a reputation in various parts of both Africa and Southeast Asia as a tropical milk-producing breed. On government farms, where she receives adequate nutrition, in addition to what is given the calf for maintenance, a cow averages 3,500 to 4,000 pounds a lactation. This is good production for a Zebu breed.

A prominent dewlap and hump and large but not droopy ears characterize the breed. The color is usually medium-dark red but may vary

Progeny-proven Red Sindhi bull. Photographed at the Livestock Experiment Station in Malir.

Red Sindhi cows. Photographed at the Livestock Experiment Station in Malir.

from light yellowish to very deep red. The horns are short, frequently stubby on the bull but somewhat longer on the female. Both the sheath and the umbilical fold are small, compared to those of many other Zebu breeds. Mature cows weigh as much as 800 pounds, bulls 1,000.

The Livestock Experiment Station at Malir, outside Karachi, is developing a high-milking strain of Red Sindhis by selecting progeny-proven sires and high-producing dams. In 1963 the herd consisted of 260 animals. The top selection of first-calf heifers in 1963 averaged 4,600 pounds for a 302-day lactation period. The highest producing cow averaged more than 6,000 pounds.

Cows are fed green chop of berseem, alfalfa, and pangola grass when pasture is not available in the area. Rainfall in the region is normally 2 to 3 inches annually. The cows receive a balanced grain ration in proportion to milk production.

In addition to supplying its own needs, the artificial insemination center at the station furnishes semen to five substations in the area. A total of 1,410 cows were bred by this method in 1962, and the practice was gradually growing. Owners with limited holdings are prejudiced against breeding a cow until she is well along in her lactation. Most cows are offered for artificial insemination six to eight months after calving.

Dhanni.—This small Zebu breed is widely used for draft in the northern part of Pakistan. It is moderately humped and varies from nearly all white with black spotted markings to nearly all black with some white markings, especially a white topline and bottomline.

Saudi Arabia

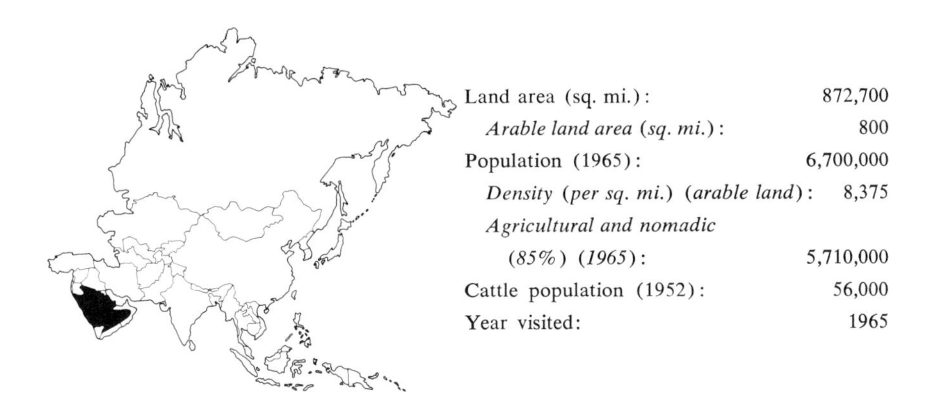

Land area (sq. mi.):	872,700
Arable land area (sq. mi.):	800
Population (1965):	6,700,000
Density (per sq. mi.) (arable land):	8,375
Agricultural and nomadic (85%) (1965):	5,710,000
Cattle population (1952):	56,000
Year visited:	1965

THE KINGDOM OF SAUDI ARABIA dates from 1932, when all of the area came under the rule of one king. The country lies in the same general latitudes as Mexico. Occupying most of the Arabian peninsula, it is bordered on the north by Jordan, Iraq, and Kuwait and in the south and southeast by Yemen, the South Yemen People's Republic, and the Trucial Sheikdoms, or Trucial Oman. The Red Sea and the Gulf of Aqaba are on the west and the Persian Gulf is on the east. Until recent years the area has been well isolated from the rest of the world. The government is an absolute monarchy in which, until recently, the provincial sheiks were all-powerful in their own domains. Slavery, although still existing, was made illegal in 1962.

The irrigated land in the oases and the small, mountainous strip by the Yemen border total only 800 square miles and are the only areas where sustained plant growth can be maintained. The rest of the country, an area nearly the same in size as the part of the United States west of the Rocky Mountains, is desert, although a somewhat more productive one than much of the African desert. Bedouins travel more than half of it to get the scant growth for their flocks and camels. The government is stable and law and order prevail. The economy is healthy because of the tremendous petroleum reserves, which generate ample foreign exchange.

CATTLE BREEDS

Most of the migrations which carried the cattle that contributed to the many types and breeds in Africa traveled across Arabia in the past. The humpless Brachyceros, or Shorthorn, cattle brought by their owners in their slow migrations to lower Egypt, probably were crossing the northern end of the Arabian peninsula shortly before the dawn of history. In later migrations, about 2000 B.C., Longhorn Zebus traveled south across Arabia, entering Africa in what is now Somalia. Much later Shorthorn Zebus followed the same route. None of these races left any recognizable traces other than the hump in present-day Arabian cattle.

The native cattle today are small and nondescript and are so varied in the common cattle characteristics that they cannot be classified as a breed or even a type. Locally they are referred to as Baladi cattle, the cattle of the country. The average cow in good condition weighs from 450 to 800 pounds. Humped animals predominate, but humpless individuals are seen. Most of the animals have very short horns, but some individuals have sizable horns. Color, conformation, and size vary widely.

It is reasonable to assume that the native cattle had their origin in remnants of the Longhorn Zebu, Shorthorn Zebu, and perhaps even the Brachyceros, which were moved across the land thousands of years ago in migrations. The cattle seen today are only the degenerate results of centuries of uncontrolled breeding of these animals and possibly later additions of other races. Deficient nutrition and selection by survival have produced a very hardy animal excellently adapted to the environmental conditions.

Typical Baladi cow. Photographed at Qatīf Oasis.

The only exotic cattle in Arabia are small herds of the Jersey, Red Danish, and Swiss Brown breeds found at the two government experiment stations.

MANAGEMENT PRACTICES

Villagers near the oases own the Baladi cattle, which are maintained primarily for their scant milk production and to some extent as a prestige symbol. The milk is obtained mainly for household use by the owner and his retainers. The Bedouins who travel the desert have found sheep and goats much better adapted to their needs; only a few keep any cattle. The cattle-sheep ratio of the whole country is approximately 1 to 200.

Cattle in town and village are maintained in fenced yards with an adjoining open shed. Feed is brought to them as green chop. Cows are milked for the owner's household; very little milk is sold. Cows are seldom used for draft since donkeys are generally employed for transportation and field work. Bulls are sold for slaughter. An owner's small group of cattle usually includes 2 bulls of varying age and 10 cows. In 1965 a good milk cow giving about ten pounds of milk a day would bring $225 as a producer but only about $100 if sold for slaughter.

Dairy herd of a village dweller.

The herds of exotic breeds on the government experimental farms are maintained in solidly constructed barns and let out for a few hours' exercise every day in an adjoining paddock. The animals are maintained in good condition, but heat stress takes its toll of production.

MARKETING

The principal livestock market is at Jidda, where cattle imported from East Africa are sold. Transactions are between the importer and the retail merchant, who has his purchase slaughtered in the municipal abattoir. Sale is by the head, in 1965 at prices equivalent to 12 to 15 cents a pound liveweight. Beef in the shops sold for 50 cents a pound without the bone, compared to 65 cents for mutton.

CATTLE DISEASES

Cattle diseases are not a serious problem in Arabia. The sparse population and the exceptionally dry climate are probably the reasons.

GOVERNMENT AND CATTLE

The interest of the Ministry of Agriculture in cattle consists principally of maintaining purebred herds of exotic milk breeds. At the Derab Farm, near Riyadh, a former royal stable has been converted to a dairy barn

Baladi cow of a settled Bedouin. Photographed at Al Hasa Oasis.

where a herd of Red Danish and another of Jersey cattle are maintained. A herd of imported Jerseys is kept at the Hufuf M'Zeira Experiment Station. Both are well-operated establishments but, beyond the milk they produce, do not appear to be accomplishing much in the way of cattle development.

By providing housing and limited facilities to maintain a few head of cattle, some effort is being devoted to settling Bedouins around the oases in an attempt to woo them away from their nomadic life with sheep and goats to a more productive existence. This, however, is not a Bedouin's way of life and whether or not he will remain settled is problematical.

OUTLOOK FOR CATTLE

It is conceivable that the government work with exotic milk breeds may lead to small, healthy dairy areas in the major oases. It is doubtful whether the European types are the best animals for this purpose, however. Some of the Zebu milk breeds such as the Red Sindhi or the Sahiwal or the excellent Shami cow of Syria, all of which are well adapted to the heat stresses encountered, would probably do better. Beyond this advancement not much can be expected in the way of cattle development in Arabia.

Syria

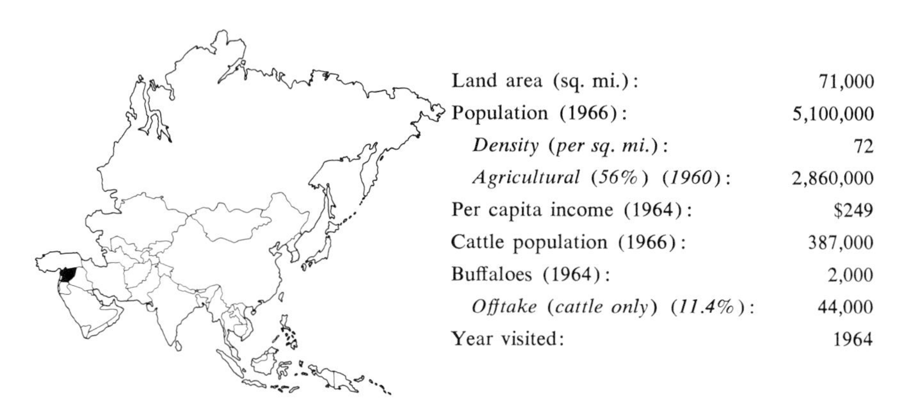

Land area (sq. mi.):	71,000
Population (1966):	5,100,000
Density (per sq. mi.):	72
Agricultural (56%) (1960):	2,860,000
Per capita income (1964):	$249
Cattle population (1966):	387,000
Buffaloes (1964):	2,000
Offtake (cattle only) (11.4%):	44,000
Year visited:	1964

THE SYRIAN ARAB REPUBLIC lies in the path of the trade and the conflicts which have involved Egypt, Asia, and Europe since the dawn of history. Located at the eastern end of the Mediterranean Sea, it is enclosed by Turkey on the north, Iraq on the east, and Jordan on the south. Lebanon and Israel lie in the southwest corner, between the border and the sea. In area Syria is two-thirds the size of Arizona and lies in the same latitude. The population is almost entirely Arabic.

All of the Middle East, including present-day Syria, which has always been the object of conquest, was under Turkish rule for the four centuries preceding World War I. Except for the few months that it broke out as an independent kingdom in 1920, Syria was under a French mandate from the League of Nations until after World War II, becoming independent in 1946. The country has been in political turmoil ever since, with frequent army coups. For a few years it was a member of the United Arab Republic, from which it seceded in 1961. Since then it has been effectively under army control. Expropriation of private enterprise, agrarian reform, and close association with the Soviet Union have been foremost in the political policies of the country.

The northern part is included in the Fertile Crescent, the historic breadbasket of the Middle East. The Euphrates River drains this area from northwest to southeast. Rainfall on the coast varies from 20 to 40 inches and falls to 10 inches or less in the east. Extensive irrigation is used in the drier parts of the north. To the south of the arable area is desert. The economy is agricultural; wheat, barley, and cotton are the principal crops. Agriculture has suffered severely from the vacuum in management which followed redistribution of land under the agrarian reform policy.

CATTLE BREEDS

The cattle seen in Syria today are nonhumped, although Zebu cattle must have traveled through this area in ancient times on their migrations to Africa. In addition to the nondescript cattle of the country, two indigenous breeds are found in Syria. In recent years small numbers of exotic milk breeds, largely the Friesian, have been imported to increase milk production.

Shami.—This breed, also called the Damascus Cow, is probably the foremost indigenous milk-producing animal outside of Europe. Its origin is not known, but it has been grown in Syria and adjacent areas for an unknown period of time. Systematic selection has not been made for milking ability, although in past generations of breeding some preference was undoubtedly given to bulls which had better-yielding daughters and to dams with exceptionally good production history. Individuals give up to 16,000 pounds of 4 per cent butterfat milk in a

Shami bull, two years old.

Rangy Shami cow.

Deep-bodied Shami cow.

lactation, and some of the small herds of 10 to 20 cows in the dairies near Damascus average close to 9,000 pounds a lactation.

The Shami is usually a dark red to brown, sometimes approaching black; a medium tan is also seen occasionally, but the color is invariably nearly solid except for small white undermarkings on some individuals. Conformation varies considerably, from a rangy type to an animal with a good, deep body and a rather fine bone structure. The tail stock is prominent, and the Shami has a slight tendency to be sway-backed. The horns are small and short and turn either up or down. Mature cows in good condition weigh about 1,000 pounds, bulls 1,250 pounds.

The breed has acquired a remarkable tolerance to the endemic diseases, and it is not subject to the heat stress which causes deterioration in imported European cattle. Given adequate feed, a Shami cow is seldom seen in poor condition and enjoys a high degree of longevity. A productive life of ten or twelve years is not uncommon for good-producing cows on adequate nutrition.

In 1964 this remarkable breed was in danger of extinction. The total population in Syria was estimated at 10,000 to 15,000 head, and there are only a few of the breed in neighboring Turkey and Lebanon. Many cows were sold off as their owners were caught in the extremities of the land-reform program. Ill-advised attempts to upgrade Shami cows by crossing them with Friesian bulls have become popular and are also a threat to the breed.

Julani.—This indigenous breed is small, the mature cow weighing approximately 800 pounds. The black-and-white color markings are reminiscent of the Holstein, to which the Julani is in no way related. The hair coat is long and shaggy, the horns very short and stubby. Except for size the Julani bears much resemblance to the Sharabi of Iraq—in the coloring; the short, stubby horns; and the general conformation.

Julani bull.

Julani cow and calf.

The Julani is seen mainly in the rougher country and is used for draft, although milk is taken for household use.

Baladi.—This term is used locally to designate the cattle of the country. They are nondescript in color and conformation, are nonhumped, and comprise a good part of the cattle population.

Exotic Breeds.—The Friesian is the principal exotic breed imported in an effort to increase the production of milk. It must be maintained under carefully controlled conditions, and even then heat stress soon takes its toll in causing debility and a decrease in lactation.

MANAGEMENT PRACTICES

The cattle of the country are pastured on the native grasses in the rougher land and on crop residues. Hay is not put up and only the mature dry growth is available for feed from January to March, when cattle rapidly deteriorate in condition.

In the Ghuta, the irrigated oasis near Damascus, the herds of Shami cows in the small dairies are well cared for, although the facilities are quite limited. Cows are taken to alfalfa pastures after the morning's milking and are tethered in a row across the field. Each cow is allowed enough rope so that she can consume the feed in an allotted area before she is taken in at night. This method of pasturing clears a strip possibly 20 feet wide across the field each day. By the time the animals reach one end of the field, there is sufficient regrowth at the other end to allow the procedure to be repeated. The interval between cycles is approximately twenty-five days. Heifers being grown out as replacements are pastured in the same manner. No more efficient method of pasturing can be devised if labor costs, which are low in Syria, can be ignored.

Shami cow herd on Ghuta method of pasturing.

These small dairies usually allow the calf all the milk a cow produces during the first forty days of her lactation. After this period of time the calf is tied in front of the cow so she can see it as she is milked. Four or five pounds of beans a day is fed during milking.

The owner of a 10- to 20-cow herd keeps his own bull. Males are usually sold for slaughter, but it is customary to keep a bull calf that is a full brother of the bull being used for breeding. If the sire is found to produce good-yielding daughters, he is sold and his young brother takes his place. Such rudimentary selection has played its part in the development of the Shami cow.

The better dairies of this type around Damascus average as much as 9,000 pounds of milk a lactation from a herd of about 15 cows. An exceptional dairyman now weighs the yield from individual cows, but production is usually gauged simply from the weights for which the producer is paid. In neither case does the yield include the milk given to the calf for the first forty days after it is born. Milk sold for 6 cents a pound in 1964, and dairying was a profitable enterprise. A good Shami cow, when one could be found for sale, cost $350.

A government center in Aleppo is trying to introduce artificial insemination, but very few cows are being brought in because of cultural objection to this method of breeding. In 1964 the bull battery consisted of 1 Friesian, 2 Red Danish, and 4 Shami bulls. The laboratory facilities are adequate and modern practices are followed in semen preparation.

MARKETING

The two major livestock markets in Syria are found in Aleppo and in Damascus. The Damascus market is held only on Friday, the Moslem holy day. Both markets are located in an open area where each owner keeps his animals—sheep, goats, cattle, camels, and an occasional buffalo—in a small group in anticipation of a sale. All transactions between the owner and either a meatshop proprietor or another farmer are by the head. Many cattle sales fall in the latter category. Nearby farmers with small holdings often come to buy an animal or two to feed further before selling them for slaughter. A rather thin animal—bull, ox, or cow—weighing perhaps 800 pounds sells for $100. Any cow in milk, which must have a calf by side, brings $125 even if practically a canner. No good Shami cows are seen at the sales. Most cattle sold in the markets are Baladis or young Shami males not wanted for breeding. Near Damascus, bulls for slaughter are sold entire; in the Aleppo market, steers are usually offered.

Cattle section of the Aleppo Livestock Market.

ABATTOIRS

In Damascus the municipal abattoir adjoins the livestock market. A large animal—a cow, a bull, a camel, or, occasionally, a buffalo—is killed by hamstringing, then throwing and cutting the throat. The carcass is then raised by a hand hoist for skinning and eviscerating. In slaughtering sheep and goats, the throat is cut and the carcass handled by hand. Slaughtering starts at 3:00 A.M. and is completed within five hours. The meatshop owner who has bought the live animal sees it delivered immediately to the slaughter floor and killed. The carcass of a sheep or a goat is thrown over the shoulder of the purchaser, who walks by the inspector with the liver and lungs in hand. If these organs pass, the inspector's assistant stamps all four quarters. Large animals are inspected and stamped before the finished carcass is removed. Because of cultural requirement all meat is sold and consumed the day it is killed. There is no refrigeration in meat stores or in most homes.

GOVERNMENT AND CATTLE

Beginnings are being made in a technological approach to the problems of agriculture generally. They are recent and small in scope, and, as far as cattle are concerned, are for the most part directed at unrealistic goals. Government plans call for the establishment of a large Friesian dairy herd in Aleppo to serve as a demonstration center and a source of bulls for upgrading native cattle.

A College of Agriculture was established at the University of Damascus in 1960. A few head of both purebred Friesian and Shami cows are maintained, with the idea of conducting crossing experiments.

A German team is operating the Experiment Station for Animal Breeding and Husbandry at Deir-el-Hajar, on the edge of the Ghuta. The station was started by the French in 1935, during their period of influence, as an agricultural demonstration center. Whatever progress was made was abandoned in 1958 at the time of the Suez crisis, and the station was then maintained by the Syrian government on not much more than a caretaker basis until the country united with Egypt. The Egyptians then undertook to reactivate the center but moved out when Syria withdrew from the United Arab Republic in 1961.

Recently arrangements were made with the University of Berlin to furnish the personnel and establish a demonstration center for animal breeding and husbandry. Six technically trained German agriculturalists supervise the operation, which is handled in a most efficient manner. This is to be accomplished in three years and the operation then turned back to the Syrian government. Small purebred herds of registered Angler, German Brown, and Friesian cows had been imported from Germany for the project. Plans include the addition of a control group of Shami cows. The current program to increase milk production by bringing in exotic breeds does not appear promising.

OUTLOOK FOR CATTLE

The cattle population of Syria decreased from 584,000 head in 1957 to 387,000 in 1966. Although there was a severe drought during this interval, the liquidation of cattle following the land-reform programs was in large measure responsible. In the atmosphere of expropriation of private businesses, as the large landowners lost their land they sold their cattle in the fear of losing them also. The small operator will continue to keep his draft animals and the dairyman his few cows, but under present conditions any expansion in cattle raising is unlikely. There is good grazing land in much of the rough country in the north that is underutilized in spite of the large number of sheep and goats feeding there. Milk is in short supply and, even under the unsettled conditions of the economy, its production is a profitable enterprise.

There is no evidence that the course charted by the government to increase production by importation of exotic breeds will be successful. Such considerations, along with the day-to-day uncertainty about the stability of the government, do not hold much promise for the future of cattle in Syria. It is to be hoped, however, that some international institution will find a way to continue the Shami breed.

Turkey

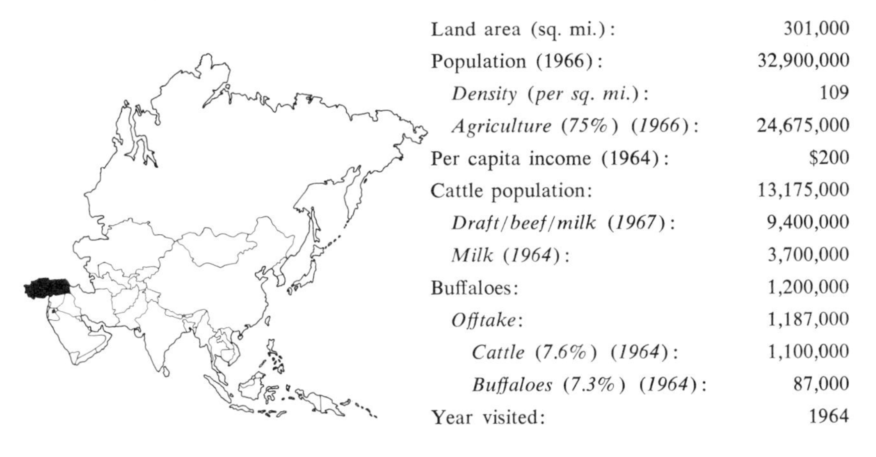

Land area (sq. mi.):	301,000
Population (1966):	32,900,000
Density (per sq. mi.):	109
Agriculture (75%) (1966):	24,675,000
Per capita income (1964):	$200
Cattle population:	13,175,000
Draft/beef/milk (1967):	9,400,000
Milk (1964):	3,700,000
Buffaloes:	1,200,000
Offtake:	1,187,000
Cattle (7.6%) (1964):	1,100,000
Buffaloes (7.3%) (1964):	87,000
Year visited:	1964

THE REPUBLIC OF TURKEY occupies the peninsula long known as Asia Minor, which lies between the Black Sea on the north and the Aegean and Mediterranean seas on the west and south, respectively. Turkey in Asia borders with the Soviet Union and Iran on the east and Iraq and Syria on the south. It accounts for 97 per cent of the land area of the country, is one-tenth larger than California and Nevada combined, and lies between the same general latitudes as the part of California north of Los Angeles. Across the Sea of Marmara, the Bosporus strait, and the Dardanelles, at the southeastern corner of Europe lies Turkey in Europe, an area a little larger than the state of Massachusetts and in similar latitudes. This area is bordered by Bulgaria on the north; that country and Greece both lie to the west.

The Ottoman Empire, which had ruled most of the Middle East for 650 years, disintegrated at the end of World War I. In 1921 Turkey emerged as the only independent state of the former empire; the various provinces of the empire were mandated to several of the allied nations until eventually they became independent. The Republic of Turkey was formed in 1923, and there was rapid advancement along cultural and economic lines under what amounted to the often severe, but benevolent, leadership of Kemal Atatürk. This momentum ceased with his death in 1938, but general political stability has existed since then except for an army coup in 1960. A new constitution was approved a year later by popular vote and civilian government was returned.

The Turkey in Europe, only 9,100 square miles, is a rough country with a central valley enjoying a fair rainfall. Turkey in Asia consists mainly of the high Anatolia plateau, 2,500 to 4,500 feet in elevation, with rugged mountains in the east on the Soviet and Iranian frontiers and lower mountains on the other sides. It is rather poor farming country. Rainfall along the coasts varies from 20 to 30 inches and decreases inland; it is rather marginal for dry-land grains on much of the farming area. About one-tenth of the cultivated land is under irrigation. The soils are fair to poor and have been denuded of minerals by centuries of cropping, with very minor replacement by fertilization. Large-scale use of cattle dung for fuel has further decreased plant nutrients. The natural grasslands are badly overgrazed, the 14,500,000 cattle population competing with nearly four times that many sheep and goats for the available feed.

CATTLE BREEDS

The native cattle of Turkey are known locally by simple terms such as Native Grey or Native Black. Indigenous breeds have been maintained and improved by selection on government farms. Exotic breeds, imported from Europe, have had a considerable influence on the cattle population, and this is increasing. These animals have been introduced primarily to increase milk productivity.

Native Grey.—This is the local name for an indigenous representative of the Grey Steppe cattle of southeastern Europe. Immigrants from the Balkan Peninsula brought them into Turkey. Their antecedents are considered to be the same as the Iskur breed of Bulgaria and what have been called the Native Cattle of northern Greece.

The Native Grey, accounting for 10 per cent of the cattle population, is a rather small animal, well adapted to arid regions. The males are often much darker grey than the females, which are light grey, sometimes nearly white. The horns are rather large and upswept, occasionally lyre-shaped toward the ends. The animal is rugged and has a good bone structure and a fair beef conformation. Primarily used for draft, its milk is taken regularly for household consumption by the owners.

The potential of the breed is demonstrated by herds on government farms, where they are maintained on a good nutrition level and have been under some selection for improvement in type. There a mature cow averages 900 pounds and bulls weigh as much as 1,400 pounds; in the hands of the village farmer, a cow seldom weighs 700 pounds and a working ox rarely weighs more than 900 pounds. The small size of the country cattle results from many generations living on a deficient diet.

Native Grey bull. Photographed at Cifteler Harasi Experiment Farm.

Native Grey cows. Photographed at Cifteler Harasi Experiment Farm.

Native Black.—The solid black draft type indigenous to the Anatolia plateau and also used for both milk and meat is locally referred to as the Native Black. The horns are thick and rather short, growing straight out from the head with a slight forward curve. The conformation is quite similar to that of the Native Grey, and the herds on the government farms show a similar improvement over the farmers' cattle. The typical village cow weighs about 600 pounds, but selection and better feed have developed a much larger animal at the government stations. The average milk yield is 1,200 pounds a lactation with a 4 per cent butterfat content, but on the government farms has been increased to 2,600 pounds.

The Native Black is the dominant breed in Turkey, comprising over 40 per cent of the total cattle, and is found generally over the central part of Turkey in Asia.

Native Black bull, 1,300 pounds. Photographed in Konya Hara.

Southern Yellow-Red.—Locally called the Southern Yellow-Red, this indigenous milk breed is seen in limited numbers in the coastal area south of the Anatolia plateau. It is very similar in appearance to the Shami breed of Syria and is said to have been introduced from that country. The color is uniform, usually a solid yellow-red to a near-brown. The medium-sized horns are upturned but often short and stubby on the bull. The bone structure is quite fine and the general conformation is rangy. Milk yields up to 6,000 pounds a lactation are obtained from the best cows, although the average farm yield is about half this amount.

The Yellow-Red accounts for 5 per cent of the national herd.

Eastern Red.—This breed predominates in the eastern part of Turkey in Asia, where it is used for draft, meat, and milk. It comprises 22 per cent

Southern Yellow-Red bull. Photographed at the Alata Teknik Sahcivarlik Okuli (Alata Technical Garden School).

Southern Yellow-Red cow. Photographed at the Alata Teknik Sahcivarlik.

of the cattle population; only the Native Black has a larger representation in Turkey.

The Eastern Red, ranging from red to walnut, is said to have resulted from the mixture of Simmental, Swiss Brown, and German Angler breeds with native cattle. The village cow averages 1,500 pounds of milk, which is increased to 5,000 pounds on the government farms by selection and adequate nutrition.

Exotic Breeds.—The Swiss Brown (German Brown) from Germany and the Austrian Swiss Brown (Montafoner) were imported after World War I to upgrade the Turkish cattle for milk production. Purebred herds are maintained at the government farms. This breed and its crosses on the native cow now account for nearly 3 per cent of the national herd.

Some years after the importation of the Swiss Brown and the Austrian Swiss Brown, the Friesian was introduced, followed by the Simmental and the Jersey. Both the Friesian and the Jersey were used rather extensively but did not become as popular as the Swiss Brown. A few Aberdeen-Angus and Hereford bulls have also been put on some of the government farms in an effort to develop an improved beef type.

Buffalo.—Buffaloes are used in the coastal areas for draft and milk. Most are of the river type and many are of the Nili breed.

Buffalo team. Photographed near Bursa.

MANAGEMENT PRACTICES

Turkey is primarily an agricultural country in which the farmer with limited holdings is the backbone. Three-fourths of the population depends directly on farming, and 85 per cent of the families concerned own their land or a major part of it, the maximum holding being twenty-five acres. The other 15 per cent of the farm families are tenants or employees of city-dwelling landowners. Nearly three-fourths of the total cultivated area belongs to small farmers who own their land; the remaining fourth is in units which are usually large enough to permit some degree of mechanization, although cattle, horses, or buffaloes are still used for draft on many of them. Only one-tenth of the cultivated land is under irrigation, although water could be made available over larger areas at an economical cost. Pastures have been overgrazed for centuries, and the minerals from the cultivated lands have been mined by crops. The practice of communal grazing, which is general throughout the country and under which anyone may feed as many animals on the common pastures as he wishes, is one basic cause of this condition. Only insignificant steps are being taken to correct these conditions.

In the country the primary use of cattle is for draft, although milk is taken for household use and to some extent for sale in the villages. Milk from sheep and goats is generally preferred.

With one exception, worn-out work animals and unwanted young account for most of the animals sold for slaughter. Near the city of Konya, in south-central Turkey, a beginning has been made in the grain fattening of cattle for slaughter. This practice was initiated by the example of the government experiment farm, Konya Hara. Many of these cattle feeders live in town and have a few bulls in a shed behind their house, as well as others similarly maintained under the care of men working for the feeders. Bulls at weights of 700 to 1,000 pounds are purchased at one to two years old from small farmers and are fed a grain ration that consists of barley with sunflower cake for added protein. Only bulls are fed, for a fattened steer sells for 3 to 4 cents a pound less than

Finished Swiss Brown–Native Black bull, two and one-half years old, 950 pounds. Photographed in Konya.

a bull. The feeder stock was being purchased for 13 to 14 cents a pound liveweight in 1964 and the finished animal sold for 19 to 20 cents. In 1966 these feeder-operators had progressed to the point at which open feed lots were beginning to be used. Several of these operations feed as many as 900 head annually. Such enterprises depend on the export market to Syria and Lebanon and to a limited extent on the higher-class trade in Ankara.

Except for a few large farms in the surrounding country, the Istanbul milkshed is in the city itself. In 1964 the 500 dairy operators were milking 10,000 cows and 2,000 buffaloes within the city limits. An owner has from a few head to a maximum of 100. The milk is sold to numerous peddlers, who hawk their wares on the street. Pasteurization, refrigeration, and the most simple sanitation procedures are rare. Cows are usually maintained in dark, filthy sheds on a dirt floor, tied so close together that they barely have room to lie down. Feed and water are brought to them. No bedding is used, and manure is removed only as a matter of necessity. A cow attendant tends 12 cows and fares about as well as his charges.

A dairy of 20 or more head usually maintains a bull—anything capable of breeding. Calves are all sold when a few days old. Cows are milked, bred, and calved, tied in practically the same position all their lives. Buffalo cows, maintained in the same manner, are not bred but are purchased in the country after calving, milked for one lactation, and then sold for slaughter.

Most cows are Friesian crosses; the buffaloes are of the Nili or Murrah breeds. A replacement Friesian cow purchased in the country when fresh costs $300 to $500, depending on a visual appraisal of her milking ability, and buffalo cows usually cost $650 or more; when sold for slaughter, either brings less than $100. Cow milk sells to the peddlers for 6.5 cents a pound, buffalo milk for 11 cents a pound. Hay costs $40 a ton, grain about 3 cents a pound. Feed, because of the cost, is the one item scrupulously handled. A base ration of hay and some straw, wheat bran, and brewers' refuse, when available, is fed along with a liquid mixture of commercial wheat flour and flax-seed meal.

The government has been planning for years to correct the Istanbul milk situation, but in 1964 little had been done except passing a law making it illegal to operate an unlicensed dairy. Only a handful of dairies, however, were so licensed; yet the others continued to produce milk.

There are a few well-run dairies near Istanbul, milking as many as 100 cows or more. These establishments have modern equipment, milk parlors, and, in some cases, refrigeration, pasteurization, and bottling facilities. At least one also produces cheese and butter. Small purebred herds of either Friesian or Swiss Brown are maintained for the production of bulls for sale as well as cows to be placed in the milking herd. Such establishments are usually owned by successful businessmen in Istanbul.

Some use of artificial insemination, initiated in 1949, is now being made. In 1964 the largest single use in Turkey of the method was at a government farm, Cifteler Harasi, which began the practice in 1961 and is using it exclusively on the herd of 350 brood cows. A number of small artificial insemination centers in the country have 3 or 4 bulls each to serve the surrounding areas. In 1968 a total of 33,000 cows were bred artificially, with a conception rate of 70 per cent.

MARKETING

The butchershop proprietor buys direct from the producers and slaughters his purchase in one of the municipal abattoirs. In 1964 the price of slaughter animals was the equivalent of approximately 11 cents a pound liveweight, sales being by the head. Very little attention was paid to quality or condition except in the case of a badly emaciated animal.

ABATTOIRS

Turkey has more than 500 municipal abattoirs and four large government-owned and operated facilities, which together probably handle slightly less than half the cattle processed. The rest are killed in private slaughterings without any inspection.

CATTLE DISEASES

Looking toward the control of cattle diseases, Turkey began her veterinary education 125 years ago. The immediate objective was the control of rinderpest, which had caused heavy losses. Through systematic control measures this disease was eradicated in 1932. Effective preventive measures are taken against anthrax, Texas fever, and both internal and external parasites.

Foot-and-mouth disease has been controlled by vaccination. Turkey is given credit for minimizing the spread of this disease from the Near East to Europe by the control and quarantine measures which were enforced. Brucellosis and tuberculosis are under control on the government farms and at state breeding stations and also in certain other areas. Most of the vaccines used are now produced in Turkey.

A number of years ago one government farm suffered two severe outbreaks of foot-and-mouth disease. Now every animal is vaccinated three times a year. They have also encountered severe losses from brucellosis and are now vaccinating all heifer calves at four to eight months of age. All cows are now tested for tuberculosis, and positive animals are slaughtered.

In the Konya area, where the cattle-feeding industry is developing, most of the farmers as well as the feeders voluntarily follow an adequate immunization program. Calves are vaccinated at six months for blackleg and anthrax and once a year thereafter for the latter. All cattle are vaccinated twice a year for foot-and-mouth disease. Most heifer calves are vaccinated for brucellosis at four to eight months of age. This program is entirely on a voluntary basis and is, therefore, only partially effective.

GOVERNMENT AND CATTLE

As is customary in eastern Europe, all matters pertaining to animal husbandry are under the direction of veterinarian personnel. In Turkey this authority is the General Directorate of Veterinary Affairs in the Ministry of Agriculture. In addition to furnishing veterinary service, the government veterinarian stationed in a district acts as a county agent of sorts, advising farmers on agricultural practices generally and on animal husbandry particularly. Directors and staff of the government farms are invariably veterinarians.

In the days of the sultans, the Turkish cavalry was the elite of the military establishment. Excellently equipped farms (*harasi*) were maintained for the breeding of horses to supply the cavalry. After the republic was founded, these farms were converted to agricultural work. The fifteen government farms and breeding stations now located throughout the country are conducted somewhat like university experimental stations in the United States. There is one research center. In addition to the usual type of agricultural work, many of the stations have sizable cattle projects under way and also serve as demonstration centers of improved methods in cattle management. Some of these farms continue to raise horses and are still called *harasi* (horse farm) as opposed to *hara*, which is simply a farm. A visitor to a *harasi* is always shown the horses first. As the visitor approaches the barn, the man in charge, following in the footsteps of the barn sergeant of cavalry days, snaps to attention in the doorway and in a thoroughly military manner reports the condition of the animals under his care, noting particularly a new foal or any ill-

Purebred Montafoner bull. Photographed at Cifteler Harasi.

Purebred Montafoner cow in foreground. Photographed at Cifteler Harasi.

nesses. After an inspection of the 40 to 60 horses in a barn has been completed, those animals in training are turned loose and walk to the exercise ground outside the barn. Here they form a circle and on command from the trainer go into their different gaits. At a final command they return to their stalls. After this exhibition the cattle are shown. Some of these government farms have done unusual work in improving the native breeds as well as in upgrading them through the use of exotic bulls.

Soker Ciftligi, near Eskisehir, is a government farm started in 1935 to utilize beet pulp from a nearby sugar factory. It was later changed to a breeding station with the objective of determining the results of crossing Swiss Brown bulls on two indigenous breeds, the Native Grey and the Native Black. The initial results indicated that the Native Grey crosses gave an animal with a higher milk yield than the Native Black crosses. The present program plans to continue breeding all progeny continuously to Swiss Brown bulls. The crossbred herd is now averaging 9,600 pounds of milk a lactation, as compared with 11,000 pounds in the purebred Swiss Brown herd.

Cifteler Harasi, east of Eskisehir, has a herd of purebred Montafoner cattle, the base of which was imported from Austria; a selected herd of Native Greys; and another of Montafoner–Native Grey crosses up to the fourth generation. Excellent records are kept on milk production and breeding. By selection within the breed for milking ability over an eight-year period, the Native Grey has been brought from an average of 3,300 pounds a lactation to 5,300 pounds. The principal basis of comparison for milking ability used is the production during the first lactation. The pure Montafoner average is 5,200 pounds for heifers bred as two-year-olds. The herd of Montafoner–Native Grey crosses, which includes F_1, F_2, and F_3 crosses, is averaging 4,000 pounds for cows of the same age.

At Karacabey Harasi, southeast of Bursa, Swiss Brown–Native Grey crosses have been compared with Friesian–Native Grey crosses. The conclusion from this work has been that the Friesian cross gives a slightly higher milk production but that the Swiss Brown cross is a hardier animal and is, therefore, preferred.

At Konya Hara, near the town of Konya, the improvement of the Native Black breed has been undertaken for both a milk-type and a beef-type animal by crossing with Aberdeen-Angus bulls. A herd of 140 head of Native Black mother cows and another of 50 purebred Aberdeen-Angus cows are maintained. Selection is being made within the Native Black breed for a milk-producing type and, based on conformation, for a beef type. The progeny of each of these types is being bred continuously to Aberdeen-Angus bulls. In 1964 this work had not proceeded far enough so that any conclusions could be drawn, but the Aberdeen-Angus cross on the Native Black cow had produced some excellent animals.

An activity that could have a more far-reaching effect on animal husbandry in Turkey than the breeding experiments of the government farms is the system of government schools which was established to combine conventional education with practical training in all phases of agriculture for teen-age children. Alata Technical Garden School, near the town of Alata on the eastern end of Turkey's Mediterranean coast, is one of the five schools of this type. Farm children in the area—boys from 14 to 17 years of age and girls from 12 to 15 years—are eligible to enter. Board and room, tuition, books, and clothing are furnished by the government for the duration of the three-year course. The present enrollment is 300 boys and 200 girls. On graduation many of the students return to their farm homes and are having their effect on improving the age-old agricultural practices.

A 15-cow herd of excellent Southern Yellow-Reds is maintained to implement the cattle part of the curriculum. Good records are kept on production and breeding, and the animals are handled in a practical, husbandlike manner. All work, including the record keeping, is done by the students, under the direction of the instructors. The facilities—sheds, yards, and feed storage—are of simple construction but are entirely adequate and have been designed to illustrate what a small farmer could provide for himself.

The cows in the school herd are pastured for nine months of the year and brought in morning and night for milking on a concrete platform with a shed roof. During the winter they are kept in stanchions in the barn. The top cow produces more than 12,000 pounds of milk a lactation; the average for the 15-cow herd is 8,300 pounds, which the students can compare with the 3,500 to 4,000 pounds the cows at home produce. At milking, a balanced grain ration is fed in proportion to the individual's milk production—one pound of supplement to three pounds of milk. This supplement consists of foodstuffs available to farmers in the area—corn, barley, and wheat bran, with sunflower cake for protein.

OUTLOOK FOR CATTLE

Before the establishment of the republic in 1923, the people of Turkey had little contact with the Western world. The only relations had been those of a diplomatic nature. Agriculture and stock raising had not advanced beyond medieval and nomadic practices. To convert to modern methods 25,000,000 persons engaged in primitive agriculture, many of them in inaccessible locations, is an undertaking requiring generations, rather than years. The headman of the village, around whom agriculture centers, is in supreme authority, and any improvement in husbandry practices must stem from him. The cattle of Turkey have to compete for the available forage with four times their number of sheep and goats, whose meat and milk the predominantly Moslem population prefers. Some areas not now fully utilized can be made productive for stock raising, but the possibilities for advancement in cattle raising for the present lie in the improvement of the native cattle by crossing and better pasture management—the establishment of grazing control, better-adapted grasses, and good watering facilities. By such means the productivity of the cattle of the country could be materially increased without any increase in number.

There is a wide gap between the small farmers with their few head of cattle and the government farms with their sophisticated breeding programs. Better bulls are furnished the cattle growers near the government farms, but the effect of this impact is small when total numbers are considered. The vocational schools, as exemplified by the one at Alata, are a big step forward in the areas they serve; but there are only five of these schools with fewer than 3,000 students, and their influence can affect only a minute part of the agricultural population.

Today Turkey is well oriented toward the Western world. The productivity of both crops and livestock on the socialist-type farms across the border in Bulgaria is much greater than that of the Turkish farmer on his small holding. As the day approaches when the production of food will have to be man's first consideration in that part of the world, this comparison will not go unnoticed. The situation does not look very favorable for private enterprise in Turkish farming.

The government must take a more grass-roots approach to the basic problems of animal husbandry before any large improvement in cattle raising can occur. In some sectors beginnings have been made—the government farms, the vocational schools, the commercial feeding for market in the Konya area. On the other hand, the deplorable conditions in the city-enclosed milkshed of Istanbul are left untouched. In the final analysis, advancement in stock raising in the future will depend on how much effort and money government will put into the necessary programs. The answer to this is involved in the unsettled internal political scene, where the major concern is with such problems as Cyprus, rather than agricultural developments to feed the people.

Burma

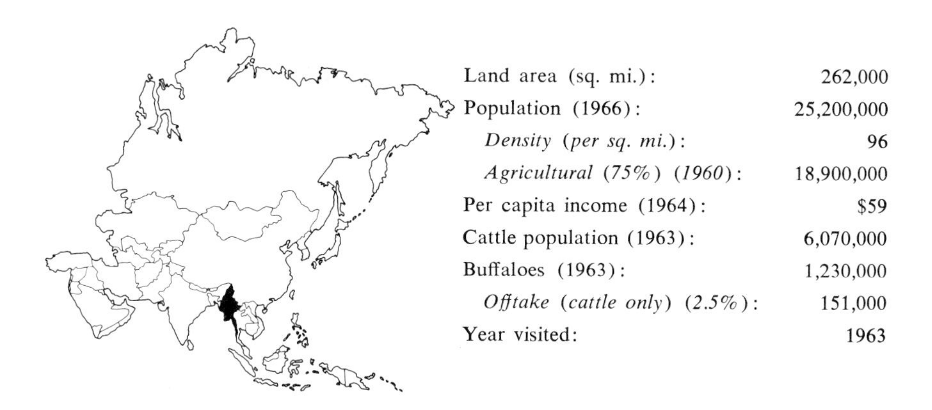

Land area (sq. mi.):	262,000
Population (1966):	25,200,000
Density (per sq. mi.):	96
Agricultural (75%) (1960):	18,900,000
Per capita income (1964):	$59
Cattle population (1963):	6,070,000
Buffaloes (1963):	1,230,000
Offtake (cattle only) (2.5%):	151,000
Year visited:	1963

THE UNION OF BURMA occupies a somewhat diamond-shaped area where the western part of the Indochinese peninsula joins Asia. There is also a narrow tail 500 miles long extending part way down the west of the Malay Peninsula. The northern part of the diamond is a wedge between India on the west and China on the east. The southeastern boundaries are with Laos and Thailand. The Bay of Bengal lies on the southwest and, at the extreme western tip of the diamond, East Pakistan touches Burma. The area is nearly as large as Texas; the latitude is in the same general range as Mexico. Burma has rich mineral deposits and jade has been mined there for centuries.

There is a story that a traveler from Venice reached Burma in 1435. The Portuguese were definitely there in 1511 and established some trade; but only some pirate outposts, which had broken off from their legitimate enterprises, remained for any length of time. English and Dutch traders came to Burma in the seventeenth century, and in the next century the English and French were competing for the Burmese trade. In the early part of the nineteenth century, the Burmese began to move in on Assam, which was part of India. This action led to three wars with Britain over a period of sixty years, culminating with Burma being annexed as a province of India in 1886. In 1937 it was placed directly under the control of a British governor and remained in this status until the Japanese occupation during World War II. After liberation in 1945 independence movements resulted in the Union of Burma (outside the Commonwealth of Nations), which was recognized by Britain in 1948.

Constitutional government did not last long. Since 1958 the military have been in control, with the exception of one brief return to civilian government, and have followed a constantly increasing socialistic policy. The 1,200-mile frontier with China has undoubtedly had its influence here. In addition the Chinese have exhibited a grass-roots approach in their propaganda efforts. Chinese schools are free but those teaching English and European languages require a tuition fee. After 1963 most contact with the Western world was broken off, and practically all private enterprise except farming was expropriated. In the final analysis, agriculture also is effectively dominated by government through controlled outlets and prices.

The Burmese people are descended from immigrants who came from China in unrecorded migrations in ancient times. Much more recently both Indians and Chinese entered the country and became sizable elements in the population before World War II. Following the war, and especially after Burmese independence, their numbers declined materially. Europeans, principally the British, who came in during the period of British control probably never exceeded 12,000 and have now departed.

The economy is primarily agricultural, with rice the main crop, along with some cotton and rubber. Annual rainfall varies from 200 inches in the narrow coastal strip on the Malay Peninsula to 80 inches around Rangoon and 24 inches in the driest inland areas.

Cattle are raised mainly for draft purposes and outnumber buffalo

Typical Burmese cattle.

6 to 1. Buffalo are used for cultivation mainly in the paddy fields in the wetter areas near the coast. The cattle seen near Rangoon are a nondescript, degenerate type of humped animal. Negative selection—the better bulls being kept as draft animals and the poorer ones being allowed to run with the cows—has contributed to this condition. The nondescript animals are varied in color, although many white and light-grey animals are seen. The hump is small and the dewlap and umbilical fold are not nearly as pendulous as those of most Zebu cattle.

The buffaloes are the swamp type seen throughout Southeast Asia. They maintain themselves in better condition than the cattle since they are more efficient users of the rough forage available.

Buffalo cow grazing on rice stubble.

The government operates a dairy and a milk plant on the outskirts of Rangoon. Sterilized milk is bottled and will keep without refrigeration for two months. Milk production, however, plays no significant part in the livestock economy.

The Rangoon slaughterhouse is an open shed, not even screened, with a concrete floor. Killing is done by Moslems. The animal is thrown, faced toward Mecca, and the throat cut. The carcass is then raised by a hand chain block to facilitate eviscerating and skinning. Sanitary provisions are rudimentary. In conformance with local custom, all meat is disposed of the day it is killed. There are no refrigeration facilities.

The Burmese are predominantly Buddhist and although they eat beef, it is not a favored article of diet. Other elements of the population probably consume most of the production. Cattle and buffaloes that are no longer useful for draft are the only animals slaughtered. Both will continue to be raised primarily as work animals in the foreseeable future.

Ceylon

Land area (sq. mi.):	25,300
Population (1964):	11,500,000
Density (per sq. mi.):	455
Agricultural (76%) (1964):	8,750,000
Per capita income (1964):	$123
Cattle population (1966):	1,851,000
Buffaloes (1966):	1,002,000
Offtake (1966):	172,000
Cattle (8.5%):	157,000
Buffaloes (1.5%):	15,000
Year visited:	1967

THE DOMINION OF CEYLON is a beautiful and climatically pleasant island directly east of the southern tip of India in the Indian Ocean. It lies in the same general latitude as Panama and is a little more than half the size of Louisiana. The mountainous highland in the interior, which enjoys a near-temperate climate, is where Ceylon's famous tea is grown.

The history of the island has been long and varied. During the sixth century B.C. the Sinhalese people from northern India began migrating to the island, bringing with them their cattle and probably their buffaloes. These people developed an advanced agricultural economy with extensive irrigation works, some of which have recently been rehabilitated; many of the works that were reclaimed by the jungle centuries ago, however, are still unused.

For 2,000 years the Sinhalese built kingdoms, tilled the soil, and developed an advanced civilization. Their progress was interrupted from time to time by foreign invaders, mainly from southern India. There were also occasions of internal conflict between the island kingdoms, but these disturbances did not greatly disrupt the economy. The Portuguese landed at the turn of the sixteenth century and held the seaports for 150 years. The Dutch took over at the middle of the seventeenth century and were in turn forced out in 1795 by the British, who had the whole island secured by 1815 and began their highly successful colonization program. Inadvertently they sowed the seeds of the major element of discord on the island today by bringing in laborers from the southern Indian Tamils. The antipathy between the Hindu Tamils and the Buddhist Sinhalese is thus of long standing and completely irreconcilable. This feeling still is so great that when independence was granted in 1948, 1,000,000 Indian Hindus, or nearly one-tenth of the population, were disenfranchised. India refused to repatriate them and in 1966 they remained a people without a country in spite of agreements between the Ceylonese and Indian governments to rehabilitate them.

Under governments that have been increasingly socialistic since independence, the economy deteriorated to the breaking point until the middle of 1965, when elections placed a more moderate element in control. What this group will accomplish in channeling the rich agricultural resources of the country into a viable economy remains to be seen. The labor unions, more than 1,000 of them, remain in control of the radical political parties and are a powerful tool for blackmail against the now-ruling faction. The withdrawal of financial aid by the Soviet Union and China after the turn toward the West in recent elections and the reactivating of United States AID could signify a changing trend.

Public corporations operate much of the business enterprise and owe their existence to the expropriation of foreign concerns by the recent leftist regimes before their ouster. The management of these corporations is inefficient, and many have continuously run in the red. The saving grace has been that the large, British-owned tea plantations were largely unaffected in this movement—even from the specter of land reforms. The plantations account for more than 60 per cent of the foreign exchange, and those in power realized there would be no funds for image building if this income disappeared.

Few places enjoy the favorable conditions for an intensified agriculture that are found in Ceylon. Climate is the principal support for the economy, since there are no other major natural resources. The southwest monsoon, from April to August, and the northeast monsoon, from November to January, bring more than 80 inches of rain annually to all the island except to the "dry zone" in the west and north, where the precipitation averages 60 inches. Here irrigation is necessary for the cultivation of paddy rice, the principal crop; large areas not now culti-
southwest monsoon, from April to August, and the northeast monsoon,
improvement of the ancient irrigation works. A country with a twelve-month growing season and a "dry zone" which gets 60 inches of rain is certainly agriculturally well endowed. (At one time Ceylon was referred to as the "granary of the East.")

The large plantations growing tea, coconut, and rubber are mechanized, but the small farmer and the local cart transportation utilize cattle and buffalo for draft purposes.

CATTLE BREEDS

Apparently there were no wild cattle on Ceylon, although many of the wild life species of India are native to the island. Indigenous wild buffaloes on the island conceivably could have been domesticated, but there is no evidence of this having happened. The two main bovine infusions were first made by the Sinhalese, who brought their cattle and probably buffaloes with them, and, much later and in smaller numbers, by the English dairy breeds which accompanied the Scottish settlers during the British colonization period. Following these animals, Zebu draft breeds were imported from India for plantation work, and more recently Indian breeds of milk cattle and buffalo were imported.

Sinhala.—What can now be considered indigenous cattle of Ceylon are distinctive, small, humped animals, varied in color but often nearly solid black, named for their original owners, the Sinhala. These cattle were well established on the island before the time of Christ. They are now a degenerate type, reduced in size to a 400-pound cow and a 500-pound bull, apparently by generations of negative selection. Evidence that a larger animal must have existed in the Sinhala's ancestry is seen in some of the cart draft bulls, weighing as much as 800 pounds and surely

Sinhala ox, 600 pounds.

Sinhala cow, 450 pounds.

selected for size and kept on better feed. The animals have a definite polled tendency, many cows being seen without horns.

The Sinhala have now been interbred to a considerable degree with the more recently introduced Cape Cattle.

Cape Cattle.—When British colonization began at the turn of the nineteenth century, the early settlers, a large proportion of whom were Scottish, brought with them their milk cattle as they traveled around the Cape of Good Hope. To distinguish these animals from the native cattle, they were called Cape Cattle and are so known today. Cape Cattle were indiscriminately interbred and are now a heterogeneous mixture of Ayrshire, Shorthorn, Friesian, and Jersey. The temperate-zone Cape Cattle became acclimated to the high country, with its moderate climate, where their owners developed the tea plantations. They interbred to some extent with the native Sinhala; but for the most part the exotic breeds were maintained separately, although they freely mixed among themselves.

Small owners now own most of the Cape Cattle, usually growing only 2 or 3 head and seldom more than 10. The animals are milked for household use and local trade or for sale to the Milk Board, the government agency controlling milk marketing.

Cape Cattle cow.

Indian Zebu Breeds.—The Kangayam breed of India was imported as the need for draft animals increased with plantation development. It is of medium size, the bull weighing as much as 1,000 pounds. A white to

Kangayam draft team of one ox and one bull.

light-grey animal with a moderate hump, the Kangayam has quite large horns that extend upward, backward, then frequently turn inward at the end. The dewlap and sheath are large.

This breed has been quite well maintained in a more or less pure state and is now widely used for cart draft. Often the bulls are not castrated and are still worked satisfactorily.

Indian Zebu cattle were also introduced in numbers by tobacco planters along the Mahaweli Ganga, in the dry zone in the east central part of the island. These animals appear to have developed locally into a distinctive Zebu type, locally called Tamankaduwa, and have been maintained in a pure state because there were few other cattle in the area. They are solid white, have a rather small hump, and are now used in this region as cart animals and to some extent for milk.

Tamankaduwa cow.

In recent years some of the Zebu milk breeds have been imported for upgrading the native cattle, particularly the Tharparkar, locally called the White Sindhi, and the Red Sindhi.

Sinhala Buffalo.—Said to be a descendant of the buffalo of the early Sinhala people, this small animal is of the river type. A mature cow does not exceed 800 pounds, a bull 1,000 pounds. The horns are smaller and lack the wide spread of the typical swamp buffalo of the Far East; neither do they have the curl of the Indian milk breeds. They are good draft

Tharparkar cow. Photographed at the Government Livestock Project, in Tamankaduwa.

Red Sindhi cow. Photographed at the Government Livestock Project, in Tamankaduwa.

animals for their size and are widely used for work in the rice paddies. Many males used for draft are not castrated.

Sinhala buffalo cow.

Murrah Buffalo.—The government livestock projects have recently imported large numbers of this breed from India. The object is to develop sizable milk production from the establishment of such herds.

Murrah buffalo cows on improved pasture. Photographed at the Government Livestock Project, in Tamankaduwa.

MANAGEMENT PRACTICES

Most of the cattle and buffalo are in the hands of small owners. They are grazed on crop residues when available, otherwise on uncultivated areas and along the roadside. Small owners milking cows will bring hand-cut feed to their milk cows. Deficient protein in the tropical grasses accounts for the poor condition in which most cattle are seen.

Cattle shed of a small dairy owner.

On the large plantations and the government livestock projects, reasonable management usually prevails and in instances can be called excellent. Some of the large plantation owners, individuals as well as corporations, have become interested in milk production, intrigued by the tax incentive which is granted for the expenditures involved. One of the most enterprising of these is Mahaberiatenne, owned by Anglo-Ceylon General Estates. This plantation is located at an elevation of 1,400 feet at the beginning of the high country. Lying in the Teldeniya Valley, in a pocket area of 60-inch rainfall with 100 inches across the ridges on either side, it enjoys natural advantages for cattle raising. Coffee, cacao, coconut, pepper, and kapok are the major crops there; but, as a further diversification, a well-planned dairy was started a number of years ago. A herd of 300 Friesian and Jersey cows is maintained, the base stock of which was imported from Australia and England. The Friesians average 8,000 pounds of milk a lactation, the Jerseys 4,000. Cows are on pasture after the morning's milking for four hours, after which, to protect them from the sun, they are brought to the sheds, where they are kept overnight. Concentrate consisting of coconut meal, rice bran, and corn (grown on the plantation) is fed in proportion to milk production. Milk is sold to the Ceylon Cold Storage Plant and is delivered to Colombo, 80 miles away, for 12 cents a pound as a premium product. It retails to the carriage trade in Colombo at 25 cents. The Milk Board pays an average price of 7 cents a pound for raw milk, which is pasteurized and sold for 11 cents a pound.

In spite of the excellent management and exceptional care of the livestock, heat stress at the comparatively low altitude appeared to be taking its toll and the cows were not in thrifty condition. The practice is to cull all producing cows after their fifth lactation because of decreased production. Heifers are bred at eighteen months, which means discarding them at seven or eight years of age.

Milk barn at Mahaberiatenne.

Artificial insemination was started by the Ministry of Agriculture in 1950. There is a central collecting station at Kundasale, which lies at an elevation of 2,500 feet at the beginning of the high country. There are also one other collecting station and a total of twenty-eight substations. Semen is used fresh, with a maximum storage period of forty-eight hours. Egg yolk citrate is used as a diluent, and the rectal method of insemination is employed. Some 10,000 cows are now being bred annually and a conception rate of 58 per cent on the first insemination is obtained. A large part of the breeding is on the government livestock projects.

MARKETING

No cattle markets as such exist in Ceylon. Cattle for slaughter were bought by dealers from the small farmers at approximately 6 cents a pound in 1967. The weight is estimated by the girth measurement in some instances; elsewhere sale is by the head. Butchers in Colombo purchase stock for slaughter from the dealers at an average price of 7 cents a pound. The government-fixed price for retail meat was 19 cents a pound; but, because of the scarcity of slaughter animals, there was considerable black marketing. Those prices were 20 cents a pound for meat with bones in or 23 cents for that without bones. Most slaughter animals except worn-out work animals are killed at two to two and one-half years of age.

ABATTOIRS

A typical municipal slaughterhouse of the Asian type is located in Colombo, but much of the beef in the country is killed by unsupervised local butchers. In 1963, 157,000 cattle and 15,000 buffaloes were killed; this would represent an offtake of about 8.5 per cent for cattle but less than 2 per cent for buffaloes.

CATTLE DISEASES

Ceylon is said to be free of rinderpest and tuberculosis, but outbreaks of foot-and-mouth disease occur. Hemorrhagic septicemia is serious, particularly during the wet season. Government farms and a few large private operators vaccinate against these diseases at the end of the dry season. Government farms also test for brucellosis and slaughter any reactors. Sampling of cattle in the private sector shows an incidence of less than 1 per cent of brucellosis, a percentage not considered serious. At Mahaberiatenne all cattle are sprayed with Cooper-Tox for four successive days every month for tick protection, but this practice is exceptional.

GOVERNMENT AND CATTLE

The major interest of the government in cattle is in the development of increased milk production. Importations of dairy products currently are requiring $20,000,000 of the country's foreign exchange and amount to more than 6 per cent of the total imports. The land could certainly harbor dairy cattle in sufficient numbers to enable the country to be self-supporting in dairy products. Many conflicting elements, such as scarcity of risk capital and, probably most important of all, getting the small operator who owns most of the cattle to change his management methods, present difficulties in realizing this potential. Gifts of cattle are solicited in the aid programs of other countries; sizable tax benefits are offered large operators for any dairy expenditures; and import licenses are freely granted to import breeding stock.

The Ministry of Agriculture has sponsored the construction of a condensed milk plant and a spray-drying plant with capacities considerably greater than the visible supply of milk necessary for their operation. New Zealand furnished the funds for the condensery and in return required that all milk powder and butterfat used for reconstituted milk be purchased from her until Ceylon is self-supporting in milk.

The Government Livestock Project, Ambawela, is near Nuwara Eliya, at an elevation of 6,000 feet, and comprises 4,000 acres just over the ridge from some of the best tea plantations on the island. The soils and terrain here are said to be unsuitable for tea cultivation but obviously can be converted to good pasture. Some 2,200 acres are now in improved and unimproved pastures, and the remaining scrub jungle is being so converted. The foundation stock on the farm was a herd of Ayrshire cows brought in to supply milk to the British army in 1943. These animals have gradually increased and recently have been augmented by a gift of 500 poor-quality Jersey heifers from New Zealand. The total cattle population has been brought to 2,260 head of Jerseys and Ayrshires. An unusual feature is that the cows are pastured from 7:30 A.M. to 3:30 P.M. instead of being stall-fed. They are brought to barns for milking.

Another large government livestock project is at Tamankaduwa, on the east-central side of the island, in the so-called "dry zone." Although rainfall averages 60 inches annually, there is practically no precipitation from June to October. The extent of this unit is 24,000 acres, and the cattle population is 6,000 head. The Murrah buffalo population is 2,500, which the project intends to increase to 4,500. Of the cattle 2,000 are Red Sindhi and Tharparkar, 1,500 various crosses, and 2,500 native Sinhala. This is considered a foundation herd to be used in a cross-breeding program with the Jersey with a view to developing a high milk-producing strain for the dry zone.

The cattle at Tamankaduwa are in noticeably better condition than those at Ambawela. This situation is the opposite of what would be expected since Ambawela is at an elevation of 6,000 feet and enjoys a nearly temperate climate with even a few degrees of frost occasionally during the winter, while Tamankaduwa is tropical throughout the year. Possibly the heavy rainfall at Ambawela, more than 100 inches, and the high humidity, averaging 75 per cent, may have some debilitating effect on the temperate climate breeds; or there may be some mineral deficiency in the soils which causes their condition.

OUTLOOK FOR CATTLE

Ceylon has the potential to become at least self-supporting in beef as well as in milk. With the ample rainfall over all the island, carrying capacity is sufficient for both the necessary draft animals and sizably increased milk and beef herds. Additional grazing areas could be made available by establishing good grasses in the reforestation areas and also under the coconut groves. Extensive areas of scrub jungle could be cleared and good pastures established, as is being demonstrated at Ambawela. Soil surveys have shown that 800,000 acres of land in the dry zone which can be brought under irrigation are better suited for use as pasture than for cropping.

Much good planning has gone into the programs of the government livestock projects, which could serve as examples of the possibilities in cattle production. An unusual feature of the program is the long-range plan to develop these projects as demonstration centers to be eventually turned over to the private sector; the government farms would then be shrunk back to the size required for research and experimental work. This seems a rather Utopian idea in a country where only yesterday expropriation of private corporations was the order of the day.

The government is inviting the expenditure of foreign capital and offering tax relief to foster expansion in cattle operations of the country. This may augur well for the future, but the past specters of expropriation and disregard for private enterprise will probably have to lie for a while longer before the new policies have an appreciable effect on the cattle industry.

India

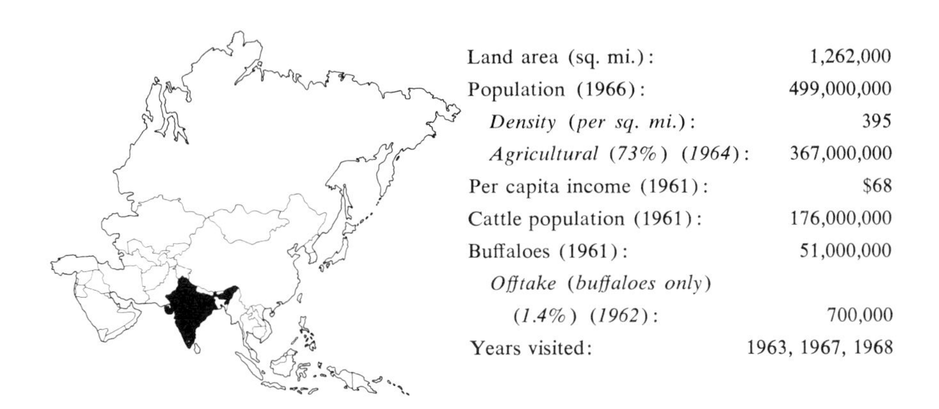

Land area (sq. mi.):	1,262,000
Population (1966):	499,000,000
Density (per sq. mi.):	395
Agricultural (73%) (1964):	367,000,000
Per capita income (1961):	$68
Cattle population (1961):	176,000,000
Buffaloes (1961):	51,000,000
Offtake (buffaloes only) (1.4%) (1962):	700,000
Years visited:	1963, 1967, 1968

THE REPUBLIC OF INDIA occupies a triangular area with the Himalaya Mountains in the north as the base and the apex lying in the Indian Ocean, far to the south. The broad plain between the Himalayas and the Vindhya Mountains drains to both the west and the east coasts. Below the Vindhyas, drainage goes east from the high, broken ridges along the west coast. The latitude approximates that of Mexico and Central America down to Panama. The land area is one-third larger than this portion of the Western Hemisphere, and the population is ten times larger. Several European countries and a few New England states, however, have greater population densities—in the Netherlands the number of people to the square mile is more than double that in India.

Landing in 1498, the Portuguese were the first Europeans in India and established trading settlements on both the east and the west coasts. With the decline of Portugal as a sea power, the Dutch began to move in during the early seventeenth century and were soon followed by the British. During the eighteenth century India was a hotbed of confusion, the native states, the English, and the French, who by this time had gained a foothold, all being involved. The situation began to be resolved during the first part of the nineteenth century, with the British in substantial control. Soon they developed complete domination of the area now occupied by India and Pakistan.

After World War I a strong nationalist movement developed and culminated in the dominions of India and Pakistan being established in 1947. The carving out of the areas occupied by a predominantly Moslem population into what is now Pakistan and leaving to the new India the predominantly Hindu states was an essential prelude to successful independence. Religious differences between the Hindu and the Moslem are irreconcilable, and unless one or the other is in complete control, any form of peaceful coexistence is obviated. The Moslem element is now approximately 10 per cent of India's population, and the Hindu makes up 6 per cent of Pakistan's. Both of these situations cause considerable unrest but not to the point of disrupting the continuity of government.

In 1950 both India and Pakistan obtained full independence but remained members of the Commonwealth of Nations.

The mineral wealth of India is surprisingly small. Only 1 per cent of the population is engaged in industry. Indian agriculture occupies nearly three-fourths of the population. Except for a relatively small arid region in the northwest, rainfall generally varies from 30 to 100 inches, concentrated during the monsoon, from June to September. The climate is generally tropical, although the northern regions lie within the temperate zone. One-fifth of the cultivated land is irrigated—more than half from the canal systems originated by the British and the remainder from wells and from reservoirs which collect surface water. The soils, however, have been depleted of humus and plant nutrients by centuries of removal of all plant growth for human and animal consumption and of manure for fuel. At least 100,000,000 head of the cattle population serve no useful purpose. These two factors alone are enough to account for the inability of India to feed herself. Efficiently farmed, half of the cultivated land of India, if supplemented by the cattle now permitted to die of old age, would supply the entire population with an adequate diet.

The combined cattle and buffalo populations in India are roughly in the same proportion to the human population as cattle are in the United States, approximately 1 bovine for every 2 persons. The religious concepts of the Hindu, however, dominate the handling of cattle in India. Buffaloes and cattle are primarily draft animals, although milk is taken from both. The flesh of neither is used for food except for a small number of buffaloes consumed by the non-Hindu elements of the population. In most states the slaughter of cattle is prohibited by law. Regulations regarding the slaughter of buffaloes are more lenient, although it is generally illegal to kill an animal under fifteen years of age or one capable of reproduction.

More controlling than legal restrictions, however, is the religious ban of the Hindu against taking the bovine life or eating its flesh. The extent to which such concepts dominate the diet is illustrated by the attitude toward eggs: formerly a Hindu would not eat an egg because of its being a potential source of life; recently, however, nonfertile eggs have been introduced as "vegetarian" eggs and the less orthodox Hindu who is economically able to do so is beginning to eat them. Cattle are permitted to wander at will in most towns and cities. They glean bare any available plant growth, and part of their sustenance consists of the handouts of vegetables given by shopkeepers as a form of charity. These city cattle are usually in better condition than the work animals seen in the country.

City-dwelling cattle in Calcutta.

CATTLE BREEDS

The Zebu cattle of India and Pakistan are a distinct type unrelated, as far as is known, to any wild cattle found today or to the progenitors of nonhumped cattle. Their distinctive characteristics are the very loose skin, the pendulous dewlap and sheath, and the pronounced hump, which is invariably transmitted to any new breed which has been developed by crossing with nonhumped cattle. As a result of countless generations of natural selection, Zebu cattle became adapted to the tropical climates and developed the ability to withstand high temperatures, ticks, and many tropical diseases which would decimate the European breeds. Humped cattle that originated in India were known in Egypt and the Near East long before the Christian Era. They have played a dominant part in the development of the cattle populations in the tropical and warmer areas of the world. During the second millennium before Christ, migrations of their owners had carried the longhorned humped cattle to Africa. The humped shorthorn followed later; descendants of both, as well as the mixtures with humpless cattle, make up the large majority of the African cattle population of today. The present-day Indian Zebu breeds include those with prominent horns and some with short horns. Although no lines of descent can be traced, the breeds now seen with prominent horns presumably had the same ancestors as the long-horned humped cattle that found their way to Africa, and the same would hold for the short-horned types.

Zebu cattle also found their way to many of the large islands of the world—to Madagascar (by way of Africa) and to the Philippines, Indonesia, and other islands of the Pacific. Early migrations took the animals to Southeast Asia, China, and the Near East. In all of these parts of the world, the Zebu now predominates, frequently to the exclusion of any nonhumped cattle except for recent introductions for experimental crossing.

In modern times the Western Hemisphere, which had no indigenous cattle, has been the source of widespread development of Zebu cattle founded on importations from India. One of the earliest of these importations was the introduction in 1870 of the Indian Kankrej breed into Brazil, where it became known as the Guzerat for the district in India where it was grown. It was followed by other Indian breeds—the Ongole (known as Nellore in Brazil), the Gir (Brazilian Gyr), and the Krishna Valley. All of these breeds are grey to near-white with the exception of the Gir, which has red or brown patches on a largely greyish-white background. Principally from developing a fixed cross of the Gir and Kankrej, a Zebu breed known as the Indo-Brazil was developed. All of these breeds, with the exception of the Krishna Valley but including the Indo-Brazil, were bred for beef characteristics; and herdbooks were established by the breed societies, which were organized for each breed.

Zebu cattle were brought to the southern United States from India, Brazil, and the West Indies, where they had been introduced toward the last of the nineteenth century. The American Brahman was developed entirely from Zebu breeds, predominantly the Guzerat, but also with some mixture of the Krishna Valley, Nellore, and Brazilian Gyr. The Santa Gertrudis breed was fixed at five-eighths Shorthorn and three-eighths Zebu of different breeds. Both are beef types, superior for this purpose to any of the original Indian Zebu breeds from which they derived.

The Food and Agriculture Organization publication on the Zebu cattle of India and Pakistan lists twenty-eight breeds.[1] They are largely confined to government farms or to the large establishments of a few

[1] Food and Agricultural Organization (United Nations), *Zebu Cattle of India and Pakistan.*

private owners, which are known as "organized farms." The cattle of the small farmer and those which roam the streets of the cities and towns defy any classification by breed. These animals comprise all but a minor part of the total cattle population. In some areas the cattle of the country show a degree of similarity, although often without uniformity in appearance, to one of the recognized Zebu breeds.

The same situation applies to the Indian buffaloes as to the cattle. Several breeds are recognized by animal husbandry men, but the buffalo that the Indian farmer works in his field and milks is just a buffalo.

Discussion of some of the more prominent breeds of Indian cattle will be found under the heading "Government and Cattle," for they are usually seen on government farms.

MANAGEMENT PRACTICES

The typical farmer is a village dweller. His cattle are run in community herds. Those animals not being worked are taken out to the stubble fields for grazing, or to such other pasture as can be found, during the day under the constant eye of a herdsman and brought back to the village at night. The upper 25 per cent of the farmers cultivate five to eight acres and have 2 buffaloes or oxen for draft and 2 cows (or buffaloes) for milking. The farmer's plow is a wooden stick, sometimes iron shod, the same implement that has been used for centuries. Oxen and buffaloes supply most of the power for cultivation. There is less than one tractor in the country for every 10,000 acres under cultivation. If the land is irrigated and it is necessary to raise water, the power is frequently supplied by cattle.

In the wetter areas buffaloes are more often used than cattle for draft purposes. On the sparse, coarse roughage, which is the usual feed, buffalo are more efficient, both for draft and milk, than cattle. The estimated average annual milk production is less than 400 pounds a cow compared to 1,100 pounds for the buffalo. In 1962, 51,000,000 cows and 24,000,000 buffaloes were kept for milk, 90 per cent of these being in the hands of farmers owning only a few head each. Cows are milked morning and night, the young calves being allowed enough nourishment to permit survival. A portion of the milk is retained for family use and the remainder sold in the village. In the cities in which bottled milk is not available, particular housewives demand that the cow be brought to their door and milked to ensure the milk's being fresh and undiluted. Both cows and buffaloes are maintained in sheds in the cities for milking. These units are unsanitary and poorly maintained, usually with from 5 to 30 animals tied in place continuously.

A major disservice to Indian agriculture is the universal use of cow dung for fuel. Every bit that can be gleaned from the areas where cattle are handled is compressed by hand into round cakes about eight inches in diameter, then plastered on any convenient vertical surface, such as a tree or the side of a building, for drying, after which it is stored in neat round piles. It is used primarily as fuel for cooking by the farmer whose family collects it and is sold for this purpose in the villages. The annual production is estimated at more than 200,000,000 tons and is an important article of commerce.

Storage pile of dried cow dung.

Every bit of plant growth is removed from the land, first by harvesting and then by grazing when everything is grubbed down to the ground by the famished cattle. Such practices have denuded the soil of humus and, in addition to causing low fertility, create a condition in which erosion becomes an increasingly serious problem. While commercial fertilizers are used to some extent and their use is increasing, this practice falls far short of providing sufficient volume to obtain reasonable crop yields. As a result the Indian farmer has some of the lowest productions known, an average of thirteen bushels of wheat an acre where half the sown area is irrigated, or 900 pounds of rice an acre, all grown on irrigated land.

Government-maintained artificial insemination centers have been established throughout the country, but the number of cattle bred by this method is small. The cultural antagonism to the method has been difficult to overcome.

Since there is no slaughter of cattle for beef, the animals usually die of natural causes in the field except for the relatively few in the homes for old cattle, discussed under "Government and Cattle." Immediately at death, the remains are consumed by vultures, a common sight in the countryside. There is no stigma involved in use of the bones, which are gathered and sold for processing into bone meal. There is also some salvage of the hides.

Vultures disposing of a deceased cow.

MARKETING

Cattle sales are held in villages throughout India. The stock offered is usually in very poor condition. Trading is mostly between farmers. Any sale of buffaloes for slaughter in the few areas where this is practiced is handled very quietly in the country by traders dealing directly with the owner.

ABATTOIRS

The only beef produced in India is that obtained from the slaughter of worn-out work buffaloes or unwanted young. The religious ban of the Hindu against killing cattle to a degree extends to include the buffalo, and slaughter operations are managed in a clandestine manner. The slaughterhouse in Delhi, operated by the municipality, issues slaughter permits, each of which permits the licensee to kill 1 mature buffalo or 2 calves a day. In 1963 there were only sixty-eight such permits in force and the daily kill averaged less than 100 head.

The slaughterhouse is under the management of a Hindu, but the work is all done by Moslems. The facility consists of a large, open shed into which the animal is led. Here it is thrown and its throat is cut while two or three men hold its head. Skinning and evisceration follow, after

Village cattle sale being held in southwestern India.

which the entrails are dragged outside for further processing on the ground. Sanitary provisions are nonexistent. Sheep and goats are processed in a similar manner in the same facility.

Slaughter starts at 9 o'clock in the evening and is completed at 4 o'clock in the morning. The Moslem owner of the animal then arrives to start bargaining with the vendors. About 700,000 buffaloes are slaughtered annually throughout the country, representing an offtake of 1.4 per cent of the buffalo population.

CATTLE DISEASES

The cattle of the country rely largely on their inherited tolerance of endemic diseases, although in 1962 there were 4,000 veterinary centers and current plans called for doubling this number. Rinderpest was the most serious disease, and the program for combating it called for the entire bovine population to be vaccinated by the end of 1964. In 1968 this disease was said to be fairly well under control. The government stations and large organized farms vaccinate for rinderpest, hemorrhagic septicemia, blackleg, and to some extent foot-and-mouth disease. The latter is generally prevalent but kills few cattle.

GOVERNMENT AND CATTLE

The Ministry of Food and Agriculture is endeavoring to improve the quality of livestock and management methods. These efforts include all the commonly recognized channels—education, including veterinary colleges; extension work; livestock breeding centers; experimental and research stations; and artificial insemination and veterinary services. Very little of this improvement gets down to the small Indian farmer living in his village with his few head of cattle and buffaloes. His way of life bears no more resemblance to that of the white-collared research worker in a government station than does that of his undernourished cattle to the excellent animals on a balanced ration at that same station. These institutions are well operated and follow modern practices in animal husbandry.

The Indian Agricultural Research Institute at New Delhi is a continuation of the Imperial Agricultural Research Institute, founded in 1905 by the British at Pusa. The activity originated in an attempt to alleviate the periodically recurring famines in the more heavily populated areas. Although concerned primarily with crops, the institute is also interested in cattle breeding and management. Both Sahiwal cattle and the Murrah buffalo have been improved by systematic selection.

In 1904, the year before the institute was established, its Sahiwal herd was founded. Selective breeding for milk production had increased the average yield per cow from 3,200 pounds per lactation in 1914 to 7,500 pounds in 1953, with individuals giving as much as 13,000 pounds. Cows are milked four times a day and fed an ample ration in proportion to their milk production. Bulls are sold to other government stations and some to individuals.

India's largest cattle enterprise is the government-operated Aarey Milk Colony, just outside Bombay. It was started in 1947 in an effort to eliminate the countless ill-kept dairies of buffalo cows that were scattered throughout the city. In this endeavor it was highly successful until recently, when the unkempt town dairy began creeping back.

The colony provides modern facilities for 15,000 buffalo cows, belonging to approximately 250 owners, who milk their own animals. The milk is sold to modern plants, operated by the municipality. The cows are maintained in well-constructed, paved sheds, open on all sides and having metal roofs. Manure washings are collected and pumped back on the meadows. Pará grass is the principal roughage and is fed green-cut for the most part. In the twelve-month growing season this yields thirty tons an acre on an air-dry basis, with eight or nine cuttings from the irrigated meadows.

From 3,500 pounds per 300-day lactation when the colony was started, the average milk yield has been increased to 4,500 pounds. This gain results from better nutrition as well as selective breeding for milk production. Cows must produce a minimum of 14 pounds of milk a day

Progeny-proven Murrah buffalo sire. Photographed at the Aarey Milk Colony.

Mature Murrah buffalo cow. Photographed at the Indian Agricultural Research Institute, New Delhi.

to be included in the colony herd. The maximum yield per cow is 10,000 pounds a lactation. In addition to the green-cut Pará grass, which is fed at a ratio of 2.5 pounds to 100 pounds of body weight, approximately twelve pounds of concentrate, consisting of crushed pulses, wheat bran, and peanut or cottonseed oil cake, are fed.

Breeding is by artificial insemination, the rectal method with fresh semen being employed. The bulls used are excellent specimens, well maintained. In some countries in which attempts have been made to employ artificial insemination in breeding buffaloes, the method has been unsuccessful, probably because of the difficulty in detecting when the cow is in heat, the buffalo cow being a notoriously shy breeder. The success of artificial insemination at Aarey is the result of using vasectomized bulls for detection.

Murrah buffaloes account for 70 per cent of the herd. Most of the rest are of the Nili breed, which in appearance is almost identical with the Murrah except for the white spot on the forehead. There are also some of the Ravi and Mehsana breeds, both closely related to the Murrah.

In 1962 a cow-breeding scheme was inaugurated at Aarey with the objective of producing higher yielding cows of the Rath and Gir breeds.

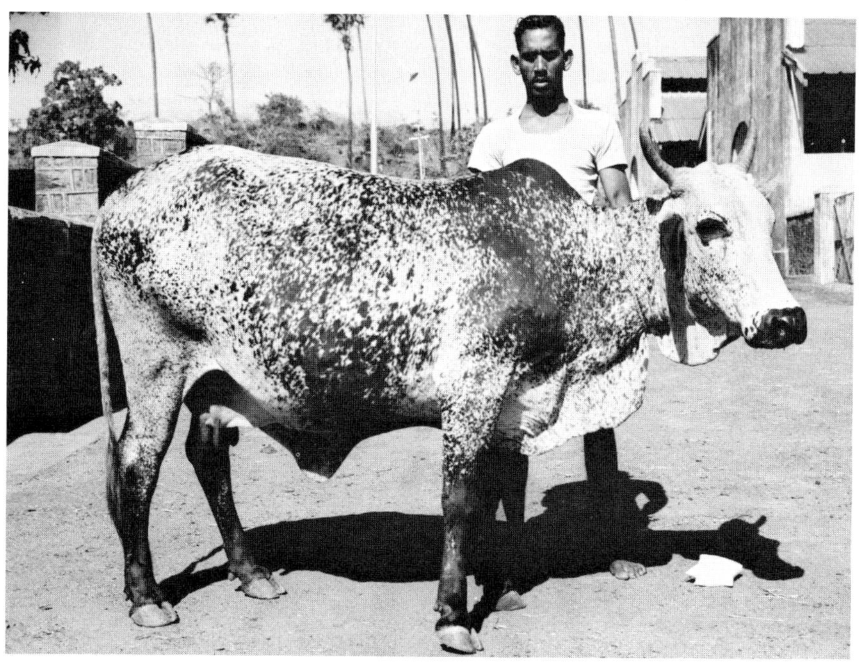

Rath cow. Photographed at the Aarey Milk Colony.

Mature Gir bull. Photographed at the Aarey Milk Colony.

Mature Gir cow. Photographed at the Aarey Milk Colony.

The Rath is a rather nondescript type, hardly warranting the term "breed." It is usually considered an unfixed mixture of Sahiwal and Red Sindhi. Red is rather a prominent color, although nearly all black and mottled black-and-white animals are seen. A mature cow weighs about 900 pounds.

The Gir is the breed which has been so well developed in Brazil. The color is white or light grey, with red or brown patches. The horn has a distinctly oval cross section as it leaves the skull and gradually becomes nearly round after several inches. The average weight of a mature cow is approximately 1,000 pounds.

The Rath cow, which normally has a milk production of about ten pounds a day, is being crossed with both Sahiwal and Jersey bulls for increased milk production. The Gir breed, the milk production of which ranges from thirteen to eighteen pounds a day in the colony herd, is being kept pure.

A small herd of Holstein-Friesians, a gift from Australia, had been maintained at the colony for one year in 1967. They were in fair condition, having been kept in isolation from the other cattle. That they were

Mature Sahiwal bull (horns blunted), 1,200 pounds. Photographed at the National Dairy Institute, in Karnal.

suffering to some extent from heat stress seemed probable because the heifers that had calved and were of excellent quality and well fed were averaging only twenty-five pounds of milk a day.

The National Dairy Institute at Karnal, in northern India, has a total herd of 1,200 head, mostly of Red Sindhi, Sahiwal, and Tharparkar cattle and Murrah and Nili buffaloes. Experimental crossing using Brown Swiss bulls imported from the United States on the Indian breeds of cattle is being conducted in an effort to develop an improved milk type of Zebu. Excellent examples of the three Indian breeds are seen there.

One of the better-milking Zebu breeds is the Sahiwal, which has an exceptionally well-developed udder for a Zebu. The breed is of medium size for humped cattle, mature cows in good condition weighing from 900 to 1,000 pounds and bulls weighing as much as 1,400 pounds. The horns are rather small and have only a slight curve. The animal is a solid, rather light red, with only occasional small, white markings. The hump is smaller than that of most Zebus.

In the government stations the Red Sindhi has been selected to a considerable extent for milk productivity. It is one of the smaller Zebu breeds, mature cows in good condition weighing 700 to 800 pounds and bulls weighing as much as 1,100 pounds. The horns are of medium size

Mature Red Sindhi bull, 1,100 pounds.

Mature Red Sindhi cow, 800 pounds. Photographed at the National Dairy Institute, in Karnal.

and slightly curved. Hair color varies from medium to dark red, with only occasional, very minor white spottings. The hump is of medium size on the female but quite large on the male.

Except for color the Tharparkar is quite similar to the Red Sindhi and in some countries into which it has been imported is called the White Sindhi. It is light grey and has dark to almost black shoulders and neck. The horns are of medium size, extending upward and frequently showing an inward curve on the female; but on the male they are short and thick.

The government maintains more than sixty farms, known as *Gosadans*, as old-age homes for infirm and nonproductive cattle so that they may live out their natural lives; more of these institutions are currently being provided. As charity, wealthy individuals have established privately financed institutions to serve much the same purpose. These farms may produce some milk for sale, the proceeds going toward the upkeep of the institution. The number of useless cattle these farms can care for is but an infinitesimal fraction of the country's 176,000,000 head; yet they illustrate the extreme nature of the Hindu cultural attitude where cattle are concerned. The government may be thinking that the *Gosadans*

Mature Tharparkar bull, 1,000 pounds. Photographed at the National Dairy Institute, in Karnal.

eventually may become the means of humane elimination of useless cattle; but, if this is the case, the policy is a matter never mentioned.

OUTLOOK FOR CATTLE

The devotional regard in which they are held by their Hindu owners is the key to the future of India's cattle. Letting them live out a natural life, from a Western viewpoint, is hardly compatible with permitting the cattle to exist in the famished condition in which they are frequently seen. Milk production of the village owner could be doubled if he would give 1 cow the feed that 2 now consume and dispose of the other. Any fundamental improvement in the cattle husbandry of India will have to start along some such line. Until this is done, the development of higher-milk-producing strains in the well-kept herds of the animal breeding stations will be only a matter of scientific interest.

The future of India's cattle, as well as of her people, is not a pleasant picture. Famine would be serious today without the donations of wheat from the United States and other countries. The day when famine will strike and such relief will no longer be available is not many years away. The population increases in India at the rate of 10,000,000 human beings annually. The increase in food production in comparison is only nominal. Modern agricultural practices could adequately feed the population, but what would become of the millions of small farmers and their families who would be left without a means of livelihood and the millions of cattle and buffaloes that would be replaced by tractors? It is conceivable that the cattle could be absorbed in the world market for beef, which faces a continuing shortage; but the Indian would starve before he would see them slaughtered. An even more immediate solution that could appreciably alleviate the situation is being ignored. If just the non-productive and low-productive cattle were eliminated, more food could be made available for human consumption, if not from the cattle themselves, from the plant growth that maintains them. Instead, more old-age homes are being built for the cattle unable to work. India is one place in the world in which cattle are eating people.

Nepal

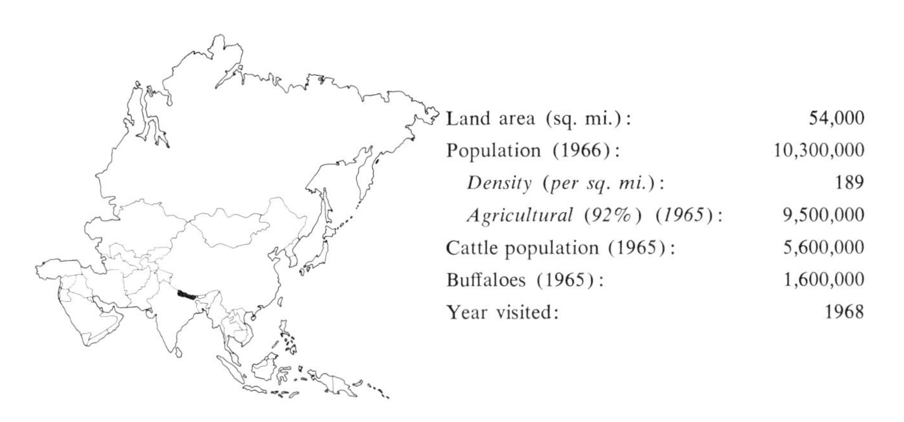

Land area (sq. mi.):	54,000
Population (1966):	10,300,000
Density (per sq. mi.):	189
Agricultural (92%) (1965):	9,500,000
Cattle population (1965):	5,600,000
Buffaloes (1965):	1,600,000
Year visited:	1968

THE KINGDOM OF NEPAL stretches along the eastern part of the northern boundary of India, lying between that country and Tibet and bordering with the Indian protectorate state of Sikkim on the east. The latitude is generally the same as that of Florida. Nepal is about half the size of Colorado, extending 500 miles from east to west, with an average width of 100 miles. Within this narrow range the extremes of elevation are such that three climatic zones, which vary from subtropical to arctic, exist, each extending completely across the country. Along the length of the Indian border in the south, the fertile Ganges plain is the region known as the Terai. It extends to the jungle area below the first mountain ranges. Next is the midlands, where most of the people live. A mountainous area with deep valleys, it ranges in altitude from 4,000 to 10,000 feet. To the north of the midlands is the high plateau country, a rugged alpine region ending in tundra, with Mount Everest and the high peaks of the Himalayas on the border with Tibet.

Nearly one-third of the land is forest, somewhat less than two-fifths is cultivated, and one-seventh is perpetually under snow. Most of the rainfall occurs during the southwest monsoon, from mid-June through September, and averages about 60 inches annually in the vicinity of Katmandu, the capital, in the midlands. Many of the short, steep watercourses from the mountains to the Terai are dry except during the rainy season.

Nepal has an ancient recorded history of local ruling rajas and various conquerors from outside, who invaded the area at different times. It was a land of many small kingdoms, separated by natural boundaries, until the latter part of the eighteenth century, when the principality of Gurkha unified most of the mountain states. During the first half of the nineteenth century, there was considerable conflict with the British rulers of India; but from the latter half on, relations were friendly.

In the 1840's, although the country nominally remained a monarchy, the ineffectual kings came to be dominated by a hereditary line of prime ministers, the Rana family, who were the factual rulers and exercised dictatorial power for more than 100 years. The kings were maintained in pampered luxury but were kept illiterate and were virtually prisoners. In 1951 the curtain fell for the Rana dynasty, although the numerous descendants still retain powerful influence in the army and the commercial activity of the country. In that year the king, who had fled to India, managed to resume control and proclaimed the country a constitutional monarchy.

Nepal has three principal groups of people—the Tibeto-Burmese, who came in from the east; the Tibetan, who immigrated from the north; and the Indo-Aryan, who entered from the south. From the central region of the Tibeto-Burmese group came the famous Gurkha soldiers who served in the British and Indian armies.

The cattle of Nepal, as varied as the terrain and the people, are interesting in both type and methods of husbandry. The official Nepalese estimate of the cattle population is 5,600,000 head of cattle and 1,600,000 buffaloes. These numbers are more than twice those of other estimates. The economy is essentially agricultural; yet throughout the midlands very little use is made of the large cattle population except for the production of manure as a fertilizer. In the Terai, however, they are used to some extent for draft.

CATTLE BREEDS

The domesticated yak is found in the alpine region below the Tibetan border. The Chowri, a cross between the yak and the native Zebu cattle, which are found throughout the higher midlands in the northeastern part of the country, are maintained at a slightly lower elevation. Nondescript Zebus, locally known as Hill Cattle, are seen throughout the rest of the midlands, where they are more numerous than the buffaloes. In the Terai the buffalo predominates, although there are also large numbers of Zebu cattle. Buffaloes in both the Terai and the midlands are raised mainly for the production of milk.

Hill Cattle.—As seen in the midlands, these cattle are a degenerate type of Zebu. They have resulted from the indiscriminate crossing of the cattle which have found their way in the past from India to this area of Nepal. Many generations of inbreeding under conditions of inadequate nutrition have resulted in a nondescript animal of widely varying size, conformation, and color.

These dwarf cattle weigh as little as 350 pounds at maturity. The largest Hill cow normally seen does not weigh more than 600 pounds and very few bulls reach 700 pounds. Frequently black or dark brown, often with a white spot on the forehead, Hill Cattle may have white markings on the body and legs. There are some grey animals, but red coloring is rare.

A hump is always present but varies widely in size. Many individuals exhibit only a very small remnant of a hump, but occasionally larger bulls are seen with the prominent lopped-over hump characteristic of some Indian breeds. The horns tend to be short, thick, and stubby, although individuals are seen with medium-sized, upswept horns. Conformation varies considerably, but the sloping rump typical of the Zebu is usually not prominent.

In some areas in which they have been isolated by extremely rugged, mountainous terrain, distinctive types of Hill Cattle displaying consider-

Hill cow on Mount Mahabharat Lekh. Note hair growing on the poll.

Kirko bull from a Chowri herd. Photographed in the Pike area.

Kirko cow and her calf, which was sired by a yak bull.

able uniformity have resulted, apparently from natural selection in the locality. An example of this is seen on Mount Mahabharat Lekh, near the town of Daman. This long, quite isolated mountain provides a grazing range at elevations of 4,000 to 14,000 feet. There the fine-boned Hill Cattle are quite uniform in conformation and other external characteristics and have a distinguishing growth of shaggy hair on the poll. Black is the predominating color, often with white markings on the legs and underline; there are also light-brown and grey animals. The hump on the male is quite sharp and prominent, and the horns are thick and rather short; the female has larger and thinner horns, which are usually upswept, and a slight hump. A mature bull weighs approximately 450 pounds, cows from 350 to 400 pounds.

An even more distinctive type of small Zebu, locally called Kirko, is seen in northeastern Nepal, at the edge of the high plateau region. Its nearly complete isolation from the cattle at lower elevations has resulted in a quite uniform type of small cattle. Characteristics are similar to those of the Mahabharat Lekh cattle, without the shaggy hair on the poll. Black-and-white patched animals are quite common and often have a white topline.

The Hill Cattle near Katmandu and the large towns are generally of

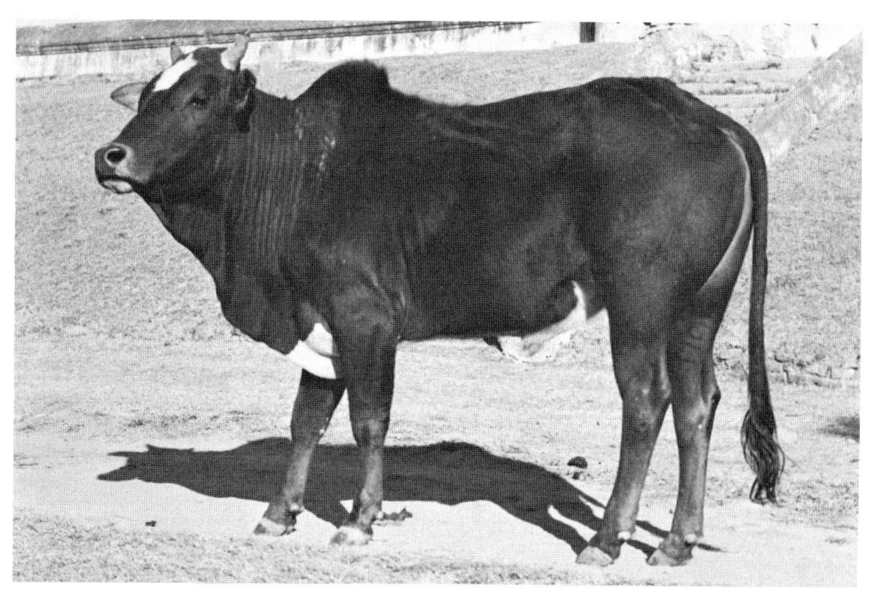

Hill bull, 450 pounds. Photographed near Katmandu.

Hill cow, 375 pounds. Photographed near Katmandu.

more nondescript character than those in most of the grazing areas lying between the jungle and alpine regions, probably because of more interbreeding with other Zebu breeds, as well as some admixture with the small number of European cattle which have been imported from time to time.

Terai Cattle.—The cattle in the Terai are a mixture of the Indian Hariana and Tharparkar breeds with the Hill Cattle, the Hariana influence usually predominating. Locally these cattle are given no particular designation but are here referred to as Terai Cattle.

Although smaller in size, they are similar in general characteristics to the Hariana cattle of the Indian countryside. Color is light grey to white, and the horns are of fair size, rather thick, and upswept in a single curve. Mature cows range from 600 to 800 pounds in weight, and the average bull is possibly 100 pounds heavier. Examples are seen of relatively pure Hariana and Tharparkar bulls which have been brought over from India for upgrading. The oxen are much used for cart draft and weigh 800 to 1,000 pounds.

Terai cattle at the edge of the jungle region.

Buffalo.—The native buffaloes of Nepal are small, ranging in weight from 750 to 900 pounds. The hair color is usually slate or dark grey. Their outstanding characteristic is the backswept horns which totally lack the tight curl of the Murrah and the other milk-buffalo breeds of India. They are seen from the Terai to elevations of 9,000 feet at the edge of the Alpine region.

Murrah buffaloes, imported from India in the recent past, are now widely used in crossing with the native animals to obtain increased milk

Native buffaloes at 9,000 feet. Photographed in northeastern Nepal.

production. Many herds now have individuals of both types. In the Terai buffaloes are maintained for milk and for work in the rice paddies, and limited use is made of them for cart draft; in the midlands they are kept entirely for milk production.

Murrah buffalo herd in the midlands.

Yak.—The yak was introduced to the alpine region of Nepal from Tibet many centuries ago. While this name is the generic term for the species, "yak" is also used to designate the male and "nak" the female. It is indigenous to the high plateau area north of the Himalayas in Tibet and also to the Mongolian plateau.

Specifically adapted to high altitudes, the yak is said to be unable to survive at elevations below 8,000 feet. The body is barrel-shaped and the legs are quite short. The hair coat is at least four inches long, which gives the characteristic shaggy appearance. Even longer hair covers the entire tail. The horns are rather large and upswept and turn backwards toward the tips. The yak is usually black or tannish grey, although animals of various other colors are seen. The male and female are practically indistinguishable in external appearance. Weight varies from 500 to a maximum of 800 pounds.

Chowri.—This is a cross between the Hill Cattle of the northern and higher midlands, locally called the Kirko, and the yak. The general appearance is somewhat similar to that of the yak, although the hair is not as long and shaggy. As used by the outsider, "Chowri" commonly refers to any cross of either sex between the yak and the Zebu, but breeders of

Yak bull, seven years old, on winter range at 9,200 feet elevation.

the Chowri distinguish between the yak–Zebu cow cross and the Zebu bull–nak cross as follows:

Parents	*Female Offspring*	*Male Offspring*
Yak-Zebu cow	Urang Chowri	Urang Yopkio
Zebu bull-nak	Dimjo Chowri	Dimjo Yopkio

Little difference exists in external characteristics between the Urang and the Dimjo Chowri, although the Dimjo probably has more the appearance of the yak. The horns of both frequently may have a forward thrust toward the tips instead of the backward thrust seen on the typical yak horn. Except for this forward thrust, the horns of the Dimjo are usually characteristic of the yak, but those of the Urang are frequently smaller. Both have long hair and a hair-covered tail. The coloring of either Chowri is more varied than that of the yak and white patches are more common, but black or grey predominates in most individuals.

The female Chowri, either the Urang or the Dimjo, is an exceptionally fertile animal and calves every year. The male (Yopkio) is said to be infertile. The breeding of the Chowri has been practiced for centuries by the mountain people to obtain an animal with a higher milk yield. Both the Urang and the Dimjo Chowri yield considerably more milk than either of the parents, although the Dimjo excels the Urang in productivity. Both Chowris are remarkably long-lived animals, having a maximum of eighteen lactations.

The average milk yield during a lactation period of 180 to 250 days, depending on pasture conditions, is about six pounds a day, although individuals have been known to give more than three times this quantity. The butterfat content ranges between 5 and 8 per cent, the milk from the Dimjo Chowri averaging 0.5 to 1 per cent higher than that of the Urang.

MANAGEMENT PRACTICES

Throughout Nepal, except in the high, alpine regions where the yak and the Chowri are husbanded, one of the major reasons for maintaining cattle is the production of manure to be used as fertilizer. Some manure in the form of hand-molded cakes is saved for use as fuel, as is done in India; but this use is not widespread and is seen mostly in the Terai. There both cattle and buffaloes are herded on the stubble fields or such waste-land grazing as is available and are brought to the owner's dwell-

Dimjo Chowri.

ing at night for housing under a thatched roof. Buffalo cows are milked regularly, but milk is taken from female cattle only incidentally for household use. Rice and other straws are used to some extent for feed and some hand-cut grass is saved as hay; but the major feed requirement is supplied by grazing except in the case of draft animals. They are fed under roof—straws, hand-cut grass, or edible leaves, which are also gathered by hand and brought in for the animals' consumption. Buffaloes are the principal draft animals for field work in the Terai, although cattle are utilized to some extent for this purpose. Oxen are the principal cart draft animals, but buffaloes are also so used in a limited way.

In the midlands most cultivation is done by men and women using hoes. It is highly exceptional to see either oxen or buffaloes doing field work. Buffaloes, both the native type and the Murrah crosses, are maintained primarily for milk production in this area.

All salvageable manure is meticulously saved and carried to the field, usually by manpower, in two shallow baskets hung from the ends of a short pole which is carried across the shoulders behind the neck.

After the jungle belt, which lies between the Terai and the midlands, is passed, there is very little level land except for small areas in the wider

Urang Chowri.

river beds. All slopes less than 30 degrees are terraced to permit cultivation. Even the comparatively flat bottom land is contoured or terraced. The cultivated areas therefore consist of irregular, narrow ribbons, sometimes only one or two yards wide. The difficulty in employing animal draft power on such plots is conceivably the reason that they are cultivated by hand. Even in the larger valleys such as Katmandu where oxen could be used for cultivation, field work is with the hand hoe.

In the midlands there is considerable plant growth on the slopes that are too steep for terracing; and here as well as in the stubble fields cattle are grazed. During the day they are under the care of herdsmen, frequently small boys or girls, and are brought into sheds or to an area on the ground floor of the owner's dwelling at night.

The alpine area at the northern limits of the midlands, particularly in northeastern Nepal, is the land of the yak and the Chowri. The yak is found at elevations as low as 10,000 feet in the winter and as high as the permanent snow line of 15,000 feet in the summer; the Chowri ranges up to 13,000 feet in summer and as low as 8,000 feet in the winter. Both follow the grass under the constant supervision of herdsmen.

Yaks are not indigenous to Nepal but have been brought in from Tibet for centuries. At higher elevations they are used mainly as pack animals and are consumed as food. The people in the high country are of Tibetan origin and of the Buddhist religion and are not averse to eating cattle flesh. A little farther down, yaks are used mainly for breeding the Chowri.

Within the altitude range of 8,000 to 13,000 feet in northeastern Nepal, there are two types of Chowri owners—the breeders, who raise Chowris for sale, and the milk producers, who keep their herds mainly for milking. While these people are village dwellers and have fixed abodes, the herders follow the grass with the season and may be gone for weeks at a time in their search for desirable pasture. The milk which is taken and not consumed as food is converted to cheese or ghee, the clarified butter much used for cooking. These products are packed to Katmandu on the backs of both men and women for distances requiring as much as fourteen days of travel.

The average milk production of a Chowri herd is about six pounds a day per cow. Herds which are the property of one family are kept separate both while grazing and while bedding down at night. Some grass is cut and used sparingly for feed when the snow cover is deep enough to make this practice necessary. Calving always takes place in the spring, probably because of the nutritive level which varies with the seasons. Bulls are kept in the cow herds the year round but conception occurs in the early summer, corresponding with the time that the new grass growth has permitted the cows to recover from the effects of the low nutrition level on matured dry grasses during the winter.

In the Chowri herds producing milk, the Dimjo tend to outnumber the Urang and are preferred because of the higher butterfat content of their milk. Both calve every year to initiate the lactation period, but very few cows are seen that are not the first cross of yak-Zebu or Zebu-nak parents. Because the mountain people love their animals, an unusual circumstance occasionally permits exceptions to be made and a second-cross calf is allowed to live. The various crosses beyond the first generation have elaborate local names; the Tolmu, for instance, is the female progeny of a yak sire and an Urang Chowri. None of these animals are of much importance, however, because of the low milk production of the female and the infertility of the male.

Tolmu Chowri. Photographed in the Pike area.

In the Chowri-raising region the Hindu element has begun to overlap the Tibetan Buddhists, and the religious ban on cattle slaughter, along with the law against killing cattle, presents a problem to the Chowri grower. This obstacle is partly alleviated by not allowing unwanted calves, both male and female, to suck so that they die a "natural" death. This solution is not universal. In the vicinity of Pike, a major Chowri-raising area sixty miles air line but eight days on foot east of Katmandu, there are said to be 5,000 hybrid males roaming the ranges, unclaimed by any of the Chowri owners. In the past there was some trading of the hybrid males across the border in Tibet for use as pack animals, but this exportation is now illegal.

In the midlands and the Terai the many unowned cattle around the towns and villages survive on whatever they can glean from the streets, open areas, and roadsides. All are amply protected by the religious attitude toward cattle as well as by the law against their slaughter. In Katmandu there is an asylum for unwanted cattle (like those in India), the Pashupati Gaushal (temple house of the cow), which is maintained as a charitable institution.

CATTLE MARKETS

There are no organized cattle markets in Nepal. Transactions are between one owner and another. Naturally, there is no sale of cattle for slaughter and only an occasional local buffalo bull or male calf is sold to the butchers. Slaughter of male buffaloes is legal, but most of the animals killed are driven up from India to Katmandu and are sold on arrival by the Indian traders to local Nepalese butchers. Some female buffaloes are sold and illegally killed.

ABATTOIRS

The largest slaughtering facility in Katmandu processes an average of 15 head of buffalo daily. It is located on the Baghmati River, a tributary of the Ganges, and consists of an unevenly paved floor, with no other facilities. Skinning is done here and the carcass is dressed and cut up and then distributed to the numerous stalls throughout town which sell meat.

Several smaller slaughtering units of similar type in the city kill only a few buffaloes daily. Buffaloes killed for consumption by the small Moslem population are handled in separate facilities in accordance with the orthodox methods required by that religion. The other large towns have facilities similar to these. In the whole Katmandu Valley a daily average of only 100 buffaloes is slaughtered.

CATTLE DISEASES

The wide climatic range of Nepal lays it open to many cattle diseases. Until a rinderpest immunization program was inaugurated in 1965, very little effort had been made toward the control and treatment of cattle diseases. The fact that very few exotic cattle have been introduced into the country, with the exception of the Zebu from India which can hardly be put in this category, has led to the development of considerable tolerance to the many common diseases which are endemic. This acquired tolerance is more pronounced in the Terai than in the hill cattle.

Rinderpest and hemorrhagic septicemia are probably the two most serious diseases. Liver fluke is encountered in swampy areas at elevations up to 6,000 feet. A limited amount of drenching is used to treat this parasite and some use is made of copper sulfate to treat snail-harboring waters, but there is no major program for control. Brucellosis is widespread but calfhood vaccination is not practiced. Tuberculosis is common and is impossible to control because of the legal and religious bans against the killing of cattle. Foot-and-mouth disease is endemic, and there is no control other than the infection of the animals in an outbreak area by application of the saliva of a diseased animal. This procedure is used in an attempt to infect all animals in an area with the disease at the same time and thus avoid continuing recurrences. After suffering several months' to a year's loss in condition, most cattle recover. There is some vaccination for blackleg on outbreaks.

The rinderpest control program, financed by the Oxford Committee for Farm Relief and administered by the Food and Agricultural Organization, calls for the vaccination of all cattle and buffaloes in the Terai and visualizes the establishment of an immune belt the entire length of the country along the Indian border. This program is the first major effort for the control of cattle diseases in Nepal. Its success, which definitely can be anticipated, may pave the way for further comprehensive efforts to control endemic cattle diseases in the country.

GOVERNMENT AND CATTLE

The elemental stage of the whole government organization naturally encompasses those departments concerned with agriculture and animal husbandry. It must be remembered that it was only in 1951 that the virtual dictatorship of the Rana regime was eliminated. That family's principal contribution to cattle development was the maintenance of a herd of exotic cattle which had been imported to provide milk for palace consumption. In 1968 one aged cow remained in the former cow barns of the palace, which are now used for experimental and developmental work. The Department of Animal Development and Veterinary Services has been in the process of establishing an artificial insemination breeding center here since 1963. Very few cattle except those belonging to the station have been so bred, however. A few Jersey and Brown Swiss bulls have been imported in past years in efforts to improve the milk production of the Hill Cattle by upgrading, but nothing of particular note has developed from these efforts.

The government's major activity involving cattle is the establishment of thirty-three veterinary hospitals throughout the country. Immunization and treatment are free when applied for; but, unless there is a serious outbreak of an easily recognized disease, such requests are few. Diseased cattle are also treated if brought to these hospitals.

The Veterinary Laboratory in Katmandu, with modern equipment for the production of vaccines, was established in 1967. The production of biologicals for the control of several of the endemic diseases was contemplated, but in early 1968 only rinderpest vaccine for the eradication program in the Terai was being produced.

OUTLOOK FOR CATTLE

The basic obstacle to cattle development is the religious attitude of the Hindus, who probably account for 90 per cent of the population in Nepal. The cow has the greatest religious significance, and both the legal and religious ban on the killing of cattle of either sex prohibit their slaughter even for the control of disease. Each animal must be permitted to live out its natural life as far as it is able to do so. Even among the cattle raised for production purposes, this regulation results in a surplus of useless males in addition to the large numbers of nonproductive cattle near the towns. Under such prohibitions, improvement by artificial selection for desired characteristics such as milk productivity is severely penalized, if not made impossible. Eradication or even control of such diseases as tuberculosis cannot be accomplished. The future for the cattle of Nepal is so bound up in these cultural aspects that it is far from bright.

With the buffalo the picture is considerably different, largely because unwanted males may be slaughtered. This practice has made possible the upgrading of native buffaloes by the introduction of the Murrah breed from India, and further progress along this line can be anticipated.

An intriguing line of cattle development lies in the possibilities of the Chowri, a most unusual hybrid whose offspring exceeds the ability of either parent in two notable aspects—milk productivity and fertility. Chowri growers maintain that a male of the seventh cross can produce offspring. This may be purely legend—actually the degree of sterility in any of the male crosses of the yak and the Zebu has never been authenticated. For centuries the breeding of the Chowri has been solely in the hands of the mountain people who raise them. The only assistance they have enjoyed from the outside world has been isolated instances of technical advice concerning the manufacture of cheese and butter. Whether anything can be accomplished in the way of developing a fixed type with the desirable attributes of the Chowri in milk production and fertility has not been investigated. Such a project would be of considerable genetic interest beyond the possible benefits to the mountain people of Nepal who have developed this remarkable hybrid.

Southeast Asia

Indonesia

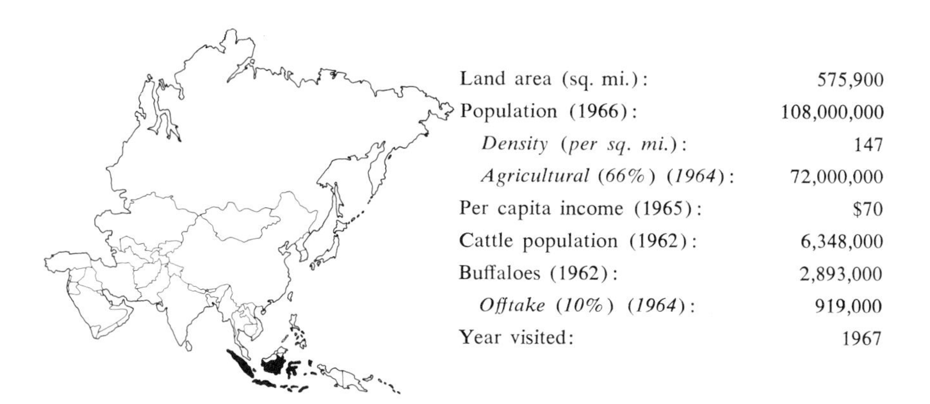

Land area (sq. mi.):	575,900
Population (1966):	108,000,000
Density (per sq. mi.):	147
Agricultural (66%) (1964):	72,000,000
Per capita income (1965):	$70
Cattle population (1962):	6,348,000
Buffaloes (1962):	2,893,000
Offtake (10%) (1964):	919,000
Year visited:	1967

THE REPUBLIC OF INDONESIA includes not only the chain of 2,500 islands that stretches for 3,000 miles along the equator from the tip of the Malay Peninsula almost to Australia but also the southern three-fourths of the island of Borneo, which lies immediately north of the center of the island group. Borneo, four-fifths the size of Texas, is largely tropical jungle, thinly populated with only eight inhabitants to the square mile. The other islands are equal in area to Texas, Louisiana, and Mississippi combined. Java harbors 65 per cent of the total population and has a density of nearly 500 persons to the square mile. West Irian, the western half of New Guinea, is currently under Indonesian control, its final status to be determined by a plebiscite in 1969.

The Portuguese were the first Europeans in what is now Indonesia, establishing a trading post on Sumatra in 1509. The Dutch and the British followed, most of their early footholds being on Java. The Dutch eventually prevailed and had their position well consolidated by the turn of the seventeenth century, although spasmodic warfare with the native rulers continued until the nineteenth century. Britain held Java for seven years during the Napoleonic Wars and the Japanese overran the islands for three years during World War II. Except for these interludes, the Dutch were in continual control for well over three centuries and until they recognized the independence of Indonesia at the end of 1949. With Dutch control gone, troubles multiplied rapidly. Sukarno, the leader in the independence movement, utilized the country's resources for personal aggrandizement and international image building and finally attempted to lead it into the Chinese Communist camp. This action culminated in an abortive communistic coup at the end of 1965, which was soon followed by a complete takeover by the military in 1966. Subsequently several hundred thousand Communists and suspected sympathizers were killed. The economy of the country had been completely demolished, and it is faced with a long, uphill road before it can reach even a subsistence level.

The people are of Malay stock and most are now of the Moslem faith. On Bali, however, there are still many Hindus. Most of the Dutch have left the islands, and there are very few Europeans in the country.

The large islands of Java and Sumatra have mountainous backbones. Even little Bali has a mountainous area. In natural resources, which include large areas of good, undeveloped agricultural land, Indonesia is by far the richest of any of the island states of the western Pacific. Its people are the poorest, having an average annual income of under $70.00, one of the lowest in the world. Agriculture is sharply divided into two classes, the large estates averaging 4,000 acres and the small farmers' holdings of 2 or 3 acres each. Oxen and water buffaloes supply the draft for working the small farms and for much cart work. A small, rugged horse is also used for hauling.

CATTLE BREEDS

Java is one of the few places in the world where representatives of the tribe Bovini still exist in a wild state. Many of the domesticated animals are direct descendants of these. There are also exotic breeds, which have been introduced during the past century.

Banteng.—This indigenous animal, although similar to cattle in general characteristics, belongs to a different branch of the Bovini, the Bibos. It is still found in the feral state in the more remote parts of Java. There were two bulls in the Surabaja zoo in 1967.

These animals are of fair size, weighing 1,100 pounds. Their horns are of medium length and are quite thick, turning upward and then inward at the tips. Although the legs are rather long, the body has good conformation, parallel topline and bottomline, and a wide, deep barrel. The male is very dark grey, bordering on black, with sharply marked white legs below the knee and a distinctive white patch in the center of the rump. The female has much finer horns, thrust rather directly upward, nearly straight or with a slight twist. Her color is tan, much like the Jersey, but with the white legs and rump like those of the male Banteng.

Banteng bull. Photographed at the Surabaja Zoo.

Bali.—The domesticated Banteng seen on the island of Bali is a counterpart of the wild animal although smaller in size. A mature bull weighs 800 pounds, the cow 600 pounds. It makes an excellent draft animal and is widely used for this purpose. When mature males are castrated, the hair coat after four months begins to change from the nearly black-grey color and in the course of a year turns to the tan color of the female. If the bull is castrated when young, his horns will be thinner and will assume the straight upthrust of the female. Both the bull and the cow have a very thin line of black hair running the length of the back.

Bali bulls have been exported to Malaysia for upgrading the small Kelantan breed. There it is claimed that the progeny is sterile; this opinion is partially correct in that the male offspring of a Bali cross on other domesticated breeds is sterile for the F_1 and F_2 crosses but is fertile on the third. The female progeny, however, is always fertile.

Madura.—This breed, native to the island of Madura, is an unplanned cross of the Banteng and the Zebu. It is smaller and has smaller horns than the Bali, the domesticated Banteng, but is similar in most other characteristics. Unless on special feed the average bull weighs about

Young Bali bull.

Mature Bali cow.

550 pounds, the cow 450 pounds. Both male and female are the tan color of the Bali cow. The legs below the knee are white or nearly white and there is a white patch on the rump, but these markings are not as distinctive as those on the Bali.

Young Madura bull.

Javanese.—Essentially the same animal as, but somewhat larger than, the Madura, this breed is generally seen on the island of Java, for which it is named. It can be considered the same breed as the Madura.

Java Ongole.—The Nellore, or Ongole, cattle of India were brought to Java during the nineteenth century for draft use on the sugar estates. They were indiscriminately bred with Javanese cattle and have produced the Java Ongole, an animal widely used for draft. It is intermediate in size between the two parent breeds, is usually light grey, and retains a distinct hump that is smaller than that of the pure Nellore. If they are to be used for draft, the males are usually not castrated.

Grati.—Several European milk breeds have been introduced from time to time for experimental crossing on native cattle. Of these crosses the

Young Madura cow.

Java Ongole draft team. The animal on the left has been castrated.

Friesian bull on the Javanese cow has been the most successful and has become known as the Grati because the Grati area, south of Surabaja, in eastern Java, has been the center of this development.

The Grati withstands the climatic stress much better than the pure Friesian, and the cross has become well established. In milk productivity it materially exceeds the native Javanese breed. The average Grati cow is smaller than the Friesian, weighing about 1,000 pounds. No fixed relationship exists in the crossing of the Friesian and Javanese, although the proportion of Friesian is usually more than one-half. Any animal predominantly Friesian in type is called a Grati and sells as such.

Grati cows of a dealer. Photographed in the Grati area.

Buffalo.—Both swamp and the smaller river buffaloes are widely used for work in the rice paddies. River buffaloes are seen mostly in Java, swamp buffaloes in Bali. Both are typical of the types seen throughout Southeast Asia.

MANAGEMENT PRACTICES

Most of the cattle and buffaloes in the country are maintained by small farmers, who utilize them primarily for draft. The cows are milked in some areas, a practice which the Veterinary Services are endeavoring to promote. There is some sale of bulls on Bali for fattening and slaughter on Java. With the lush plant growth resulting from the twelve-month growing season and ample rainfall, feed for these small cattle holdings is not a problem. The animals are grazed in small, rough areas not susceptible to cultivation and along the roadside, tethered and frequently unattended. The Indonesian takes good care of his animals and they are usually seen in very fair condition.

Bali cattle have a well-deserved reputation as a superior beef animal in Southeast Asia and are exported to Hong Kong and also to Singapore and Malaysia when political conditions permit. An organized feeding program is common with animals grown out for export. The three cattle

River buffalo cow. Photographed on Java.

Swamp buffalo cow. Photographed on Bali.

dealers on Bali have the island area divided among themselves and purchase bulls at four years of age or older from the small farmers. These animals are placed in the hands of the "feeders," farmers located in the higher parts of the island, and castrated. They are then fed better on the available forage than the typical work animal, although the former are also used for draft as occasion requires. The feeder-farmer usually has 2 or 3 such animals in his charge. The feeding period is a minimum of one year and frequently stretches to two. The dealer pays the feeder 3.3 cents a day for the time the animal is in his care if the finished weight is 935 pounds or more but only 1.3 cents if it is under this weight. Often some rice bran is fed as supplement so that the animal will reach the premium weight. This practice produces weights of as much as 1,100 pounds. The best animals grade the equivalent of high-good by USDA standards. Vaccination for foot-and-mouth disease is required fourteen to thirty days before shipment to Hong Kong.

A bull as received for growing out is the typical color of the Bali male (dark grey to black). Within twelve months after castration, he changes to the tan color of the female.

On Madura at the end of the rice harvest there is an age-old festival

Bali bull, five years old, 850 pounds, just placed in the hands of a feeder-farmer.

Finished Bali steer, eight years old, 1,100 pounds.

which centers around the bull race. The bulls are lavishly decorated for this occasion, and the participating teams are paraded in front of the spectators before the race. Each animal has a brightly colored umbrella fastened over his head for the parade, and his horns are covered by an extension of the brilliant harness. Ceremonial musicians and various attendants also enter the parade. The driver rides on a wide tongue, the front end of which is fastened to a yoke across the necks of the team, the rear end dragging the ground. Control is maintained by a rope, passed through the center membrane of the nose of each bull and held by the driver. The course is 110 meters over grassy turf. Two at a time, the teams race in an elimination contest which is continued until one of a starting line of thirty to forty teams has finally won. The same kind of elimination by competition of pairs is held for the original losers, one team finally placing as the winner among them. These bulls are bred from selected Madura stock to race and are specially fed and trained until their first contest, when they are about four years of age.

Milk production in the Grati area is usually an adjunct to the small farmer's main pursuit of crop growing. If he has Grati cows, he may milk as many as 12 head; if he is producing only from his Javanese cows, which are maintained principally for draft, he probably milks only 2 or 3.

In an effort to increase milk production, the Veterinary Services have initiated farmer co-operatives to handle the milk production at a central point and are attempting to provide artificial insemination service by a Grati or a Friesian bull. Such a co-operative is formed of 150 to 200 members, milking a total of 800 cows and producing 1,500 pounds of milk a day. In 1967 milk was being bought by the pasteurizing plant, where there was one, at 2 cents a pound and was selling at retail for 4 cents. If no plant was located in the vicinity, the milk was finding its way to the retail trade at 3.7 cents a pound.

MARKETING

Cattle markets are held on fixed days of the week in the large villages. Cattle are bought by butchers for local slaughter and by dealers for transport to slaughterhouses in the cities. Farmers also trade to some extent among themselves. To save feeding costs, a draft animal no longer required for work is often sold, and a replacement is purchased when the need again arises—a practice generated by the runaway inflation to which Indonesia is subject. On an estimated weight basis in early 1967, prices were about 3.3 cents a pound for slaughter animals. Nothing

Team of racing bulls prepared for the opening parade.

illustrates the serious state of Indonesia's economy more forcefully than a 900-pound bull, seven or eight years old and in fair condition, returning only the equivalent of $30.00 to the owner.

Madura bull race.

Cattle market at Blimbing.

The better-quality cattle exported to Hong Kong and Singapore from Bali fare somewhat better. These sales are handled directly through exporters by the dealers who have their cattle farmed out for fattening. All returns pass through the government so that it can take advantage of the foreign exchange involved. A 1,000-pound steer in good condition brings about 7 cents a pound in a negotiated sale to the exporters.

ABATTOIRS

The larger cities have municipal slaughterhouses. Moslems kill all the animals in their usual manner, described under "Abattoirs" in the discussion of Sudan. Veterinarians inspect the carcasses. Holding yards are provided at the Surabaja slaughterhouse, where at times animals are held several days before being killed. Processing is done by employees of the butcher, the municipality simply furnishing the facilities. The plant at Surabaja kills 5,000 cattle and buffaloes each month; that at Denpasar, on Bali, kills 500 head.

A canning plant at Denpasar is well equipped for turning out tinned meat products. It formerly killed 600 head of cattle a month but was shut down in 1965 because foreign exchange was not available for the purchase of the sheet tin used for the manufacture of cans.

CATTLE DISEASES

Foot-and-mouth disease of one of the milder types is endemic. The Veterinary Services' effort to promote vaccination by the farmers is handicapped by the relatively high cost and the shortage of vaccine. The Animal Virus Disease Institute at Surabaja is equipped to manufacture the common vaccines but funds are lacking for operation; 400,000 doses of foot-and-mouth vaccine were manufactured in 1962, but this had dropped to 50,000 in 1965. Cattle for export have first call on what is produced to meet the requirement for export.

GOVERNMENT AND CATTLE

The Veterinary Services of the Department of Agriculture are making constructive efforts to improve the cattle industry of the country with the limited means at their disposal. The dairy co-operative movement they sponsored has been mentioned. Four artificial insemination centers have been established in East Java, the largest of which is breeding 50 cows a month. The handicaps in communication and transportation make the work exceedingly difficult. Only the exceptional farmer is interested in the program, and cows are usually brought to the center too late for breeding.

Ingenious practices are employed in efforts to increase the number of cattle in the country. A cow is placed in the hands of a farmer who needs one, with the agreement that when he raises 2 calves to one year of age and turns them over to the Veterinary Services, the cow becomes his property. A heifer which the Services receives under this program is placed in the hands of another farmer; a bull is usually sold for slaughter.

Because of lack of funds, the experiment stations established by the Dutch and later maintained by the Department of Agriculture of the Indonesian government have had to gear their operations to money-making programs in order to survive. The Dwidharma Experiment Station, at Pekarangan, has been reduced to twenty-five acres. The initial program was to develop a stud farm to breed Friesian cattle so that the dairy industry could be expanded by using the Grati cow. Frozen semen from Australia and Holland was to be utilized. The station is still managing to hold on to a few Friesian cows, but the main activity now is the commercial production of hogs and poultry to provide maintenance funds.

Another experiment station, the Lalanglinggah Farm, also on Bali, is surviving by selling milk from the purebred Friesian herd. Improved pastures were the primary objective here to show the small farmer what could be accomplished in increasing production by seeding adapted grass and legumes. The effort is being continued; but the funds for the work must generate from the sale of milk, which must be trucked thirty miles to market, a two-hour trip because of the condition of the roads. (Throughout Indonesia many roads have one side abandoned, any maintenance that can be expended being devoted to keeping a one-way track open.)

OUTLOOK FOR CATTLE

The tremendous resources of Indonesia include an unused agricultural potential greater than that of any country in the Far East. Even on heavily populated Java there is much undeveloped land suitable for grazing. The average population density of the rest of the islands is fewer than eighty persons to the square mile—about the same as South Carolina. A large part of the land area is above 3,000 feet in elevation, where climatic conditions approach the summer of temperate zones. Even European breeds could be satisfactorily maintained here if carefully handled.

The country is blessed, however, with an indigenous breed, the Bali, which is ideally adapted to the environment in any part of the islands and could become a superb tropical beef animal. This part of the world has a market for beef that is practically unlimited. Even today, without selection and under the native type of husbandry, the Bali ox is recognized as a superior beef animal in the Hong Kong and Singapore markets. The basically Malayan population is not indolent—the greatest need of the people is merely an opportunity to work. The terraced rice paddies on Bali which have been beautifully maintained for centuries on hillsides with slopes up to 30 degrees are ample evidence of their industry. Labor would be no problem in any cattle-raising project that could be financed.

To the extent possible within the limits of the present economy, the possibilities in cattle are realized, as exemplified by the development of the Grati cow and the feeding practices used by the Bali cattle dealer. If the country can work herself out of the economic shambles now existing, cattle raising could develop into one of Indonesia's major industries.

Malaysia

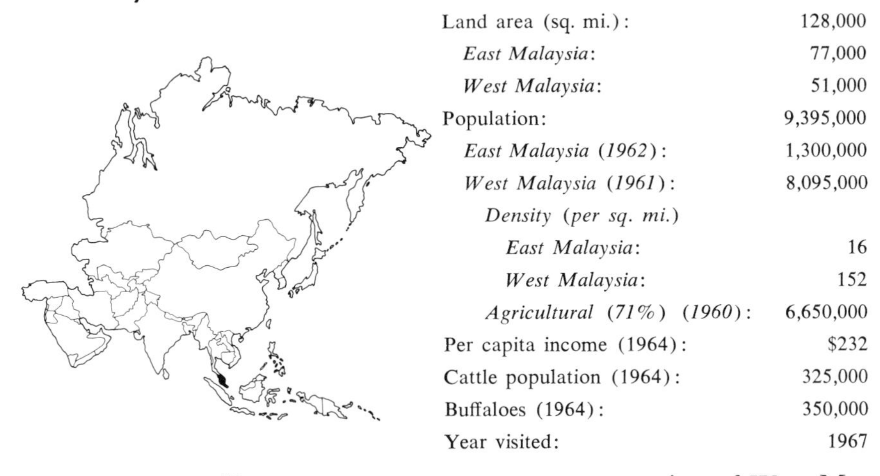

Land area (sq. mi.):	128,000
East Malaysia:	77,000
West Malaysia:	51,000
Population:	9,395,000
East Malaysia (1962):	1,300,000
West Malaysia (1961):	8,095,000
Density (per sq. mi.)	
East Malaysia:	16
West Malaysia:	152
Agricultural (71%) (1960):	6,650,000
Per capita income (1964):	$232
Cattle population (1964):	325,000
Buffaloes (1964):	350,000
Year visited:	1967

THE FEDERATION OF MALAYSIA consists of West Malaysia (the former Federation of Malaya), which occupies the Malay Peninsula, and East Malaysia (the former British crown colonies of Sarawak and North Borneo, now Sabah), which covers a strip of land along the northern and northwestern coasts of the island of Borneo. East Malaysia lies directly east of West Malaysia, across the South China Sea, and in area equals Louisiana combined with half of Mississippi. Coastal plains rise to a mountainous rain-forest region inland. West Malaysia is bordered by Thailand on the north. It is nine-tenths the size of Florida, is divided by a mountainous core running north and south, with elevations as high as 7,200 feet. A wide coastal plain lies on the east and a narrow one on the west. Both East and West Malaysia lie just north of the equator.

Portuguese traders were first in the area in the early sixteenth century. The Dutch established themselves on the southern end of the Malay Peninsula toward the middle of the seventeenth century, and the British began to take over at the end of the eighteenth century. Their colonization program was well consolidated within the next 100 years, although the final acquisition, that of the northern states, which were acquired from Thailand, was not effected until 1909.

When Britain tried to bring order back to the peninsula following the brutal Japanese occupation from 1942 to 1945, the seeds of independence were beginning to spiral. A vicious nine-year interlude followed. The Malayan insurgents were in large part commando Malayan troops which had been trained by the British to fight the Japanese during the latter's period of occupation. At the end of World War II these Malayans turned against the British army, which they outnumbered at the time. There is a strong feeling that this rebellion would not have happened if, at the end of the war, the Malayan jungle fighters had been treated as well as their British counterparts in the matter of back pay for service during the occupation. The Malayan was given 100 Malayan dollars, the British soldier, back pay for three and one-half years. At the war's end the dissatisfied natives were fertile ground for Communist indoctrination, and they continued to fight. Only meager support came from the Communist world, primarily because of lack of channels through which such support could be funneled. The final resolution came in 1957—independence for the Federation of Malaya within the British Commonwealth. Six years later the Federation of Malaysia was formed when the Borneo colonies and Singapore joined; after two years Singapore withdrew and is now an independent entity.

The Borneo area, although susceptible to major agricultural development at some future day, is now a relatively unimportant part of the federation: Malaysia today is largely western Malaysia. Even with the extensive rubber, coconut, and palm oil plantations, tropical jungle still covers four-fifths of the peninsula; and a considerable part of this is susceptible to future agricultural development. The rainfall pattern is excellent, with 30 inches falling in the December-January rainy season and another 50 inches being distributed fairly well throughout the rest of the year. With a population density of 152 persons to the square mile, the land is not overcrowded, and the voluntarily segregated mixture of Moslem Malayans, Buddhist Chinese, and Hindu Indians provides what so far seems to be a workable system of checks and balances on political excesses. Malayans constitute about half of the population, Chinese 37 per cent, and Indians 11 per cent; various other nationalities provide the remaining 2 per cent. Law and order prevail in an atmosphere of apparent permanence.

The climate, tropical in the low areas, becomes temperate in the high country except that there is no winter as commonly experienced in more northern latitudes.

Agriculture (rubber) and mining (tin) are the bases of the economy. Malaysia produces one-third of the world's supply of each of these commodities; the rubber plantations, however, are now gradually being converted to palm oil culture as world-wide use of synthetic rubber increases. Although Malaysia is not a cattle country, the draft ox and the buffalo are essential elements in the life of the farmer with small holdings. Wild gaur, one of the few remaining types of wild Bovini, are said to grow in the northern mountains.

CATTLE BREEDS

Two distinctive types of cattle and both the swamp and the river buffalo, all of which have been in the country for many generations, are maintained in the federation. All of these animals, with the exception of the indigenous Kelantan cattle, were probably introduced into the country during the nineteenth century. In recent years several exotic breeds have been brought in with the objective of upgrading the native cattle.

Local Indian Dairy.—Referred to as L.I.D., this is a degenerate mixture of various types of Zebu cattle brought from India at different times, initially for draft purposes. It varies widely in color, conformation, shape of horns, and size of hump. For generations the rubber tappers have kept this animal on poor nutrition, and one of their mature cows weighs only about 600 pounds. In an animal husbandry station of the Ministry of

L.I.D. cow. Photographed at the Central Animal Husbandry Station, in Kluang.

L.I.D. cow of a rubber tapper.

Agriculture, however, a cow in good condition easily weighs 850 pounds, the difference being caused by the level of nutrition. Work oxen, selected for size when young and then well maintained, weigh about 1,100 pounds.

L.I.D. cattle of small farmers and rubber tappers exhibit an unusual polled tendency for Zebu cattle, 10 per cent of the cows being either without horns or showing only bare stubs.

Kelantan.—A distinct enough type to be called a breed, the Kelantan can be considered indigenous, its antecedents apparently having been lost in the past. Individuals are seen which, except for size, somewhat resemble the Chinese Yellow. The color varies from light tan to dark brown, usually solid or nearly so, with some white markings; a medium reddish-brown predominates, however. A small vestige of a hump is frequently seen on the female, although it often is not noticeable; in the male the hump is more pronounced, though not prominent.

A mature cow seldom exceeds 500 pounds in weight and usually is lighter. The horns are small and upturned, often with a rather sharp curl. Polled animals are common. The breed is found largely in the state

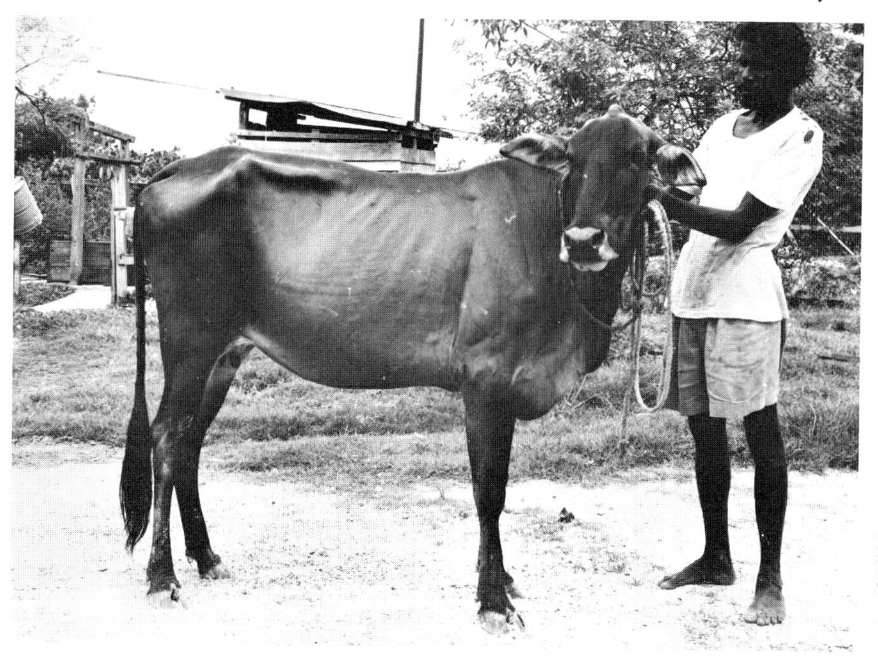

Polled L.I.D. cow of a rubber tapper.

of Kelantan, for which it is named; it is also grown in the northern part of West Malaysia and in southern Thailand, just across the northern border. This hardy, well-adapted animal is exceptionally fertile, most cows calving every year in spite of the low level of nutrition they are on. The Kelantan's compact conformation, natural hardiness, and fertility seem to show definite possibilities for its development by selection as a beef type.

Kelantan cattle are held mostly by small farmers and rubber tappers as an adjunct to their regular work. Generally very little use is made of them except for the occasional sale for slaughter. Mainly they serve as the "peasant's purse"—a quick source of cash in an emergency. In the rice paddy areas buffaloes are preferred for draft, although in some locations Kelantans, both bulls and cows, are worked.

Until the sport was outlawed in 1959, the Kelantan was trained to fight in contests between two bulls. Potential fighters were pampered with the best feed and care. In northern Kelantan, bulls are still raised and trained for fighting and are then sold in Thailand, where the sport persists.

Buffalo.—Buffaloes constitute more than half of the total cattle popu-

Young Kelantan bull, naturally polled.

Mature Kelantan cow.

lation. Mostly of the swamp type, they are used for draft and are slaughtered only as worn-out work animals. Draft buffaloes are essential to the agriculture of the country and are widely used in some sections for cartage. They graze the rice paddies after harvest and are fed on cut grass during the growing season.

Near towns and larger villages the Indian Sikhs maintain dairies of Murrah and Nili buffaloes. These animals are used entirely for milk and are well cared for. In the morning they are driven to such grazing as can be found, pastured under the eye of a herdsman, and returned to the village at night. The estimated number of milk buffaloes is 5,000 head.

MANAGEMENT PRACTICES

Most cattle of the country except draft and dairy animals shift largely for themselves. Many rubber tappers on the plantations own 1 or 2 head of cattle, which graze on whatever plant growth can be found and on crop residues. Buffaloes and oxen used for draft fare better, and feed is brought to them.

Dairying is handled almost exclusively by Indians on the outskirts of the larger towns. They have herds of 5 to 20 head of L.I.D. cows, or a somewhat larger number if the owner is milking buffaloes. Much of the

Cow yard and herd of an Indian dairy near Kuala Lumpur.

milk is distributed in bulk from cans, either by the owner on a bicycle or by an intermediate who purchases it for retail sale. As an auxiliary to the dairy operation, the Indian often keeps a team of work oxen. They are rented out by the day, the compensation being $2.50 to $3.00 for a 10-hour day for the team and working driver.

The average dairy is a down-at-the-heels enterprise consisting of a crude corral in which the cattle are held at night, a makeshift shed for protection from the sun and for milking, and a little better hut for the owner's domicile. The herd is taken out in the day by a herder for such grazing along the roadside or in stubble fields as can be found. A limited supplement is fed by making a gruel of broken rice, bran, and a little palm-oil cake. While the owner stirs the mixture each cow in milk is permitted to drink a limited quantity of it.

Artificial insemination was introduced in 1955 but has not made much progress. There are three centers for the collection of semen. Egg yolk citrate is used as a diluent and the rectal method of insemination employed. Some frozen semen has been imported from Australia, but the results have not been satisfactory. The number of cows bred annually can be counted in the hundreds in spite of the facts that communications and roads are better than in many other countries and the veterinarians doing the work are knowledgeable and conscientious in their efforts. Cultural antagonism toward the method is probably the reason for the lack of progress.

MARKETING

An official cattle market was established in 1955 for the sale of cattle for export on a quota basis. Most sales, however, are by small owners to traders, who pick up the animals at the owner's premises and transport them to the local abattoir. Depending on the distance from market, a 500-pound ox at five years of age in 1967 sold for $70.00 to $80.00, or 14 to 17 cents a pound liveweight, and passed from the trader to the butcher at the abattoir for about 20 cents a pound. Beef, which is considerably less expensive than mutton or pork, cost about 70 cents a pound in the shops.

Legal slaughter is estimated at 38,000 head of cattle and 35,000 buffaloes annually, or 12 per cent for cattle and 13 per cent for buffaloes. More cattle than buffaloes are probably killed locally by one-animal butchers. These figures represent quite a sizable offtake since many of the animals sold are more than five years of age and the killing of females, unless certified as nonproducers, is prohibited.

CATTLE DISEASES

The Malay Peninsula below Thailand is said by the Veterinary Division of the Ministry of Agriculture to be free of rinderpest and foot-and-mouth disease. No cases of anthrax or blackleg were known in 1966. Strict quarantine enforcement and the limitation of imports to only those animals from clean areas is responsible.

Hemorrhagic septicemia is serious, but extensive vaccination is materially reducing the incidence. The government farms vaccinate against this disease before the beginning of the rainy season and every four months until it is over. Internal parasites are ever-present.

The claim is made that brucellosis is unknown except in the case of an occasional imported animal.

GOVERNMENT AND CATTLE

The Ministry of Agriculture is placing considerable emphasis on cattle improvement, particularly in regard to milk production. The Central Animal Husbandry Station at Kluang, with 2,000 acres of developed pasture and 3,000 acres of undeveloped jungle, is a well-equipped, well-managed establishment. No flies are seen around the barns, which is highly exceptional in a tropical environment. In 1967 the cattle population was 1,200 head, including a Red Sindhi purebred herd, an L.I.D.

The type of Red Sindhi bull used at Kluang.

herd, and another of Red Sindhi–L.I.D. crosses. Crossing experiments have also been conducted with L.I.D. cows and Jersey or Friesian bulls, and the conclusion reached was that the Red Sindhi gives better over-all results on the L.I.D. cow. Good L.I.D. cows will have an average production of 1,000 pounds a lactation. The Red Sindhi crosses now average 2,000 pounds, with individuals giving as much as 5,000 pounds. The program calls for continuous breeding back to Red Sindhi bulls.

Improved pastures and meadows for cutting green fodder to be brought to the milking sheds have been established, using Napier and guinea grasses. Cows are run on pasture after the morning's milking until the heat of the day and are then returned to the sheds. The yield of the Napier grass on six or seven cuttings was said to be thirty-seven tons an acre annually on an air-dried basis.

Crossing the Kelantan cow of the rubber tapper with exotic breeds by employing artificial breeding is also under investigation. Hereford crosses on the Kelantan have been tried in an attempt to produce a better meat animal but results have not been encouraging. A Jersey cross has given indications of better growth.

At the Paroi Experiment Station a purebred herd of Red Sindhi cattle imported from Pakistan is maintained for the production of bulls to be distributed in the country. Johne's disease was discovered in the herd in 1959 but was subsequently eliminated.

Other efforts of the government to increase and improve livestock production are the introduction of the *pawah* scheme in the rice paddy areas, compulsory castration of scrub bulls, and the use of subsidized bulls and artificial insemination. Under the *pawah* scheme the government places a buffalo cow in the hands of a bailee and when he delivers a two-year-old buffalo heifer to the government, he gets the title to the cow. The program is then continued by placing the heifer under the care of another farmer. This scheme is spreading to cattle and, with its self-maintained momentum, is having some effect in increasing the cattle population in the plantation areas where there is much unutilized plant growth. A rubber tapper who lacks the capital to purchase a cow can manage, with the help of his children, to provide a fair living for an animal in the small stubble fields and along the roadside.

Castration of scrub bulls, while now only mandatory in a few states, can do much to improve the quality of the random cattle of the 1- or 2-cow owner. This is especially true if one of the subsidized bulls from the experiment station, either a Red Sindhi or a Red Sindhi first-cross, is used.

OUTLOOK FOR CATTLE

With the ample rainfall and its good pattern, a twelve-month growing season, and the large areas of undeveloped land, Malaysia's agricultural development, including the cattle sector, could expand considerably. The two native types of cattle, the L.I.D. and the Kelantan, are well acclimated. Upgrading them by using another tropical breed, the Red Sindhi, holds much promise. Improvement by selection within these breeds, which has not been undertaken, is another possibility. While neither the L.I.D. nor the Kelantan seems to offer much potential as a milk-producing type, the small Kelantan cow, with her compact conformation, hardiness, and fertility, should not be overlooked as a possible beef type. The newly independent government is continuing on an exceptionally even keel; and, all things considered, there should be a future for both the cattle and the buffaloes of the country.

South Vietnam

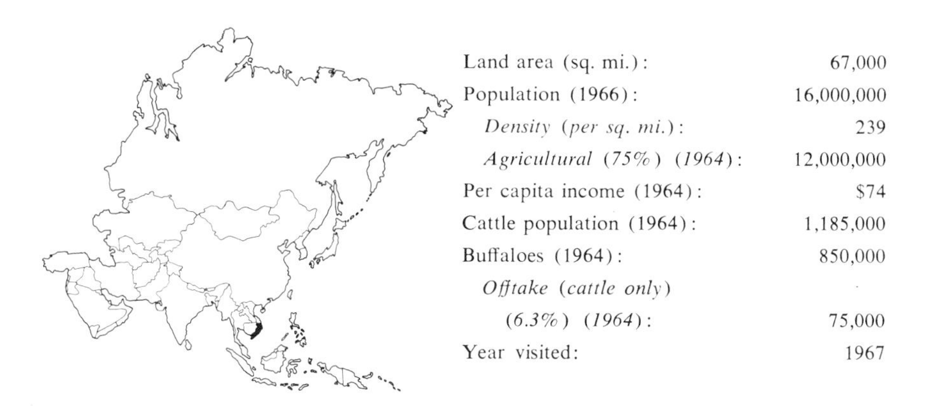

Land area (sq. mi.):	67,000
Population (1966):	16,000,000
Density (per sq. mi.):	239
Agricultural (75%) (1964):	12,000,000
Per capita income (1964):	$74
Cattle population (1964):	1,185,000
Buffaloes (1964):	850,000
Offtake (cattle only) (6.3%) (1964):	75,000
Year visited:	1967

THE REPUBLIC OF VIETNAM (South Vietnam) with its neighbors to the west, Laos and Cambodia, and to the north, North Vietnam, comprised French Indochina before World War II. The country is a narrow strip of land, 650 miles long and an average of 100 miles wide, lying along the eastern side of the Indochinese peninsula, directly across the South China Sea from the Philippine Islands and Borneo. Drainage is from the hilly country in the west and north onto the coastal plains and to the sea. In area the country is one-ninth larger than Florida, with a population nearly three times greater than that state's.

The recorded history of what is now Vietnam goes back more than 2,000 years, many of which have been troubled. The Chinese were the first aggressors but were driven out approximately 1,000 years ago. The Portuguese in the sixteenth century were the first Europeans to land and were followed by both Dutch and English traders in the seventeenth century. Later came sizable numbers of French missionaries. Eventually France subjugated all of what is now Vietnam, Laos, and Cambodia and maintained control until the Japanese entered after the beginning of World War II; Vichy France then acquiesced to the Japanese occupation. When the Free French tried to resume their former position in Vietnam at the end of the war, they encountered organized Communist opposition in the north and a strong independence movement in the south. The Indochinese War ensued and lasted from 1946 until 1954, at the end of which South Vietnam's independence was established. Then followed internal conflict and the growing escalation of guerrilla war with the Viet Cong. Later other countries became involved in an all-out war the principal burden of which fell on the United States.

The climate is tropical, with a fair rainfall pattern averaging 85 inches. The rainy season is from April to November, with most precipitation occurring from July to September. The highest producing area is the Mekong Delta with its paddy rice for which irrigation is necessary. Generally the irrigation works do not provide water during the dry season and, therefore, usually only one crop is grown. In the tropical climate two or more crops could easily be obtained if water were made available during the dry winter. The forest-covered highlands in the west and north are not nearly as productive and are more sparsely populated.

More than 80 per cent of the population is Vietnamese and most of the remainder are related people of the Indochinese peninsula. In the western and northern mountains live primitive tribes numbering about 1,000,000 persons. The 400,000 Chinese in South Vietnam are engaged mainly in trade. The European element in the population is small, mostly French, and has decreased since the days of French supremacy. It is now somewhat less than 40,000, excluding the military.

The economy is normally agricultural; but the recent buildup of the

Chinese Yellow ox. Photographed at the Saigon abattoir.

population centers by American military, the United States AID, and the many construction and supporting organizations of both have completely disrupted this. A few years ago South Vietnam was one of the major rice exporting countries of the world; in 1966–67 it was necessary to import more rice than was formerly exported. Lack of security in the farming areas and poor price incentives for the farmer have been largely responsible for this change.

For centuries cattle and buffaloes have been used for draft. The wild banteng, the same as that on Java, is native to the western highlands along the Cambodian border but was never domesticated. Across the border, the guar, the largest of the existing wild cattle, and the kuprey, another wild species, intermediate in size between the banteng and the guar, were seen by an expedition sponsored by the National Academy of Sciences in 1963. All these animals are doomed to disappear within a few years because of random slaughter by natives and by Viet Cong troops who are forced to live off the land.

CATTLE BREEDS

Cattle are principally of two types—the Chinese Yellow and a nondescript Zebu, which frequently shows a strong influence of the Chinese Yellow. Except for a few dairy animals, the buffalo is the swamp type generally seen in Southeast Asia.

Chinese Yellow.—To a limited extent this breed has been maintained in a relatively pure state. Although smaller in size, it apparently is basically the same animal as that found in Taiwan. This difference in size could have been occasioned by generations of the breed feeding on a lower nutrition level. A cow in poor condition, as it is normally seen, weighs about 650 pounds; an ox at seven to eight years of age weighs about 800 pounds.

Nondescript Zebu.—Most cattle in the country are humped and have the characteristic long ears and the heavy dewlap of the Zebu. They are unplanned crosses between various Zebu types or between the Zebu and the Chinese Yellow.

Swamp Buffalo.—This is the typical draft animal used to work the rice paddies throughout the Far East and neighboring islands.

Nili Buffalo.—A small number of these milk buffaloes have been imported for use in a few dairies near Saigon.

Zebu ox showing Chinese Yellow influence. Photographed at the Saigon abattoir.

Swamp buffalo ox.

Nili buffalo cow at a Saigon dairy.

MANAGEMENT PRACTICES

Cattle and buffaloes are primarily draft animals when in the hands of the small village-dwelling farmer. They are maintained on stubble fields or whatever miscellaneous plant growth can be found. As a general practice, only worn-out work animals are sold for slaughter.

The few dairies near Saigon, mostly operated by Indians, have a total of approximately 1,000 milk cows and 100 Nili and Murrah buffaloes. Some operations are using Red Sindhi bulls in an effort to increase the milk yield of the Zebu-type cows in their herds. Both cattle and buffaloes are maintained in open thatch-roofed sheds, and forage is brought to them. Some supplement in the form of rice bran, bean pods, and brewers' grain is fed. The largest dairy, Roussema, milks 80 head of native humped cattle and crosses of these by a Red Sindhi bull, along with 18 head of Murrah and Nili buffaloes. Most farmers making milk have from 4 to 12 cows or buffaloes. Milk is sold mainly to restaurants and costs 8.5 cents a pound.

MARKETING

Slaughter animals are customarily sold by the farmer to traders or, occasionally, direct to butchers. Most of the trade passes through the hands of a trader before sale to the butcher. Sale in the country is by the head, but transactions at the Saigon abattoir between trader and butcher are on a liveweight basis. Some attention is paid to quality. A canner by

Imported Red Sindhi bull of the Roussema Dairy.

USDA standards was bringing 28 cents a pound liveweight in 1967. A six-year-old ox which would grade low-good, the highest quality available, was selling for 34 cents. Worn-out buffalo draft animals sell for 24 cents. Beef retails for 60 to 95 cents a pound, depending on quality.

ABATTOIRS

The Saigon abattoir is a municipal facility killing 120 cattle, 80 buffaloes, and 1,200 hogs a day. Cattle for slaughter remain in adjacent holding sheds for as long as seven days. The animal is stunned with a hammer and its throat is cut as the animal falls. The operation is conducted on the floor, after which the carcass is hoisted for skinning. Butcher-owners of slaughter cattle handle the dressing and the carcass preparation before delivery to their shops. Veterinarians inspect the carcasses, but any they condemn just disappear.

CATTLE DISEASES

Disease problems are not handled systematically. The cattle of South Vietnam depend principally on their own developed resistance to disease for survival. Foot-and-mouth disease is endemic. Vaccines for rinderpest, which has been rampant for several years, and hemorrhagic septicemia are prepared in a government laboratory and administered by the Veterinary Services, but there is no adequate control.

GOVERNMENT AND CATTLE

The National Directorate of Animal Husbandry of the Ministry of Agriculture is making an attempt to rehabilitate the former French experiment stations, a difficult task during a war. Ben-Cat Experimental Dairy Farm, thirty-five miles from Saigon, was established in 1956 as an Australian aid project. Improved pastures were developed, and modern dairying facilities—including milking machines, cooling equipment, and pasteurizer—were provided. One hundred-fourteen Jersey cows and ten bulls were imported from Australia to demonstrate modern dairy farm methods in the hope that Ben-Cat could become the center of a dairy community of small farmers, each having a herd with as many as ten cows. It was all destroyed by the Viet Cong in 1967.

OUTLOOK FOR CATTLE

South Vietnam needs a draft animal and a milk cow. The Zebu, the Chinese Yellow, and the swamp buffalo are well adapted to the environment and are useful for these purposes. The cattle could be improved by artificial insemination, but this is hardly practical under the conditions of ownership which prevail—small farmers with only a few animals. The buffalo is a very efficient work animal, superior to cattle in every way except in the salvage value for meat. In the highlands meat-type animals could be grown, but any major development along this line will have to come after the war ends.

For milk production Nili and Murrah buffaloes or milk-type Zebus such as Red Sindhis would give better results than the European breeds. The few commercial dairymen near Saigon are proceeding along this line, but the limited experiment-station work is with exotic breeds ill-adapted to the tropical climate. Time will show that the Indian with his buffaloes and Zebu cows was on the right track.

Thailand

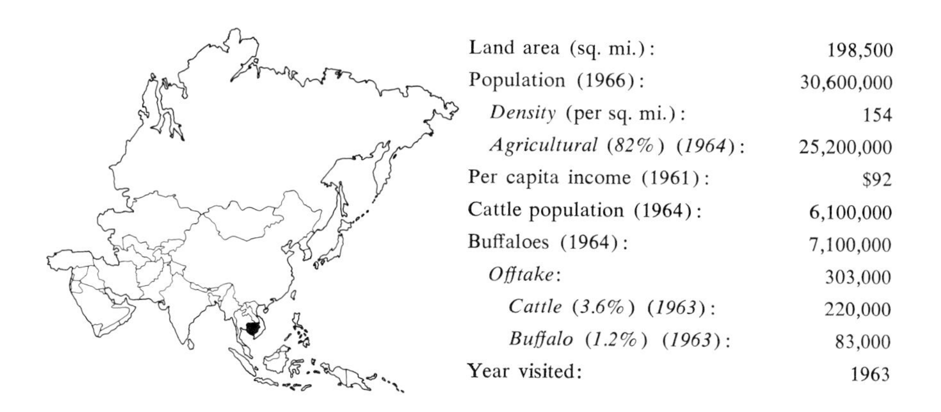

Land area (sq. mi.):	198,500
Population (1966):	30,600,000
Density (per sq. mi.):	154
Agricultural (82%) (1964):	25,200,000
Per capita income (1961):	$92
Cattle population (1964):	6,100,000
Buffaloes (1964):	7,100,000
Offtake:	303,000
Cattle (3.6%) (1963):	220,000
Buffalo (1.2%) (1963):	83,000
Year visited:	1963

THE KINGDOM OF THAILAND (Siam until 1949) occupies the east-central part of the Indochinese peninsula and lies in the same general latitude as Central America and southern Mexico. Included within its boundaries is a long, narrow strip of the Malay Peninsula north of Malaysia, with the Andaman Sea on the west and the Gulf of Siam on the east. The other borders are with Burma on the west and northwest, Laos on the north and northeast, and Cambodia on the south and southeast. The land area is nearly as large as that of Florida, Alabama, Mississippi, and Louisiana combined; the population is more than double that of these states.

The ancestors of the Thai began migrating to Siam 1,000 years ago from their independent kingdom in southern China. The population is now largely Thai, although a substantial number of Chinese live in the cities. Some Moslems from Malaysia live in the south.

The Portuguese at the beginning of the sixteenth century were the first Europeans to establish contact with the area. They were supplanted by the Dutch in the seventeenth century; the French and the British also began establishing trade relations, and there was some trade with the Japanese. Siam fought off the British when they became dominant in the area and maintained its independence throughout the colonizing days of the European powers. It was an absolute monarchy until a constitutional form of government was adopted in 1932. During World War II the country, defenseless before the oncoming Japanese, declared war on the United States and Britain. When freed from domination by Japan after the war, internal politics obviated any continuity in government and military coups were frequent until 1958. Since then law and order, with the army usually having a strong hand in government, have been maintained except for some Communist terrorist action following Thailand's token participation in the war in Vietnam.

The economy is basically agricultural, with paddy rice being the primary crop. A north-south mountainous ridge divides the country into a western area, which drains from north to south to the Gulf of Siam, and an eastern area, which drains generally east to the Mekong River, the border with Cambodia. Rainfall averages 50 inches, varying from 30 inches to 100 inches in the tropical monsoon climate; southwest winds from May to October and northeast winds from October to February bring rain before the hot, dry period, from March to May. Much of the mountain area is forest, and the long, thin strip of land stretching down to Malaysia is mostly rain forest.

Both buffaloes and cattle are used almost exclusively for draft, cultivation, and farm transport. Buffaloes slightly outnumber cattle and are a more efficient animal in working the paddies. Farming is of the village type, with small operators cultivating five to ten acres and usually owning 3 or 4 head of buffaloes and cattle.

Buffaloes are the swamp type found throughout Southeast Asia. They are highly resistant to the endemic diseases, withstand well the tropical climate, and are efficient users of the coarse, low-protein feed available.

The cattle of Thailand are mostly a nondescript, humped animal,

Swamp buffalo.

varied in color and usually having rather short, thick, and upturned horns. Only an exceptional mature animal weighs as much as 800 pounds. Although tolerant of the endemic diseases and resistant to heat stress, they do not maintain themselves as well on the low-quality feed as the buffaloes.

Thai cattle in the country.

ABATTOIRS

A Western-type abattoir, built by United States AID in Bangkok, was provided with modern Danish equipment and has cooling facilities. In 1963 the plant was killing 300 head of buffaloes and cattle, 1,000 hogs, and 5,000 chickens a day. Meat for the European hotels is aged a week to ten days, an uncommon practice in Southeast Asia. The average liveweight of cattle slaughtered is 550 pounds, that of buffaloes, 1,000 pounds. Neither female cattle nor buffaloes can be legally killed unless certified by a veterinarian as being incapable of reproduction. The kill consists largely of worn-out work animals. Useless females are slaughtered clandestinely in the country for village consumption.

CATTLE DISEASES

Many cattle diseases are endemic, the most serious of which are foot-and-mouth disease and anthrax. They are treated by vaccination in areas where outbreaks occur. Anaplasmosis is prevalent throughout the country to the extent that the native cattle develop a calfhood-acquired immunity and are not otherwise affected; exotic breeds, however, are soon decimated by this disease unless extreme precautions are taken.

GOVERNMENT AND CATTLE

At the experiment station in the Pakchong area, cross breeding has been undertaken to develop a climatically adapted milk animal. In 1963 this procedure consisted of breeding Red Sindhi and Swiss Brown bulls to native cows. The Red Sindhi cross was outperforming the Swiss Brown, as would be expected because of the Zebu's adaptation to the tropical environment.

The husbandry practices at the station provide an informative demonstration of the difference in the ability between buffaloes and cattle and between Zebus and European cattle to maintain themselves in a hot climate on coarse feed. At the end of the dry season, only mature tropical grasses are left in the pastures. Buffaloes, purebred Swiss Brown, Red Sindhi, and native cows, along with both Swiss Brown and Red Sindhi crosses on native cows, are kept on the same pasture where ample mature growth is left standing. The buffaloes are in excellent condition; the native cows, Red Sindhi, and Red Sindhi crosses are in fair condition. The purebred Swiss Browns can only be described as emaciated; the condition of the Swiss Brown crosses is somewhat better but not nearly as good as that of the Red Sindhi crosses.

An attempt was being made to employ artificial insemination, utilizing fresh semen, in the breeding program but results had not been very satisfactory. The best conception rate had resulted when fresh coconut milk was used as a diluent.

Twenty miles up the road from this government station, the Danes have established a demonstration dairy with an imported herd of 100 purebred Red Danish cows. Anaplasmosis had decimated the first importations, and the herd was then maintained under practically clinical conditions with thorough prophylactic treatment against endemic diseases. Milk production averaged thirty-five pounds a head. A balanced grain ration was fed and the cows were maintained on improved sprinkler-irrigated pastures. The Thai farmer could never hope to duplicate these practices, however. Although an interesting demonstration of how exotic cattle can be maintained in productive condition in a highly unfavorable environment, the Danish dairy did not have the practical aspects of the government experiment station with its crossbred native cows.

OUTLOOK FOR CATTLE

Thailand is predominantly Buddhist, so beef is not an important article of diet. Milk is in greater demand—and scarce. The effort to develop an acclimated milk animal could eventually lead to the establishment of sizable numbers of small dairies near the villages. There should be room here for the Murrah or the Nili buffalo of India, too. Other than for milk the major role of both cattle and buffaloes in the economy will be the same in the future as it has been in the past—to furnish the draft power for agriculture.

Far East

China (Mainland)

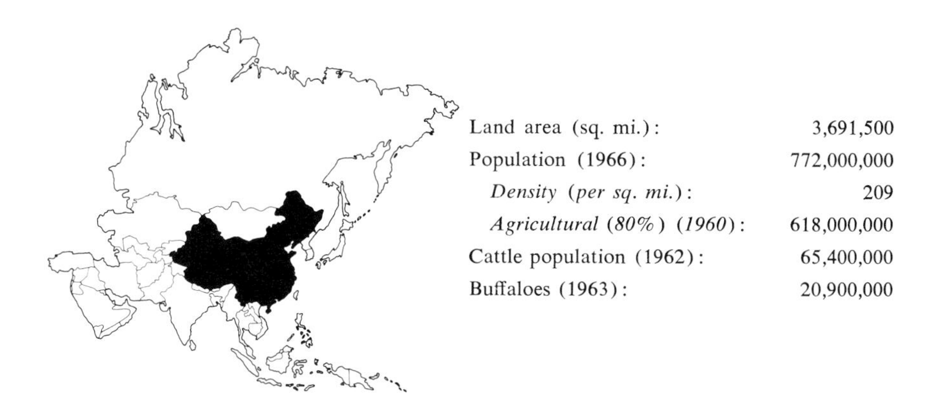

Land area (sq. mi.):	3,691,500
Population (1966):	772,000,000
Density (per sq. mi.):	209
Agricultural (80%) (1960):	618,000,000
Cattle population (1962):	65,400,000
Buffaloes (1963):	20,900,000

THE PEOPLE'S REPUBLIC OF CHINA was proclaimed in 1949 by the Chinese Communists after the four years of civil strife which followed World War II. In area the country is equivalent to the United States (Alaska excluded) and the northern three-fourths of Mexico and lies within the same latitudes. Highly irregular in shape, China borders the Soviet Union and the Mongolian People's Republic in the north; in the south the borders are with North Vietnam, Laos, Burma, Bhutan, Tibet, Nepal, and India. North Korea and the waters of the Yellow Sea, the East China Sea, and the South China Sea are on the east; and Pakistan, Afghanistan, the Soviet Union, and the disputed regions of Jammu and Kashmir are on the west.

Much of China is rough, mountainous country, and there is considerable desert area in the west. The extreme south is tropical, but in the north and on the high Tibetan plateau the growing season is short and winters are severe. Estimates of the cultivated land vary from slightly more than 10 per cent to just under 20 per cent. In the mountainous and arid regions, the country is sparsely populated; but in the southeast and the better agricultural areas, there are localities where the population density is as high as that of any other place in the world.

The antiquity of Chinese culture approaches that of Mesopotamia and, at China's peak, its accomplishments were certainly as great. The Chinese civilization which eventually developed, however, appears to have included a basic incompatibility with the Western culture which emerged from the ancient Mediterranean empires. Until the advent of the current Communist regime, no influence of the Western world, political or military, made any notable impact on the Chinese way of life. The current Communist control of all human activity has effected some drastic changes, but it is still to be determined how permanent these alterations will be and the direction they will follow.

China has traditionally been a country of small farmers; in 1943 their individual holdings were three to six acres in extent. Following World War II, the first step of the Communist government affecting agriculture was to distribute the few large estates in small holdings to the peasants. Inducing all the farmers to form small co-operative units for more efficient utilization of labor and whatever tools were available soon followed. The next step was uniting these co-operatives into the Soviet type of collective farm, with the original landowners maintaining a theoretical vested interest in the enterprise, in accordance with their contribution of land. The amount of this interest also determined the landowner's share of the income. In 1958, when these farms failed to meet established agricultural goals, the collectives, with an average membership of 200 families, were united into large communes in which the members approached 5,000 families. The initial vested rights of the individual farmer in his few acres was gone, and the state was in complete control of all agricultural activity.

Before the socialization of agriculture, the Chinese farmer depended entirely on his cattle or buffaloes for farm draft. In many parts of the country, dairy products were an important part of the diet. The heavy concentration of livestock is in the cultivated areas. The grasslands—some 40 per cent of the total land area—are utilized mostly for nomadic grazing or subsistence livestock raising and contribute little to the general food supply. There is an immense potential here for an increase in productivity if modern management methods were adapted.

Practically all agriculture except the nomadic raising of livestock is now under control of the agricultural communes, of which there were 26,000 in 1959. A great deal of government effort has been devoted to increasing agricultural production. Much of this work has been consumed in propaganda and planning, but there have been material accomplishments as well. From practically nothing twenty years ago, probably 4 per cent of the farming is now mechanized—official estimates claim as much as 15 per cent. There have been major additions to irrigation works, particularly in water storage, and sizable areas of virgin land have been brought into production. Chemical fertilizers are being produced, still in a small way; but even their limited use is an important addition to the meticulous saving of all animal and human wastes, which has been the Chinese practice from time immemorial.

The agricultural development which has occurred since World War II was largely engineered by technicians and scientists of the Soviet Union during the years before the political break with China. The state-type farm, which occupies a major part in the Soviet agricultural scheme, did not develop in China. A few were organized but their contribution to the total agricultural production has been minor. Incorporated in the commune as it now exists is a feature unknown to either the Soviet collective or the state farm, but rather somewhat like the diversified activities of the Israeli *kibbutz*. The communes have holdings in commercial and small industry enterprises and also enter into educational, cultural, and military activities. China experienced the same growing pains of socialized agriculture as did the Soviet Union, and her progress has been even slower. The nucleus of technical personnel essential to fulfill the management function was much smaller in China, and the education and training of persons for this work was very much slower. From a long-range point of view, however, there can be no question but that the present commune type of agricultural organization will produce much larger quantities of food for the Chinese people than the small farms of the peasant ever could have done.

The following discussion of Chinese cattle is based on the publication *The Livestock of China*, by Dr. Ralph W. Phillips and his associates, who visited China in 1943 to observe the livestock there.[1] (Annually repeated efforts of the author to obtain an invitation from the Chinese government to observe the cattle there were unavailing.)

CATTLE BREEDS

Descendants of three branches of the Bovini tribe were present in China in 1943—the Yellow Cattle from the Bovine, the buffaloes from the Bubalina, and the yaks from the Poephagus. The yaks were indigenous to the areas in which they were found, but where the cattle and buffaloes originated is not known.

Chinese Yellow.—Before the Communist regime assumed control, all cattle in China were called Yellow Cattle (and possibly still are), as distinguished from the buffaloes and the yaks. There appeared to be at least three common types, or possibly breeds, of cattle in the country, however. The most important of these was the Chinese Yellow, which at times still finds its way into the Hong Kong abattoir and is seen on Taiwan, where it has been carried from the mainland during the past 300 years. This breed was found in central and southern China and used

[1] Ralph W. Phillips, Ray G. Johnson, and Raymond T. Moyer, *The Livestock of China*.

Chinese Yellow bull. Photographed in east-central China. Courtesy H. Epstein

Chinese Yellow cow. Photographed in east-central China. Courtesy H. Epstein

primarily for draft. It varied considerably in size but was essentially the same animal as that described in the chapter on China (Taiwan).

Mongolian Cattle.—The characteristics of Mongolian Cattle were possibly not sufficiently fixed to warrant classification as a breed. They were definitely nonhumped and were small—mature animals described as weighing approximately 600 pounds but having much variation in size. Their color was quite varied—nearly solid black, brown, or yellow, but often with some white spotting. The horns were slender and thrust up and forward. These animals have been used for both draft and milk and are said to be better milk producers than the Chinese Yellow.

They were found scattered over most of northern and western China.

Tibetan Cattle.—This is a dwarf type of nonhumped cattle found on the Tibetan plateau. Color varies widely. In 1943 they were used as pack and draft animals and milk was taken, the average production being given as 260 pounds for a 105-day lactation period.

Buffalo.—In southeastern China the swamp buffalo was the principal draft animal as it is in many areas of Asia where paddy rice is grown. It was probably brought to China from India or the Indochinese peninsula

Mongolian bull. Photographed in northern China. Courtesy H. Epstein

Mongolian cow. Photographed in northern China. Courtesy H. Epstein

many centuries ago. The size was more variable than in most other areas, being said to range from 800 to 1,400 pounds, probably as influenced by the nutrition level. The color was black or slate, although occasionally

Tibetan cow. Courtesy H. Epstein

tan and, in rare cases, white animals were seen. The horns were typically flat, wide, and backswept with a nearly rectangular cross section, as they are generally seen on these animals in the Far East. The buffalo was

Buffalo bull. Photographed in southeastern China. Courtesy H. Epstein

Light-colored buffalo cow. Photographed in southeastern China. Courtesy H. Epstein

Black buffalo cow. Courtesy H. Epstein

used primarily for draft in the paddy areas, and infrequently milk was taken.

Yak.—This animal is indigenous to both the Tibetan plateau in the southwest of China and the Mongolian plateau in the north, and is the same animal as described in the chapter on Nepal. It has been domesticated for many centuries in both areas and, in the remoter parts, is still present in the wild state.

The domesticated female weighs 500 pounds and stands forty inches high; males are much larger. Color varies widely—black, dark brown, black with large white patches, bluish roan, and sometimes grey and white. The yak looks shaggy because of its long hair; even the tail has a bushy appearance. The horns are of medium length, upthrust, and thin; but polled animals are not infrequent. In the Mongolian People's Republic horned bulls are said to have been castrated and a polled variety developed. Yaks have been widely used as pack animals for draft and also for human transport. Milk is taken for household use, but the yield is very small.

Pien Niu.—This cross between the yak and either the Yellow or the Mongolian Cattle is found in the regions populated by the yak. The usual cross is the cattle bull on the yak cow, but the reverse cross is also a Pien

Niu. The hybrid is a more efficient draft and milk animal, about 20 per cent larger in size than either of the parent breeds. The local belief that males of the cross are sterile is so widespread that it probably can be considered factual. Females, however, breed regularly and are bred back to either cattle or the yak. (Males of the yak-Zebu cross in Nepal are sterile and the female fertile.) Back crosses, beyond the second generation at least, are considered to produce inferior offspring. The hybrid vigor of the first cross is most desired in the Pein Niu and is widely used as a draft and pack animal as well as for milk. (This development also parallels the practice in Nepal, where the Chowri, the yak-Zebu first cross, is raised for milk but continued crossing has been found undesirable.)

The varied color of the Pien Niu is much like that of the yak. The hair is not as long as that of the yak but is longer than that of either the Yellow or the Mongolian Cattle. In general the cross is intermediate in most characteristics between the two parents.

Pien Niu cows. Courtesy Ralph W. Phillips

Exotic Breeds.—In the period immediately following World War II, very few exotic cattle had been introduced into China. In 1936 there were about 300 head of the European milk breeds, mostly Holstein-Friesian, but practically all of them were killed by the Japanese during their period of occupation in World War II. These purebred animals were maintained in both university herds and commercial dairies. There was considerable interest at the time in upgrading the Chinese Yellow by the use of Holstein bulls; this breeding, carried to the third generation, gave eminently satisfactory results, with the yield of milk becoming markedly improved, although the butterfat content decreased. The animals did well in the Chungking area of south-central China but had better than average care in the university herds.

MANAGEMENT PRACTICES

All cattle—the cattle types, buffaloes, yaks, and Pien Niu—were mostly held by small farmers and by people engaged in transport. The primary use of all types of cattle was for draft purposes (in the case of the yak and the Pien Niu, largely for pack animals), and secondarily for milk. Generally only worn-out work and otherwise useless or unwanted animals were slaughtered for food.

Except for the small dairies near the cities, cattle for the most part were grazed under the care of herdsmen. In some areas a night guard was necessary for protection from theft. No hay was put up for winter feed, although rice straw and millet stalks were sometimes used. During the heavy work season, animals being utilized for draft were fed some grain if conditions permitted. Dairy animals were usually maintained under cover and feed—hand-cut grass or other roughage during the growing season and rice straw or millet stalks in the winter—was brought to them. A supplemental grain ration was generally, though sparingly, fed, the kind depending on availability in the area—corn, horse beans, barley, millet, various brans, and oil cakes. Some of the better-managed dairies fed supplemental grain in proportion to a cow's production, about three-fourths pound a day for each pound of milk. Some milk was taken from buffaloes in the areas where they were held, but this practice was not widespread.

In the areas of the heaviest population concentration, mainly in the southeast part of the country, milk was not often used as food except in the cities. In the rest of the country, many cattle were milked regularly, and the milk, butter, and casein were important articles of diet. Before World War II there were a few modern dairies with inspection and pasteurization facilities near some of the larger cities. Watering milk, however, was common practice, and it was customary to boil milk before using.

The Chinese in most respects took good care of their cattle, at least to the extent the availability of feed permitted. The small farmer would house his 1 or 2 animals in his own dwelling, perhaps sleeping above them for warmth, or maintain them in an adjacent shed. Manure was scrupulously saved for use as fertilizer. The slatted floor to permit the accumulation of manure underneath and pits with drainage from the stalls for the collection of urine were sometimes employed. The major obstacle in cattle raising was the shortage of winter feed. Arable land was at too much of a premium for the production of food for human consumption to permit growing roughage crops to be put up as hay or silage for winter feed. All straw was saved and used but there was not enough of this in much of the area where livestock was raised to supply the need. As a result many cattle had to survive the winter by grazing on whatever mature dry growth could be found. A heavy snow or prolonged cold spell caused serious losses. In mountainous regions it was the practice to take the animals not being used to the high country for summer grazing and to return them to the village for the winter.

Not much attention was paid to breeding practices. In the areas in which cattle were grazed, bulls ran with the cows the year round. There was probably some degree of selection here since draft animals are ordinarily castrated, but it is impossible to say in what direction such selection of breeding animals was headed. A bull might have been left whole because of his color or he could have been picked as a draft animal and castrated because of his size. The village owner of 1 or 2 cows took them to a bull that was maintained in the village and paid a fee for the breeding service. It was a rather common practice to feed a bull two eggs at each service, and a cow was sometimes given wine after service because this practice was considered an aid to conception.

MARKETING

There were regular cattle markets in the growing areas, usually one for each county. Cattle sales peaked in the spring, when the offering was mostly cows and calves, and again in the fall, when more bulls and oxen were sold. In the growing areas such markets were held every few days. On a representative day as many as 100 head might be sold.

Sales were arranged by a middleman who also served as a tax collector. He conducted the bargaining between the buyer and the seller, collecting about 5 per cent of the amount involved as tax and commission. There was no fixed percentage on the sale. The middleman paid an annual tax for doing business and pocketed whatever remained of the 5 per cent from the transactions he had negotiated during the year. Because the bargaining was usually conducted "in the sleeve," often the buyer did not know what he was paying and the seller did not know what he was getting until after the deal had been made. Some private sales outside the regular markets usually involved larger transactions.

The above discussion on cattle in China is based on observations made twenty-six years ago. During that period the human population has increased from 440,000,000 to 770,000,000, according to available estimates. Total cattle and buffalo population was estimated at 86,300,000 for 1962–63. On the basis of these numbers, China would

Draft ox. Photographed in northern China. Courtesy H. Epstein

be the fourth largest holder of cattle in the world, exceeded only by India, the United States of America, and the Soviet Union. Little of what is being done with livestock is known to the Western world. Whatever his fortunes have been, however, man has always kept his cattle with him since they were first domesticated; and it would be surprising indeed if the cattle of China today are greatly changed from those seen by Dr. Phillips and his associates in 1943.

China (Taiwan)

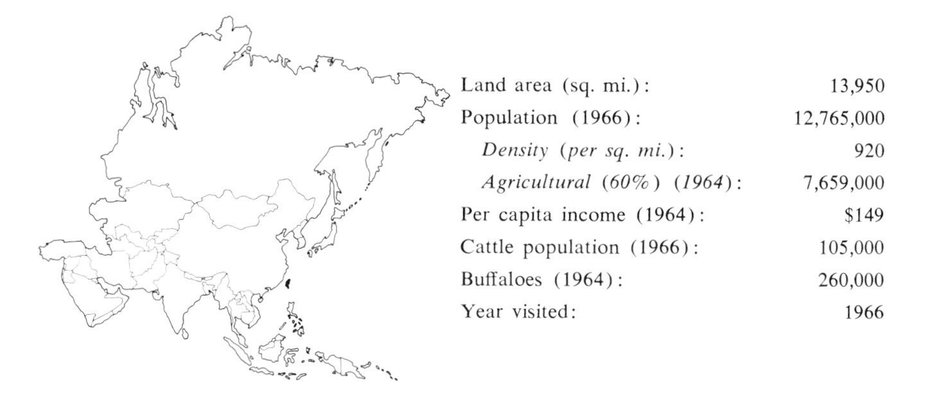

Land area (sq. mi.):	13,950
Population (1966):	12,765,000
Density (per sq. mi.):	920
Agricultural (60%) (1964):	7,659,000
Per capita income (1964):	$149
Cattle population (1966):	105,000
Buffaloes (1964):	260,000
Year visited:	1966

THE REPUBLIC OF CHINA occupies the island in the western Pacific, southeast of mainland China, formerly called Formosa (meaning "Beautiful"). Named by the Portuguese, who, coming in the sixteenth century, were the first of several waves of foreign occupiers, including both the Spanish and the Dutch, the island was Chinese territory until ceded to Japan when the Sino-Japanese War ended, at the close of the nineteenth century. At the end of World War II the island was returned to China. When the Chinese Communists overran the mainland, the Nationalists took refuge on Formosa and have remained there. "Formosa" is much more descriptive of the lush coastal plains and the high mountains they surround than "Taiwan," the Chinese word for "High Plateau," which is now used. Not quite as large as the southern third of Louisiana, Taiwan is now the homeland of more than 12,000,000 persons, the majority of whom are descendants of emigrants from the Chinese mainland who had moved in over the past three centuries; but the upper crust of the population stems from the 1,500,000 who came over en masse in 1949, when the government of China moved to Taipei, the capital.

Taiwan exhibits one of the most productive examples of intensive farming in the world. All of the 2,600,000 acres capable of cultivation grow two to four crops a year because the farmers make full use of the annual rainfall of 70 inches or more, supplemented by irrigation during dry months, and stretch the growing season from twelve to fourteen months. This unique practice is accomplished by interplanting such crops as soybeans, tobacco, and sweet potatoes between the rows of the two rice crops so that after the rice is harvested, the intermediate crop is already established. Rice and sugar are the two main crops.

The political climate is something to marvel at. Under constant threat of invasion by the Communist Chinese on the mainland (only ninety miles off the western coast), Taiwan enjoys as peaceful an atmosphere as any other place in the Far East. The overshadowing military power of the United States, coupled with what United States AID has accomplished, is obviously responsible. Taiwan is the outstanding example of what fantastic expenditures for development of an army, industry, and agriculture in a foreign country can produce in the way of tangible results while gaining the cooperation and the good will of the recipients. There is some grumbling about keeping the cream of the country's manhood, 600,000 men, in the army; but with the countless hordes of Communist China just off the coast, intensely jealous of what Taiwan enjoys and outnumbering them 60 to 1, Taiwan's action can hardly be called unduly militaristic. The thought is still general that the day will come in the not-distant future when the mainland will be regained. This dream does not seem very realistic, with fewer than 13,000,000 people on Taiwan and 700,000,000 on the mainland; but there is no question that the idea persists at all levels of the people.

CATTLE BREEDS

The backbone of the highly developed agriculture in Taiwan is two remarkable draft animals, the Chinese Yellow and the swamp buffalo. They are worked by the industrious Chinese farmer to obtain a marvelous utilization of the land. In recent years exotic dairy cattle and some of the beef breeds have been introduced. While of only minor economic importance at present, these animals appear to be on the increase.

Chinese Yellow.—This is the local name for the indigenous cattle of the mainland. The distinctive breed of China, the Chinese Yellow is said to have been first brought from mainland China in the sixth century, when a few Chinese first moved to the island. Only in the last three centuries has the breed been brought to Taiwan in large numbers.

Of medium size, an average cow in good condition weighs 900 pounds, a bull 1,200 pounds. The very uniform rather yellow-tan color is somewhat similar to that of the Jersey. The shoulders and forequarters are particularly well muscled; the hindquarters are rather narrow but heavily boned. The horns are characteristically short and thick. Though the Chinese Yellow is definitely a humped animal, its hump is smaller than, and not as sharply defined as, that of the Indian Zebu breeds; in

Chinese Yellow draft ox. Note the decorations on the halter.

Chinese Yellow cow.

conformation, however, the Chinese Yellow is in no way suggestive of the Indian cattle. Males are used for draft, for both field work and transport. The Chinese Yellow is a very docile and willing worker.

The progenitors of the Chinese Yellow were an ancient mixture of the humped cattle of India and a nonhumped race, which probably met at some time in the unrecorded past on migrations of the Zebu into Southeast Asia. These nonhumped cattle quite possibly trace to the progenitors of the Mongolian and the Tibetan cattle, both of which are humpless, seen by Dr. Phillips on Mainland China in 1945.

Black Cattle.—In southern Taiwan small cattle of a type which does not appear to be interbred with the Chinese Yellow are seen. They are uniformly black and have a pronounced, though small, hump. The bull and the cow are approximately the same size, a mature animal weighing not over 500 pounds. Conformation is inferior to that of the Chinese Yellow. Both sexes are used for draft but do not equal the Chinese Yellow for this purpose because of their small size.

This animal is quite similar in appearance to the small, black cattle that occasionally find their way to the Manila abattoir from the northern Philippines.

Small black humped ox of southern Taiwan.

Exotic Breeds.—Red Sindhi and Santa Gertrudis cattle have been imported with the objective of upgrading the Chinese Yellow to a better beef type. This activity has been confined to experiment-station work. The limited number of dairy cattle are Holstein-Friesians, the base stock of which was imported from the United States, Australia, and Japan.

Buffalo.—The swamp buffaloes of Southeast Asia are used for draft, only the worn-out and the useless being sold for slaughter. Some Murrah buffaloes from India have been imported to try to improve milk production, but there has been no sizable development along this line.

Swamp buffalo bull.

MANAGEMENT PRACTICES

The Chinese farmer loves his cattle, and he and his family tend them carefully in the small shed behind their dwelling in the village. The animals are well fed and bedded and are maintained in excellent condition. Everything that grows is utilized. When they harvest peanuts and sweet potatoes, the farmers bring in all the vines to feed their animals. Sugar cane leaves are stripped and used for feed, bedding, and fuel. Any feed not eaten is taken back to the land, along with the manure and the used bedding. Cows are kept mainly in the back country and are used by the owner for draft as well as for breeding. Milk is not generally taken. It is illegal to kill a work animal less than fourteen years of age.

Buffalo teams are usually hitched two in line for heavy hauling, and often a team of a Chinese Yellow bull and a buffalo is so seen hauling sugar cane from field to mill. Buffaloes are given much the same attention as draft cattle and are maintained in good condition.

A persistent effort is being made to develop a dairy industry. The population of all milk breeds in 1966 was nearly 6,000 head, almost entirely of the Holstein-Friesian breed. The Kaohsiung Dairy, the largest

Two-in-line buffalo draft team.

in the country, is located near the town of that name, in southern Taiwan, and is operated by a commercial firm which sells bottled milk and uses milk in its candy manufacturing business. The herd consists of 180 excellent Holstein-Friesian cows, 70 of which were imported from the United States and 20 from Japan. The other 90 were raised at the dairy. The farm itself occupies five acres. The young stock is raised on a nearby farm of twelve and a half acres. Depending on what is in season, feed is purchased from small neighboring farms—sweet potato vines, cane tops, rutabagas, cut grasses, rice bran, or corn. The diet is supplemented as needed with green chop from the farm's own small meadows of elephant grass, which is cut every twenty days. Silage is also put up by hand for use when green chop, purchased or grown, is not available.

Cows are milked in barns holding 60 head each and turned into a yard with a large shed roof for protection from the sun. Milking is three times a day, one man handling 20 cows. The average yield for a lactation is 8,000 pounds. Breeding is by artificial insemination with frozen semen imported from the United States, although two good Holstein bulls are maintained for emergency use.

Artificial insemination was introduced on Taiwan sometime during the period of Japanese occupation. Not much progress in its use has been made except in the dairy herds. The only center for the collection and distribution of semen is at the Central Research Station, at Hsinhua. Three Holstein bulls, two Santa Gertrudis, and one buffalo are kept in the stud. If liquid nitrogen is not available, the dangerous practice of storing frozen semen in liquid oxygen is resorted to. Carrier pigeons sometimes deliver ampules of semen to outlying districts. The Japanese introduced the speculum method of insemination, which in Japan is still adhered to; but three years ago Taiwan changed to the rectal method. Most of the dairy cows in the country are now being bred artificially—some 1,500 head annually.

MARKETING

Cattle and buffaloes for slaughter are purchased in the country by traders, called *tsaichs*, who truck them to small holding yards behind their dwellings in Taipei. These animals are invariably no longer useful for work or milking. An old buffalo in very poor condition and weighing 1,000 pounds brings $175; a Chinese Yellow bull in fair condition, weighing 900 pounds, sells for $200; and in 1966 steers two and one-half to three years of age, that were sold before the 300-mile haul to Taipei by the experiment station at Hengchun, were bringing 20 cents a pound.

Chinese Yellow bull that has been sold for slaughter.

The country cattle markets, held at central towns in the cultivated areas on fixed days of the week, consist almost entirely of transactions between farmers for draft animals, both cattle and buffaloes. A cart, with the front and rear wheels tied together so that they will not turn, is provided for testing the pulling ability and disposition of an animal. Only an occasional beast no longer useful for work is sold for slaughter. At the Kangshan market, held three times a week, a young Chinese Yellow draft bull in good condition and weighing 800 pounds sells for $275; a buffalo, also in good shape and weighing 1,400 pounds, brings about the same price. The value placed on a good draft animal by the local farmers can be gauged by the wage paid for farm labor, $24 a month.

Cart for testing pulling ability. Note that the front and rear wheels are roped together.

ABATTOIRS

The Taipei City Slaughterhouse is the only one in Taipei that kills cattle, although two others handle hogs. The facility is definitely unsanitary and inspection only perfunctory. All meat is delivered warm the day it is killed. The trader has his animal processed here and sells the carcass, or frequently only parts of it, to the butchershop owner. A modern new abattoir was being constructed in 1966 near Kaohsiung, the large seaport in the south, to have a capacity for 400 hogs but only 6 cattle or buffaloes daily. Beef (including buffalo meat) is not an important item in the diet of the Chinese.

CATTLE DISEASES

Disease and parasites are not a serious problem with the indigenous cattle and buffaloes, both of which have developed a reasonable degree of tolerance to them. If kept in close confinement and sprayed for ticks and biting insects, exotic cattle apparently do well but present a serious problem when pastured. Heat stress has a serious debilitating effect. Anaplasmosis and piroplasmosis are endemic and take a serious toll on exotic cattle. Unless these animals carry some Zebu influence, they deteriorate rapidly in condition without exceptional care. Foot-and-mouth disease and rinderpest are not present on Taiwan. The few large dairies vaccinate calves under six months of age against brucellosis. Most draft and milk cattle are sprayed for protection against ticks every ten to fifteen days.

GOVERNMENT AND CATTLE

Every square rod of level land on Taiwan is intensively farmed. Where the coastal plain merges with the central mountain ridge that runs the length of the island, the intervening rough, hilly land is little utilized. Experiment stations are being established to demonstrate what can be done to develop such areas for cattle raising by using improved pastures of pangola grass, elephant grass, and centrosema and other legumes.

The Central Research Station of the Livestock Research Institute at Hsinhua, near Tainan in southwestern Taiwan, is working on this type of development of marginal land and experimental breeding. The station was originally established in 1910 by the Japanese during their period of occupancy, but apparently not much was accomplished. Red Sindhi and Holstein bulls are being crossed on Chinese Yellow cows in an effort to develop a milk animal adapted to the climate. The Red Sindhi, well acclimated to tropical conditions, is producing a cross that appears to be superior to the Holstein. This illustrates the ability of the Zebu, lacking in the European breeds, to withstand the heat stress of the tropics. Regrettably, very little has yet been done to improve the Chinese Yellow cattle by selection within the breed.

At the Hengchun experiment station, at the extreme southern tip of the island, development of marginal land for beef cattle production has been undertaken. The farm consists of 2,750 acres of rough, hilly country, most of it too rugged for cropping. This region has the heaviest rainfall on the island, more than 80 inches, but it is in a poor pattern and there is a seven-month dry season. Three hundred eighty-five acres have been put into improved pangola grass and centrosema pastures and carry 265 cattle the year round.

Santa Gertrudis and Red Sindhi bulls are being used on Chinese Yellow cows. At two and one-half years of age, when the Santa Gertrudis–cross steers are marketed, the average weight is 620 pounds; the Red Sindhi at three years of age weighs 595 pounds. These weights

First cross of a Red Sindhi bull on a Chinese Yellow cow.

First cross of a Holstein-Friesian bull on a Chinese Yellow cow.

may appear light, but the native Chinese Yellow cattle in the region where the foundation stock for the herd was obtained are a smaller animal than that seen on the rest of Taiwan, probably because of under-nourishment for many generations during the long dry season.

An improved draft animal, the first cross of a Kankrej bull, imported from India, on a Chinese Yellow cow is seen at Hengchun. The Kankrej (or Guzerat) is considered one of the best draft animals in India.

OUTLOOK FOR CATTLE

The program started in 1959 for developing marginal lands for cattle raising is an intensive adaptation of the basic principle followed in many countries where large numbers of cattle are produced and often grown out in the less productive areas. It has demonstrated that good beef cattle can be grown out on some of the least favorable marginal lands between the coastal plain and the mountains and that dairy breeds can be satisfactorily maintained on the same type of terrain where there is a better rainfall pattern. The major obstacle to development of these marginal lands for use by cattle is the cultural attitude of the Chinese regarding land ownership. Land titles are a jealously guarded right, even having some religious aspects to the Chinese. The country is in the final stage of a successful land-reform program. Between 1949 and 1961 tenant-operated lands decreased from 39 to 14 per cent, a change

First cross of a Santa Gertrudis bull on a Chinese Yellow cow. Photographed at the Hengchun Experiment Station.

First cross of a Red Sindhi bull on a Chinese Yellow cow. Photographed at the Hengchun Experiment Station.

effected without undue hardship on the large landowners, who were reimbursed partially by a share of the crops raised or by government land bonds and partially by stock in dividend-paying government-owned corporations. The present ownership pattern is now three acres to each farm family, most of whom have a vested right in their holdings. In the marginal land areas contemplated for cattle development, these small farms would have to be consolidated to set up workable units; and here the peasants' land rights would become involved.

In spite of the peaceful façade generally in evidence, Taiwan exists as a military government and will probably continue to do so for the foreseeable future. It is reasonable to anticipate that population pressure and the facts that no more cultivatable land is available and that the land which is cultivated is utilized to the limit of its productivity will force the government to develop the ways and means of raising cattle in these little-used areas.

Hong Kong

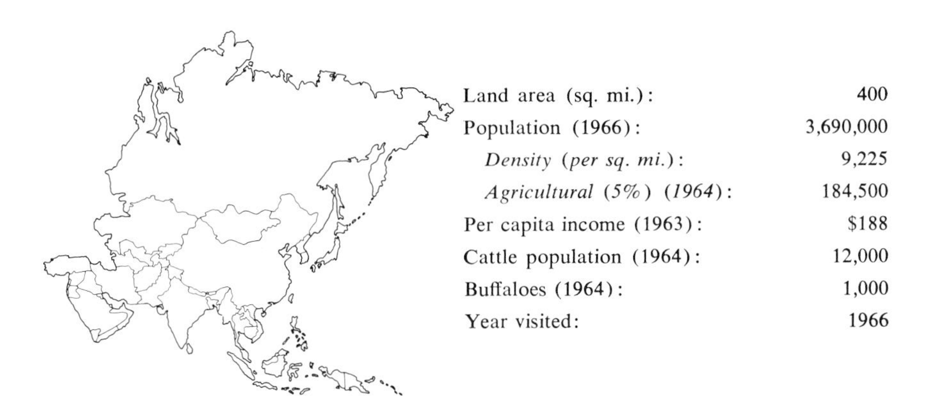

Land area (sq. mi.):	400
Population (1966):	3,690,000
Density (per sq. mi.):	9,225
Agricultural (5%) (1964):	184,500
Per capita income (1963):	$188
Cattle population (1964):	12,000
Buffaloes (1964):	1,000
Year visited:	1966

THE ISLAND OF HONG KONG, covering 29 square miles, was ceded by China to Britain in 1841. Nineteen years later Kowloon, 4 square miles of land on the Chinese mainland, was acquired. An adjoining 365 square miles of land, now known as the New Territories, were subsequently obtained on a lease basis, which expires in 1997. These areas, lying in the same latitude as Havana, Cuba, make up the British crown colony of Hong Kong, one of the world's greatest ports. The colony has the extreme population density of 9,225 inhabitants to the square mile.

Hong Kong was the first major acquisition of Japan in the early days of World War II. Restored to the British after the peace, it was the focus for a heavy influx of refugees from Communist China. The present population is nearly three times the prewar level—and more than 99 per cent Chinese. In 1958 there were fewer than 20,000 Europeans, more than three-fourths of whom were British.

Most of the land surface is rough and hilly granite outcrop, and only fifty-three square miles can be cultivated. This area is intensively farmed, some of it growing as many as six crops of vegetables. Cattle play a minor part in the economy; yet the number imported for slaughter exceed the kill in many small countries, and a unique dairy supplies most of the fresh milk.

MANAGEMENT PRACTICES

Some cattle are used for draft in the New Territories, where they are pastured in the rocky, hilly areas, which cannot be cultivated. They are small, very dark brown animals with a moderate hump. An average cow weighs 750 pounds, a bull 900 pounds. Except for color and size, they are quite similar to the Chinese Yellow, the breed from which they are probably derived.

The dairy of the Dairy Farm Ice and Cold Storage Company is lo-

Dark-brown bull. Photographed in the New Territories.

Dark-brown cow. Photographed in the New Territories.

cated on 250 acres of hilly land behind the city of Victoria, on Hong Kong island. Two thousand head of dairy stock, including thirteen hundred cows in milk, are maintained. All the land except that occupied by buildings is sowed to elephant grass on slopes as steep as 30 degrees. So well is this managed, however, that there is no sign of erosion, even though most of the 84-inch average annual rainfall falls between May

Hillside meadows of elephant grass. Photographed at the Dairy Farm Ice and Cold Storage Company.

and August. The nine milking units each handle an average of 145 cows. Land is at such a premium that one unit is two stories tall, 90 cows being handled on each level, and the management uses a twelve-story apartment building to house the farm workers. One milking unit has a herringbone milk parlor; the others are vacuum pipeline units serving stanchion stalls. Cows are kept in the barns except for a short daily sunshine-and-exercise period in a concrete paved yard. Even the young stock are kept on concrete in age groups, each with its own exercise yard.

Established in 1886 with the importation of 100 Holstein-Friesian cows from the West Coast of the United States, the farm has had its ups and downs. The herd had been radically reduced at the end of World War II. Various infusions of other milk breeds have occurred under different managements. About 30 per cent of the milking cows now are

Two-story dairy barn.

Ayrshires and Jerseys, the remainder being a mixture of Holstein-Friesians from the United States and Australia and Friesians from England and Holland.

All of the stock, from the week-old calves to the herd sires, is in good, thrifty condition, an excellent demonstration of how an exotic breed kept under carefully controlled conditions can be maintained over the years in a hot, unfavorable environment.

Breeding is accomplished by artificial insemination, using fresh semen which is kept for only a morning and an evening breeding—a maximum of twenty-seven hours. This procedure is said to have markedly improved the conception rate over the former practice of using semen stored for several days.

Selection for milk production has been followed consistently. Production averages 7,500 pounds for a 305-day lactation period. Milk, pasteurized and bottled, sells for 11 cents for each one-half imperial pint, the standard-size bottle. Except for that used by the carriage trade, most of Hong Kong's milk supply is imported as a powdered, condensed, or reconstituted product.

All male calves, except bulls being saved for breeding stock, are sold for veal, as are the culled females. Most of the heifer calves are saved for

In the background, a twelve-story apartment building housing 290 families.

replacements, however. All animals, even the purebred bulls, are de-horned.

All the roughage fed comes from 200 acres of elephant grass hillsides, reaped by hand at the proper cutting stage, trucked to the various barns, and usually fed to the animals without being chopped. Any form of machine harvesting is impossible on the steep and uneven meadows. Concentrates—barley, corn or maize, sorghum grain, and various kinds of oil cake—come from Cambodia, Thailand, Indonesia, and even Taiwan. The cows get some rice straw, brought in from Red China. Except at one operation which pumps sludge from one dairy unit onto the adjacent hillsides, manure is put back on the land by hand.

The good management everywhere in evidence at the Dairy Farm includes rigid disease-control measures. These precautions are aided by the complete isolation of the herd from other cattle. Although the farm has experienced outbreaks of foot-and-mouth disease in the past, it is now controlled by vaccination. Prophylactic treatment is also given for rinderpest and leptospirosis, and calfhood vaccination for brucellosis is practiced.

ABATTOIRS

The annual kill in Hong Kong is 160,000 cattle. The principal facility is aptly known locally as a slaughterhouse instead of as an abattoir, the usual British term. The government-maintained plant at Kennedy Town, on Hong Kong island, displays a sign "Smithfield" at the entrance, although the facility more nearly resembles a Nigerian slaughter slab than the impeccable establishment of this name in London. Built in 1894, it is badly overcrowded and the equipment is obsolete. A somewhat larger facility in Kowloon, also government operated, presents much the same picture. Both plants provide inspection for diseased animals, however.

These two government-operated facilities, run on a custom basis, are the principal sources of locally killed meat, although there are a few small private plants which are much better run. Two modern plants under construction were expected to be in operation in 1967.

Except for a few discarded dairy and work animals from the New Territories, the cattle for slaughter are imported, as are the hogs. The animals are brought by ship from the surrounding areas—wherever they can be obtained. In 1966 most of the imports, of which 60 per cent were buffaloes, were from Indonesia, Thailand, and Cambodia, with a few worn-out work animals coming from mainland China. All are brought in by importers, who sell direct to the butchers, who handle the killing and dressing themselves in the government plants, paying a fee for the use of the facilities. The dressed meat and offals, which include literally everything except the contents of the digestive tract, are sold by the butchers to the retail shop owners. Most of the cattle seen at the Kennedy Town holding yard are either the Chinese Yellow or the Bali cattle from Indonesia. The buffaloes are the swamp type from adjacent countries. Delivery of dressed meat is by truck as part of the slaughterhouse service.

The average price paid for live animals, based on estimated weight, was 30 cents a pound in 1966. The price for buffaloes was about 10 per cent less. Sales are transacted by bargaining on a per-head basis. The dressed carcasses as sold by the butcher bring about 55 cents a pound, buffalo carcasses again bringing 10 per cent less. Offals, a very appreciable item in the Chinese trade, are not included in this price. Most meat in the first-class restaurants is imported frozen, largely from China, and is of good quality. Formerly it came from New Zealand and Australia, but the stiff competition from mainland China has forced these countries out of the market.

Mainland China would seem to be the logical source of slaughter cattle for Hong Kong but does not appear to have any number of animals available. The long ocean trip in ships inadequately equipped for handling live cattle results in a serious loss in condition of the stock brought from Indonesia and the Indochinese peninsula. The grass-fattened Bali steers leave Bali in excellent shape but arrive in Hong Kong in a very gaunt condition from which they have no opportunity to recover before slaughter. The countries supplying slaughter cattle to Hong Kong need foreign exchange so desperately that they are forced to sell at whatever price is necessary to undercut the cost of New Zealand and Australian frozen and chilled beef. If the economy of the countries now supplying slaughter cattle improves and a source does not develop in mainland China, the normal trend will be toward purchasing chilled and frozen beef from the countries in a position to export it.

Japan

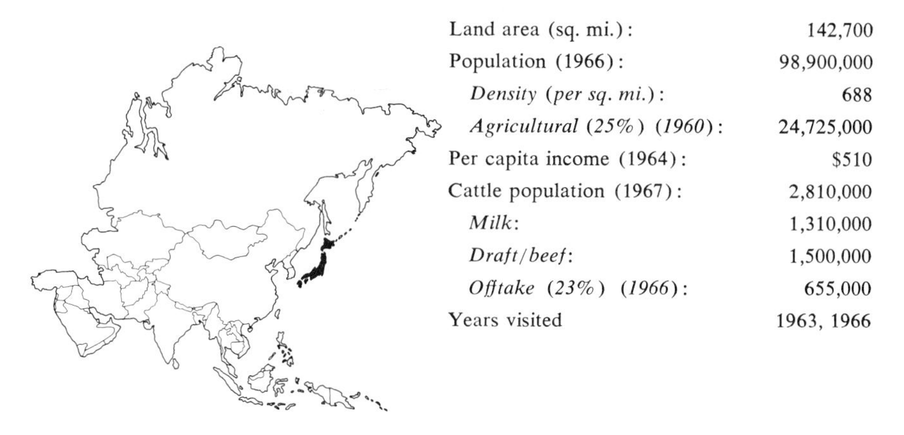

Land area (sq. mi.):	142,700
Population (1966):	98,900,000
Density (per sq. mi.):	688
Agricultural (25%) (1960):	24,725,000
Per capita income (1964):	$510
Cattle population (1967):	2,810,000
Milk:	1,310,000
Draft/beef:	1,500,000
Offtake (23%) (1966):	655,000
Years visited	1963, 1966

JAPAN is an arc of mountainous islands off the eastern coasts of Korea and the Soviet Union, lying within the same latitudes as those reaching from northern Florida to Maine. Four large islands and more than 1,000 small, outlying islands comprise the system. The total area is nine-tenths that of California, with a population that is half that of the United States. Nearly all the people live on or near the coastal plains of the large islands, where the arable land is only 15 per cent of the total area. The population density here exceeds that of the Nile Valley in Egypt. All the rest of the country is mountainous and rough terrain, much of it in forest.

Although some traces of the aborigines who inhabited the northern islands can be discovered in the present population, Japan was populated by migrations from the mainland. Very little mixture with other peoples or races has taken place. Very few foreigners other than diplomatic representatives, members of the United States military, and a small number of business people live in the country.

The Portuguese were probably the first Europeans to reach Japan, a shipwrecked crew taking refuge on one of the small islands in 1542. Christian missionaries soon followed, and considerable trade with the Portuguese was established. Later there was trading with the Dutch. These contacts were the first breaks in the centuries of total isolation and had little effect on the Japanese economy that was still dominated by a feudal system of government.

The contact of the United States with Japan, beginning in the middle of the nineteenth century, was the spark that initiated the development which finally led to modern Japan. Although this advancement was slow in gaining momentum, the economic growth since the end of the occupation that followed World War II has been fantastic. The people of both town and country are industrious to a degree found in very few countries today. This attribute, coupled with no serious drain on the country's resources for national defense purposes, accounts for the remarkable progress which has been made in the Western-type industrial development during the past twenty years. The benefit which Japan enjoys from the United States' assuming the responsibility for her national integrity is not generally appreciated.

In spite of the high degree of industrialization, one-fourth of the population is still engaged in agricultural pursuits. The rainfall is good for farming—40 inches or more in much of the country—and the drainage system from the mountains to the arable lands permits a widespread use of irrigation. Intensified farming is as meticulously practiced here as in any other place in the world. The average farm is two acres, and only 5 per cent are as large as ten acres. On the lowlands where irrigation can be practiced, frequently two crops are grown annually—always rice, with a second crop of grain or vegetables. This area accounts for half of the cultivated land. On the uplands, barley, wheat, and vegetables are grown.

Historically, cattle have been draft animals. Beef is a minor part of the Japanese diet, averaging less than three and one-half pounds annually for each person. It is, however, the base of the best-known national dish, sukiyaki. In the past few years there has been a remarkable increase in dairy cattle, principally the American Holstein-Friesian. Before 1966 the number of dairy cattle was approaching that of the draft-beef type. With the advent of the small hand tractor in recent years, there has been a sharp drop in the number of draft-beef cattle. This decrease has occasioned a shortage of slaughter animals and, coupled with a relatively low price for milk, is resulting in some liquidation of low-producing cows in the dairy herds.

CATTLE BREEDS

The native cattle of Japan are called Wagyu (Japanese Cattle), as distinguished from exotic breeds. The Japanese recognize three breeds of the Wagyu: the Japanese Black, the Japanese Brown, and the Japanese Polled.[1] None of these breeds are indigenous but, as is probably true of much of the animal life in Japan, their progenitors were brought to the islands in very early times. The logical origin of such importations would

1 Shoji Uesake, *Wagyu Cattle Breeding in Japan.*

have been from Korea, portions of which have been Japanese possessions at various times in the past. The early Japanese Black is described as a small, black animal, and pictures of it show a similarity to the Korean Black, described under "Cattle Breeds" in the chapter on Korea.

During the last decades of the nineteenth century, European breeds were imported to increase the size and draft ability of the early Japanese Cattle. There was no planned program for crossing when this was done. The principal breeds utilized at that time were the Devon, the Shorthorn, the Jersey, and the Guernsey. Later the government sponsored a program for cattle improvement, and Simmental, Swiss Brown, and Ayrshire were imported. The results of all this miscellaneous interbreeding were considered unsatisfactory, and in 1918 a program was initiated for selection in accordance with recognized standards accompanied by the registration of individuals that conformed to standards established for the Wagyu.

Each of the three Japanese breeds was differentiated from the somewhat mixed population. Although to an unknown degree they carry the influence of the European breeds which have been mentioned, this is probably quite minor since representatives of the exotic breeds introduced were few. The selection criteria, in addition to the other requirements, must have emphasized the black color for the Japanese Black, the brown color for the Japanese Brown, and the polled characteristic for the Japanese Polled, which is also black.

Following this development of the Wagyu, dairy breeds were imported and maintained in a pure state for milk production. More recently beef breeds have been introduced to a very limited extent.

Japanese Black.—This is the predominant breed of Wagyu, accounting for 80 per cent of the native cattle. It is a medium-sized animal, a mature cow weighing 950 pounds, a bull 1,400 pounds. Although developed primarily as a draft animal for work in the rice paddies, it has a very good beef conformation. It is a solid, dull black, sometimes slightly brownish, with short, upturned horns. An excellent draft animal, it fleshes well on feeding but is a poor milk producer, 2,500 pounds a lactation being considered high. Selection for type could conceivably produce an animal with the characteristics of the Japanese Black from the more primitive Korean Black.

Japanese Brown.—This breed accounts for nearly 20 per cent of the Wagyu population. It is about the same size as the Japanese Black and except for color is also reminiscent of the Korean Black. It is solid light

Japanese Black cow.

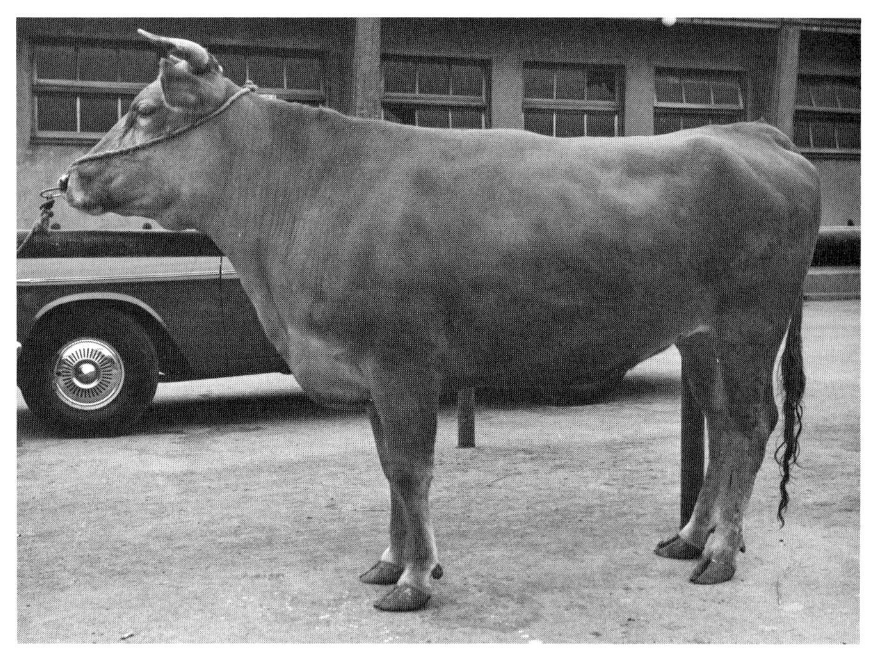

Japanese Brown cow.

brown with a slightly reddish tinge. The horns are rather short and usually have a forward thrust.

The Japanese Brown is a good draft animal but does not have the beef-conformation of the Japanese Black, being heavier in the forequarters and not as heavy in the hindquarters.

Japanese Polled.—In 1960 Japan was said to have fewer than 10,000 representatives of this breed, which is of little commercial importance. It has been selected for its polled characteristic but otherwise is very similar to the Japanese Black.

Exotic Breeds.—The predominant dairy breed is the Holstein-Friesian, with the Jersey a poor second. These breeds have been maintained pure from imported stock, and there has been practically no upgrading by the use of bulls of these breeds on the Wagyu cattle. The total milk herd of the country included 1,000,000 females in 1963 and had increased to 1,310,000 by 1967 in spite of the heavy culling of inferior animals in the past few years.

Several of the European beef breeds have been imported recently, sometimes in an effort to establish them as a source of meat and less often in another attempt to upgrade the Wagyu cattle. The Shorthorn, Hereford, Aberdeen Angus, Swiss Brown, and Devon breeds have all been imported during the past ten years in efforts to establish an improved beef type. This movement has had no important impact on cattle raising in the country.

MANAGEMENT PRACTICES

The small gasoline-powered walking tractor is replacing cattle for draft, and a majority of the Japanese farmers are now working their land by this means and doing a phenomenal job. The average farmer not only produces sufficient rice and either barley or wheat to feed his family but markets enough surplus to meet his annual cash requirements. In the irrigated, intensively farmed areas, he lives in a village; in the more hilly areas, he lives in a house on his land, often only a few hundred yards from his neighbor. In either case he frequently has a cow, which he maintains in a well-kept shed nearby or attached to his dwelling. Food is brought to her by the wife and children and she is considered one of the family.

About every fifth farmer operates a cattle business as a sideline to crop raising. He keeps a cow and raises calves to eight months of age, when they are sold to a feeder-type farmer. This feeder does not breed cattle but buys 1 or 2 head of young stock to feed until they are two to two and one-half years old. If the calf is a steer, he markets it at this time; if it is a cow of good quality, he sells it to another farmer, who will continue to feed and fatten it for another year or so until it will qualify as Kobe beef. The feeder maintains this animal in a darkened shed and takes it to a small adjoining yard each day for a few hours of sun and exercise. The feed is carefully prepared chopped rice straw, dry grass, and some grain. During the fattening period a Kobe cow gets a full feed of barley, with a proper protein supplement.

Kobe beef is a product of national pride. The center of its production is a fertile farming area adjacent to the towns of Kobe and Matsuzaka, about 300 miles west of Tokyo. The standard practice is to feed a good mature but virgin cow a well-balanced fattening ration for a year or longer so that when she is slaughtered her muscles will contain considerably more marbling than does USDA prime beef. This quality of beef is demanded by Japanese custom for sukiyaki, a dish prepared at the table by broiling thin slices of meat over a minute charcoal brazier and serving it with various vegetables, similarly prepared. Only about 10 per cent of a dressed carcass weighing 800 pounds is considered desirable for sukiyaki in a first-class restaurant. These cuts retailed at a price of about $5.00 a pound in 1967.

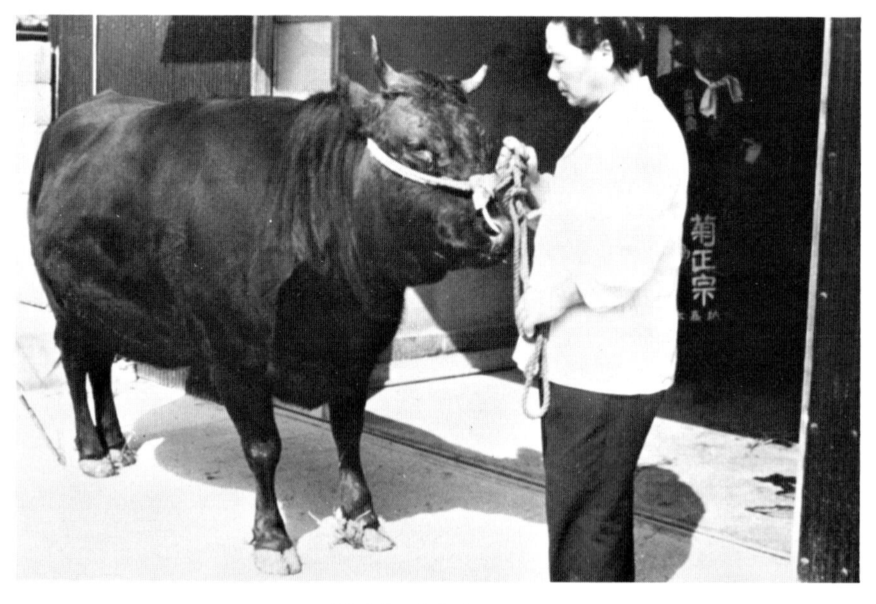

Finished Kobe cow.

In addition to the small farmer who fattens one Kobe cow, there are regular feeders handling 10 or 12 head and occasionally up to 60 in what is called a cow apartment. Such a feeder has arrangements with local farmers to carry virgin Wagyu heifers on a good growing ration until they are about four years of age. They are then brought to the feeder's own premises and put on a full feed of rolled barley, rice bran, and soybean meal, with some chopped rice straw and hay. This combination is served as a hot mash, direct from a steamer, three times a day. The grain intake for each cow amounts to about 26 pounds a day. Each animal is exercised daily, then massaged with wine; it is given a quart of beer every other day. One attendant cares for only 2 or 3 cows.

Kobe cow receiving her ration of beer.

At a national competition held annually, the best fat Kobe cow is chosen as the champion. The animals entered carry considerably more fat than championship steers at United States stock shows.

Artificial insemination has been practiced in Japan for about thirty years. The program is administered by the government. In 1960 there were 2,675 centers and nearly 26,000 licensed inseminators, of which half were practicing. More than 95 per cent of the dairy herd and 85 per cent of the draft-beef cattle are bred in this manner. The speculum method is used.

Winner of the 1962 Kobe cow championship and 1963 prospect. (The apron trophy is for 1962.)

The Japanese have developed a revolutionary method of semen preparation. The artificial vagina is used in the normal manner to collect the semen, which is then diluted. By means of a pipette small globules of the diluted semen, about one-fourth inch in diameter, are placed in depressions just large enough to hold them in a block of solid carbon dioxide. These globules freeze immediately, and the frozen pellets are placed in small plastic boxes, each of which holds 250 pellets. The boxes are then placed in a liquid-nitrogen container for storage. A standard 25-liter liquid-nitrogen unit provides storage for 25,000 pellets, which are thawed in a small quantity of a normal buffer solution immediately before insertion.

MARKETING

Cattle marketing in Japan is rather involved. The farmer sells his animal either direct to a middleman or he places it in the hands of an agricultural co-operative. This middleman or the co-op sells to a "big dealer," who arranges for transportation to the slaughterhouse. Here another transfer of ownership takes place, the wholesaler who operates at the slaughterhouse being the buyer. Title is held by this wholesaler through the slaughterhouse, which charges 1 per cent of its cost for processing the animal. The wholesaler sells the carcass and by-products either to retail butchers or to mass users such as hotels, large restaurants, medical manufacturers (for glands), and hide processors.

These various middlemen—dealer, big dealer, and wholesaler—operate on very narrow margins because of the pressure exerted by the co-operatives. In late 1966, when prices had risen more than 25 per cent over the previous year, a 900-pound steer which would grade a USDA choice brought about $650, or approximately 70 cents a pound liveweight and $1.25 a pound for the dressed carcass. A fat Kobe cow brought more than $1.00 a pound liveweight. The cost of raising cattle, the shortage of meat, and the strict quota limiting importations of beef were responsible for these high prices. This situation has prevailed in spite of the larger numbers of cattle being offered. Many of the cattle being killed in the Tokyo abattoir were very fair Holstein-Friesian cows.

ABATTOIRS

Most of Japan's locally killed beef is processed in a Western-type abattoir. The Shibaura Slaughterhouse, which handles three-fourths of the cattle killed in Tokyo, processed 71,000 head of cattle in 1965. Cattle are trucked in, this method of transportation having largely replaced rail shipments. Barns are available for stall-feeding if the wholesalers feel that a few days on feed are desirable. Animals are led individually to the killing floor. A hammer with a short, protruding tube to penetrate the skull (the same as the one used in Korea) is the killing instrument. The animal's head is held in a halter while the blow is struck. From this point on, the carcass is handled on overhead rail. Inspection of both the live animal and the internal organs is meticulous, and sanitation is exemplary. Most meat is sold warm the day it is killed from a large display floor where half carcasses are hung and where the wholesaler does his bargaining with the retail butchers. Chilling facilities are available, and beef is hung for four or five days for the trade demanding it. There is no recognized pattern for grade or price in the trading that goes on, although there is a wide variation in price with the quality of individual carcasses, many of which would grade choice by USDA standards. Two thousand horses and 400,000 hogs are also slaughtered annually.

CATTLE DISEASES

Japan has no serious cattle disease problems. The segregated manner in which cattle are handled—very few head being in the hands of any one owner and practically no mixing of the stock of different owners—along with the widespread use of artificial insemination, reduces materially the transmission of infection. Disease-control measures are strict and rigidly enforced. Some tick infestation exists in the south and in some areas brucellosis is present, but both are being eradicated.

GOVERNMENT AND CATTLE

All activities involving livestock come under the control of the Ministry of Agriculture and Forestry. Rapid changes in the economy have had a pronounced effect on cattle raising. When the walking tractor was introduced, the population of the Wagyu draft-beef animal, which had always been the prized possession of the Japanese farmer, decreased nearly 50 per cent—from 2,700,000 to 1,500,000—between 1956 and 1967. During much of this time there was an equally rapid increase in the number of dairy cattle, so the total population did not change materially; but there were substantial changes in the supply of beef.

The ministry is endeavoring to increase the production of both milk and beef. In early 1966 the rising price of slaughter cattle and the relatively low price for milk was beginning to cut into the dairy herds. This situation was corrected by granting what amounted to a subsidy on milk production, and by the end of the year the dairy herd showed a slight increase.

Programs sponsored to increase the slaughter cattle population have not been so practical. A recent program of the ministry for the importation of the exotic beef breeds for crossing on the Wagyu had resulted in only 50 head being brought in by the middle of 1967. With the good beef-type Wagyu to start with, such a crossing program is questionable. One hundred breeding centers where Wagyu cattle would be raised and calves for feeding put in the individual farmers' hands at reasonable prices are planned. Low-interest loans were authorized to farmers to finance feeding and to the farmer–co-operatives for the purchase of breeding stock. Subsidies have been offered for the establishment of improved pastures. None of these incentives seems to have been particularly effective. The population of Wagyu cattle is continuing to decrease, being 5 per cent less the first of 1967 than a year earlier.

OUTLOOK FOR CATTLE

A land not quite as large as California, with 100,000,000 inhabitants—but with only one arable acre for every 6 of them—is obviously no cattle country. Some of the mountainous areas could be developed for pasture, but such additional cattle as could be raised here would not significantly increase the supply. Promotion of the European beef breeds, which is quite active at the moment, is certainly no answer—the Japanese farmer with his good Wagyu cow, if he still has one, has all the livestock he has feed for.

Dairy cattle are another matter. There is a real need for milk; yet dairy herds were being heavily culled in 1966 because of the high price being paid for slaughter animals. Chilled and frozen beef can be imported, but fresh milk requires a cow at hand. It would seem logical that the long-range trend for cattle in Japan would be toward a good dual-purpose animal such as the European Friesian. However, cultural attitudes and national pride do not change quickly; and, in spite of the trend toward using the walking tractor, many farmers are going to keep their Wagyu cows, and Kobe beef will still be produced in Japan.

Philippine Islands

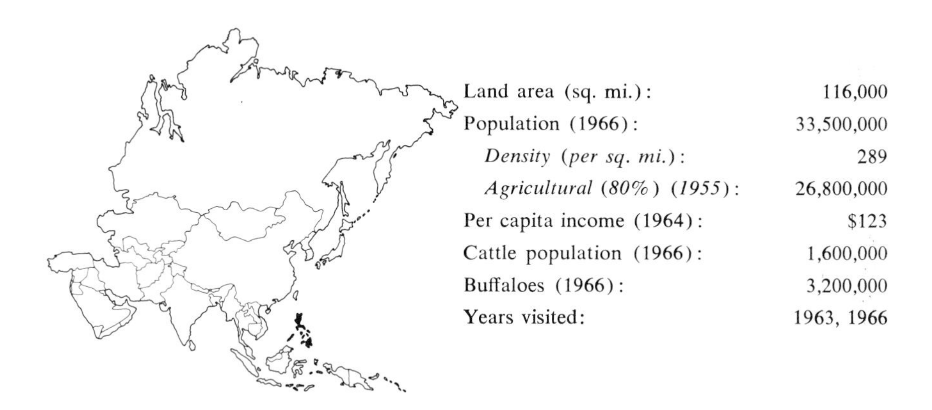

Land area (sq. mi.):	116,000
Population (1966):	33,500,000
Density (per sq. mi.):	289
Agricultural (80%) (1955):	26,800,000
Per capita income (1964):	$123
Cattle population (1966):	1,600,000
Buffaloes (1966):	3,200,000
Years visited:	1963, 1966

THE REPUBLIC OF THE PHILIPPINES occupies the group of 7,000 beautiful islands which lie across the South China Sea from the southeastern tip of Asia and extend 1,100 miles north and south between latitudes 5 and 21 degrees north. Fewer than 500 of the islands have an area of more than one square mile, and most of the population lives on only 11 of them. The total land area is equal to that of Florida and Georgia; the population is three times that of these states. The people are mostly of the Malay race, although various other elements—particularly Chinese and some Spaniards and Americans—have entered the population in the past.

The first Europeans on the islands were the Spaniards in 1521 under the Portuguese expatriate Fernando Magellan, but Portugal contested the Spanish possession for the next fifty years. Spain started settlements in 1565, and the Spanish type of colonization then prevailed for three and one-third centuries. The seeds of democratic government were sprouting before the United States took possession of the islands after the Spanish-American War, at the end of the nineteenth century. The Japanese brutally occupied the periphery of the important islands during World War II but never overcame the strong underground resistance that persisted throughout much of the country. Independence was declared in 1946, after the war had ended. The government since has known troubled times but has maintained its continuity. A Communist-dominated rebellion was successfully overcome after the last war by the Filipinos themselves. Law and order have been maintained even though government has often been riddled with nepotism and graft.

The average man of some education today is well aware of the abuse of power by those in authority; and, although he accepts the situation for the time being, he is not happy with it.

A fundamental difference between the Philippines and many other newly independent countries is the Filipino himself. Predominantly Malayan, he disproves the often accepted concept that natives of a hot climate—where food is easy to come by and protection from weather rudimentary—are congenitally opposed to work which they see no need to perform. All the average Filipino wants is an opportunity to work. While the worldwide movement of people to the cities is much in evidence, 80 per cent of the population is still engaged in agricultural pursuits. The lowland coastal areas of the ten largest islands have a hot, humid climate and accommodate most of the agriculture. The inland plateaus and mountainous country are less extensively exploited and are largely covered by forest and bush, which are meagerly developed for either livestock or cultivation but could be brought into production. Rainfall is usually ample and fairly well distributed except in central Luzon, where the rainy season is limited to June to early November.

The acute shortage of cattle in the country stems from the Japanese occupation during World War II. A total cattle population (including buffaloes) of 4,400,000 in 1940 was reduced to about 2,200,000 by the end of the war—not so much the result of actual slaughter by the occupation forces as by the Filipinos themselves, who were utilizing their animals for subsistence and trying to keep them from falling into the hands of the invaders. The fact that by 1966 the total number of cattle and buffaloes had been built back to 4,800,000 head is a notable feat. Although this number slightly exceeds the prewar level, the improved economic status of the population warrants still more cattle.

Mindanao, the second largest island and the best adapted to cattle raising, has a potential for extensive development along this line. In area it is about half the size of Florida, with a population nearly the same as that state's.

CATTLE BREEDS

The Chinese were trading with the people of the Philippines for several centuries before the Europeans arrived, and the Indonesians from adjoining islands had preceded them. No definite evidence exists concerning whether either of these groups brought cattle to the islands. There is a small indigenous buffalo, the Tamaraw, on the island of Mindoro but there were no other native representatives of the tribe Bovini. The Tamaraw, related to the Indian buffalo, was never domesticated and is now said to be in danger of extinction. The ancestors of the present-day cattle and buffaloes were for the most part introduced during the Spanish colonial days.

Representatives of numerous breeds of exotic cattle have been brought to the Philippines over the years, initially just to increase numbers but in later years with the idea of upgrading by using bulls of improved breeds. These animals fall into two groups: Oriental, as they are known locally, designating Zebu breeds from India, principally the Nellore, the Red Sindhi, the Sahiwal, and the Hariana; and Western, represented by the Santa Gertrudis, American Brahman, Brown Swiss, Holstein-Friesian, and Shorthorn breeds. Whatever their origin, it is obvious that Zebus or animals carrying some Zebu genes are better adapted to life in the Philippines. The European breeds suffer from heat stress in the hot, humid climate and do not have the same tolerance of endemic diseases.

Batanes Black.—Referred to here as Batanes Black, this small animal has no specific local name. Quite uniform in appearance, the breed is a fine-boned, heavy-chested, and slightly humped animal with short, stubby horns. The dewlap of the solid black animal is very prominent but there is practically no sheath.

Batanes Black bull.

Young Batangas bull, 700 pounds.

Mature Batangas cow, 600 pounds.

Finding its way from the Batan Islands, the northernmost outpost of the Philippines, to the Manila abattoir, the bull as its arrives for slaughter does not weigh more than 600 pounds.

Batangas.—These cattle are seen in the province of that name in the southwest corner of the island of Luzon. They are nonhumped, small, compact, rather fine-boned, and almost invariably a solid, rather dark tan color, somewhat similar to that of the Jersey. The horns are characteristically very small, stubby on the bull and sharply curled on the cow. Under the conditions in which she is kept, the average cow weighs 600 pounds; steers five or six years of age, if kept on the best grass available, may weigh as much as 750 pounds.

These cattle are said to be descendants of those brought to the island by the Spaniards in the sixteenth century. Under the care of the small farmers, the Batangas could have been segregated to a considerable extent over the years and thus maintained in something approaching a pure state. Although the Batangas is a smaller animal (which could be the result of generations of poor nutrition), in its nearly solid color, barrel-shaped body, and general conformation it resembles the Asturiana de los Valles breed as seen in Spain today.

There is some controversy concerning whether the cattle known locally as Batangas should be called a breed. Many cattle sold in Manila from Batangas Province have been brought there from some of the other islands for fattening on grass, and these animals show a hump. The fact that the true Batangas is nonhumped, however, distinguishes it from other types of Philippine cattle.

Nondescript Cattle.—The majority of the cattle in the Philippines are a nondescript, usually humped animal. They vary widely in size, conformation, color, and shape of horns. Except for the Batangas and some of the recently introduced exotic breeds, the cattle population is predominantly Zebu. Many show characteristics of the Indian Nellore breed, of which there are some local concentrations.

Exotic Breeds.—Interest is growing in exotic breeds, which are being imported for upgrading native cattle for both better milk and better beef animals. The most widely used milk breeds are the Jersey, the Brown Swiss, and Holstein-Friesian. The beef breeds have included the Indo-Brazil, the Nellore, the Santa Gertrudis, and some Charolais. Heat stress causes serious loss of condition in all of these that are not a Zebu type. Nothing noteworthy has resulted from the crossing endeavors using

Nondescript Philippine bull.

milk-breed bulls, but some success in getting a faster gaining animal has been attained by crossing Santa Gertrudis bulls on native cattle.

Buffalo.—Two types of buffalo are seen in the Philippines: the smaller one, which is locally called carabao, and the Murrah buffalo of India, which has been brought into the Philippines at different times in the past with the objective of increasing milk production. Both are utilized for draft in the paddy fields, but the Murrah or Murrah crosses are more useful as milk animals.

MANAGEMENT PRACTICES

The average Filipino farmer tills nine acres. He owns a carabao for working his land and perhaps a cow, which is also frequently used for draft. These animals are usually maintained in fair condition, grazing on crop residues, on field borders, or along the roadside, usually on tether. They are brought to the vicinity of the village dwelling at night for protection but otherwise receive little attention.

Interest in cattle raising is growing throughout the larger islands. Sugar and coconuts have always been the principal export crops, but the large landowners with ample financial resources are seeking a diversifi-

Murrah buffalo.

Philippine carabao.

cation into cattle. This is also true of smaller operators who can manage projects involving herds of a few hundred head.

Some of the large plantations have started advanced development projects. The 7,000-acre Hacienda Bigaa, on southern Luzon, is part of a large sugar-cane complex and is an outstanding example of an operation using modern cattle management methods. The present program is to determine which of several exotic breeds can be best utilized to improve the native cattle. Purebred herds of Santa Gertrudis, Charolais, and Indo-Brazil (imported from Brazil) are maintained. Bulls of these three breeds are used on three separate herds of native cows to determine which cross gives the most satisfactory results. The program was started in 1957 and now has third-cross calves on the ground. The Indo-Brazil cross is preferred on the basis of results so far obtained.

Improved pastures—largely Pará grass and the tropical legume, centrosema—have been established and are rotated on a four-day cycle. This short period was said to be necessary to prevent damage from tramping because of the frequent heavy rains. These pastures are fenced and the cattle run free, an exceptional practice in the Philippines. The carrying capacity is 3 head a hectare, or 1 head to 0.8 acre, which keeps the cattle in excellent condition.

A small Western-type feed lot is maintained for finishing steers. Three hundred head of three-year-old steers are fed out annually. The feeding period is 120 days on nine pounds a day of a concentrate ration containing 12 per cent protein. Gains of 1.5 pounds a day are obtained during the dry season, when only mature growth hay is available as roughage, and 2.6 pounds a day in the wet season when green-cut grass-legume fodder is fed. Finished three-year-old steers which would grade good, or a few low-choice, by USDA standards weigh approximately 800 pounds. Sold directly to the butcher, they brought 14 cents a pound liveweight in 1966.

On Mindanao, the most southerly of the large islands, which normally enjoys a better and more evenly distributed rainfall than Luzon and is generally a better cattle country, smaller cane growers are showing what modern methods can accomplish in raising cattle. One of these, the Valley Dairy, twenty-five miles from the City of Davao, has established a combined dairy and beef cattle operation. The dairy herd consists of 21 Zebu-type cows, selected for milk-producing ability, which have been crossed with Jersey and Brown Swiss bulls. Production averaged sixteen pounds a day a head.

The beef cattle portion of the Valley Dairy is the principal part of the business. Of the 500 acres of the hacienda, 350 acres are arable; of the arable land, the 125 acres in improved Pará grass and centrosema pasture maintain the 400-head beef herd, 200 of which are mother cows. The remaining 225 acres are cropped. The cows and calves, after weaning, are maintained in separate pastures. At one year of age the steer calves are placed in the fattening paddocks. Six 8-acre pastures of Pará grass and centrosema, established beneath coconut palms, are rotated progressively every week and carry 87 head of steers from one to three years old. The cow herd is being bred up by the use of Santa Gertrudis bulls. Three-year-old native Zebu steers on these fattening pastures are marketed when weighing about 725 pounds and dress 47 per cent. The Santa Gertrudis first-cross steers at two and one-half years weighed 900 pounds and dressed over 50 per cent.

MARKETING

The most serious deterrent to progress in cattle production is the present marketing system, which is controlled by the livestock dealers. Manila is the largest outlet for cattle. A few large operators in the vicinity are in a position to market their cattle at a fair price, but most slaughter cattle (which are worn-out work animals, both carabaos and cattle) are bought at local sales in the country from small farmers by agents of the dealers.

Sales in the country are customarily by the head, the buyer mentally appraising what the animal will dress out and making an offer which will ensure him a gross return of at least one-third more than he pays. The dealer must furnish the transportation, but that amount is not a large part of his cost. In Manila the dealer resells to the butcher, the transaction being on either a per-head or on a cold-dressed-weight basis. Prices paid by the butcher are approximately one-third more than that paid the grower in the country. The butcher-owner pays the municipal abattoir the slaughtering fee and the carcass is delivered to his shop.

Batangas steers which have been grown out to a 750-pound liveweight bring $115 to $130 in the country, or approximately 15 cents a pound. The butcher pays the dealer $160 for this animal after it has suffered a weight loss of 10 per cent during transportation and a further loss because of a holding period of perhaps several days while eating rice straw and awaiting slaughter. Zebu types of cattle shipped from the other islands bring considerably lower prices; carabaos, mostly worn-out work animals and only an occasional infertile female, bring even less. Low prices are maintained by a viciously guarded consortium of dealers who are past masters in applying the "squeeze" to the cattle growers or country dealers outside the ring. If a seller with several head of cattle ships his animals to Manila by boat, a buyer comes on board and makes a low offer. When this amount is refused, another buyer appears and makes a lower offer. This process is continued until the seller gets the idea and finally lets his cattle go for whatever price he can then get. There is no procedure by which he can take his cattle through the abattoir and sell to the butchershop owner. In addition to such means of coercion on the seller, underweighing, downgrading, and delayed payments are all practiced as a matter of principle. The effect all this has as a deterrent to cattle production is obvious.

Transportation is the second major problem in marketing cattle. The principal slaughtering facilities are on Luzon. From the other islands, cattle must be transported to Manila by small vessels with poor, or no, provision for feed and water. Shrinkage is excessive and animals arrive in poor condition. The most logical solution would be to do the processing on the islands such as Mindanao on which cattle are raised and ship frozen or chilled meats to the points of consumption. This procedure, however, would necessitate a breakdown of the cultural requirement that all meat should be purchased from the butchershop the day it is killed and eaten fresh.

ABATTOIRS

The few modern abattoirs are all short of animals, particularly cattle, for slaughter. A large plant in Manila with a capacity of 125 head of cattle an hour was closed in 1966 because no cattle were available. The plant was built by Chinese interests a few years ago to process live animals brought from Australia. The drought in that country and its effect on cattle prices, together with the drop in the Philippine foreign exchange rate, forced the shutdown. In 1966 the municipal abattoir in Manila was killing 30 to 40 cattle and 130 carabaos a day, which is but a fraction of its capacity. This is an up-to-date establishment with good holding yards, where rice straw is fed while the animals await slaughter. It is operated by the Veterinary Inspection Board, a branch of the office of the mayor of Manila.

In most places in the country, the slaughter facility is a concrete slab with primitive drainage facilities. Sanitation is very poor and is especially bad in periods of water shortage. Processing is handled by the *matanzeros*, who do the killing and dressing for the butchershop owners who have purchased the live animals and then receive the dressed meat and offals. A fee is paid to the municipality for use of the facility. To kill the animal, a knife is inserted at the base of the brain, a practice apparently carried down from the Spanish.

CATTLE DISEASES

Parasites, heat stress, and endemic diseases have the debilitating effect on exotic cattle in the Philippine Islands that is common in the tropics. The indigenous nondescript cattle and the Batangas get along quite well with a minimum of attention, although serious outbreaks of various diseases occur. The liver fluke today is perhaps the most serious cause of loss in cattle. Rinderpest was endemic until 1937, when it was practically eliminated. Foot-and-mouth disease of a comparatively mild type is present. It is said to have been introduced by imports from Indonesia in 1960. There is now good inspection and quarantine of imported breeding stock. Both anaplasmosis and piroplasmosis are present and are also thought to have been introduced by imported cattle in past years.

The only regulation on disease control outside of those applying to imports is the requirement that all animals be vaccinated every six months against hemorrhagic septicemia. This service is furnished free by veterinarians of the Bureau of Animal Industry. A small number of advanced growers spray or dip as often as four times a year for ticks, which in some areas are carriers of red water disease. A few large herds are calfhood vaccinated against brucellosis and some are tested regularly for both brucellosis and tuberculosis. Compulsory vaccination is attempted on outbreaks of anthrax and, occasionally, rinderpest, but this practice is not rigidly enforced.

Efforts to maintain the European breeds pure have, for the most part, been unsatisfactory. These animals, particularly the milk breeds, do not stand up well under the heat stress of the tropical climate. Exotic Zebu breeds do well, as do the American Brahman and the Santa Gertrudis.

GOVERNMENT AND CATTLE

The Bureau of Animal Industry is making a determined effort to build a going cattle industry on a meager budget and can point to sizable progress in recent years. A few of the very large landowners are doing substantial improvement work on their own, and some small operators as an adjunct to their farming operations are getting into dairying and are raising cattle for beef.

In an effort to build up the total cattle population, laws were passed making it illegal to kill a buffalo of either sex under twenty years of age (recently reduced to fifteen years) unless the animal were certified as being incapable of reproduction. While such a certificate is not too hard to get, at a price, if a farmer is badly enough in need of funds, it shows the extreme effort being made to build up cattle and carabao numbers.

The Bureau of Animal Industry established an artificial insemination center in 1964. A bull stud with ten imported Jersey and Holstein-Friesian bulls was set up at an experimental farm near Manila, and semen is processed and frozen using liquid nitrogen, or liquid oxygen when nitrogen is not available. The rectal method of insemination was being used. In early 1966 the program had not progressed much beyond a training program for technicians. The laboratory is adequate and procedures are in accordance with modern practice. Considerable advancement has been made in the Philippines in the artificial insemination of hogs, and the same will probably follow with cattle.

The Philippine College of Agriculture is making a start at the development of the Batangas as a beef breed. A small lot of locally purchased Batangas steer calves, nine to twelve months old and weighing approximately 200 pounds, was being fed on a balanced ration with the intention of finishing them at weights of 330, 500, 600, and 825 pounds. This experiment had been under way for only three months in early 1966, but the effect of the good nutrition on this well-adapted animal with its good conformation holds great promise if carried through to the logical conclusion of selective breeding for an improved beef type.

OUTLOOK FOR CATTLE

The Philippines have the basic elements of a limited, but healthy, cattle industry. Unutilized or poorly utilized areas with good rainfall could produce feed for several times the present cattle population. This could be accomplished on lands not suitable for cropping and without interfering with the current program to increase rice production. The people are willing to work, in fact, only want the opportunity to do so and are available to supply the labor for any type of cattle development that might be initiated. The local market would readily absorb large increases in both milk products and beef. In spite of such favorable factors, the handicaps to the development of the cattle potential of the islands are major and immediate. Inadequate transportation and marketing practice abuses have been mentioned. Capital requirements are difficult to obtain. Although there is reasonable political stability, wherever private enterprise comes in contact with bureaucracy there is the ever-present graft which permeates the economy. Encouraging signs, however, include the work of the Bureau of Animal Industry and the efforts of a few large landholders in cattle development. The day will eventually come when the islands will be self-supporting in both meat and dairy products.

South Korea

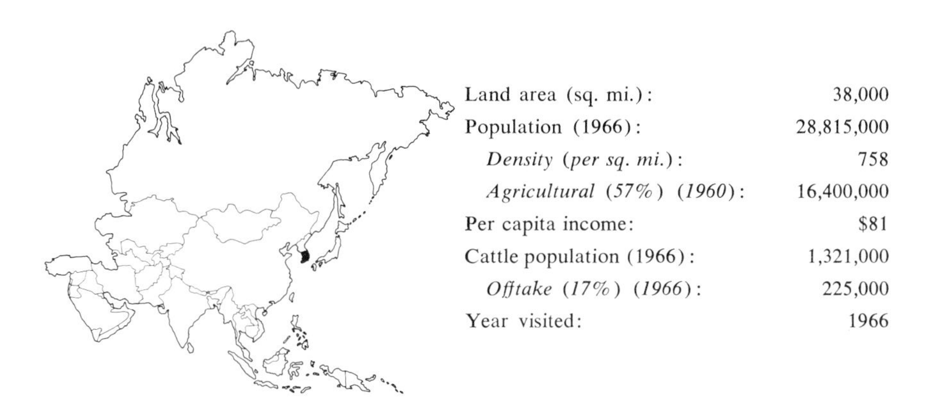

Land area (sq. mi.):	38,000
Population (1966):	28,815,000
Density (per sq. mi.):	758
Agricultural (57%) (1960):	16,400,000
Per capita income:	$81
Cattle population (1966):	1,321,000
Offtake (17%) (1966):	225,000
Year visited:	1966

THE REPUBLIC OF KOREA (South Korea) occupies the southern part of the Korean peninsula, which juts out from China toward the southern end of the Japanese Islands. The Sea of Japan is on the west, the Yellow Sea on the east. South Korea lies in the same latitude as Virginia; in area it is a little smaller than that state and has a population twelve times as large. It is a rough, mountainous land, with the principal drainage flowing to the western and southern coasts. South Korea has several island possessions, the largest being Cheju, off the southern coast. The people are of Mongol stock with some mixture of Chinese.

For many centuries China and, later, Japan have cast covetous eyes toward the whole peninsula. Parts of it at various times in the past were conquered by China. It was annexed by Japan in 1910 and remained in her possession until the end of World War II, when Korea was arbitrarily divided at the thirty-eighth parallel between the United States occupation forces in the south and the Soviet Union in the north. This division became the border between North and South Korea.

The Republic of Korea was established in 1948 and has since remained independent. Armed aggression by North Korea in 1950 initiated the Korean War, which lasted until 1953. There have been political upheavals, but a military coup in 1961 returned the government to elected civilian control two years later. Although many aspects of military government still remain, there is a feeling that law and some form of order are going to be maintained. Almost isolated from the rest of the world for centuries until annexed by Japan, Korea was without the opportunity to acquire the attributes of Western civilization until the end of the Korean War. The Japanese during their thirty-five years of occupancy exploited the land and did whatever they could to demoralize the existing culture.

With the war's end there was a tremendous increase in population, occasioned by both the refugees from North Korea who streamed into the country and a birth rate of near 3 per cent annually. From 12,000,000 people in 1953 the population had increased to nearly 29,000,000 in 1966. The population density is even greater than that of Japan and nearly equal to that of Taiwan. The production of food staples is only 85 per cent of the consumption; yet sizable rice exports are made to Japan, so pressing is the need for foreign exchange. Under such conditions, it is not surprising that a draft animal is the only type of cattle that exists on the peninsula except for some recent small importations of exotic milk and beef breeds.

CATTLE BREEDS

The cattle of the country are known locally as Korean Cattle. For the most part they are quite uniform in appearance, somewhat reminiscent of the Chinese Yellow of Taiwan but without a hump. Since the hump is a characteristic that cattle do not lose, it is doubtful whether the two types are related in spite of the proximity of the areas where they are found. On Cheju the breed is called Brown Cattle, apparently to distinguish it from the other indigenous type, a small, black, nonhumped animal, here called the Korean Black for identification. This animal is rather common on Cheju but is seen only occasionally on the mainland. Although a smaller animal and generally rougher in appearance, the breed is somewhat similar to the Black Wagyu of Japan. The Korean Cattle comprise by far the largest part of the cattle population; most are owned by small farmers, each with a one-half- to three-acre farm, where he uses either a cow or a bull for draft.

Korean Cattle.—This distinctive, indigenous breed has emerged as a good draft animal in Korea and shows no evidence of having been mixed to any appreciable degree with other cattle types. This lack of interbreeding would naturally follow from the physical as well as political, isolation of the country.

The color is a uniform dun, usually solid over the whole body and of a rather bright hue if the animal has been on good nutrition. The small, short horns grow up and forward on the female; on the bull they grow more outward and are thick and short. The average cow weighs 800 pounds, a work ox 900 to 1,000 pounds; on good feed a cow may weigh up to 1,000 pounds and a bull 1,500.

Korean bull, five years old, 1,600 pounds. Photographed at the Hwa San Branch Experiment Station.

Korean cow, seven years old, 1,000 pounds. Photographed at the Hwa San Branch Experiment Station.

Korean cow and calf roughed through the winter. Photographed on Cheju.

Korean Black.—This breed appears to be indigenous to Cheju, where it is bred separately from the Korean Cattle. Both breeds are seen on the island, although the Korean Cattle are more numerous.

A small, nonhumped animal, solid dull black in color, the Korean Black appears to be something of a degenerate type. The horns are small, curving forward, and the bone structure is very fine. Depending on the feed available to her, the cow weighs 650 to 800 pounds. Individuals run remarkably true to type.

Exotic Breeds.—In the endeavor to establish a dairying industry, the Koreans recently imported Holstein-Friesian cattle from the United States; they now number several thousand head and are maintained pure, with very little crossing with local cattle.

Small numbers of American Brahman, Santa Gertrudis, Hereford, and Aberdeen-Angus bulls have been imported for experimental breed-

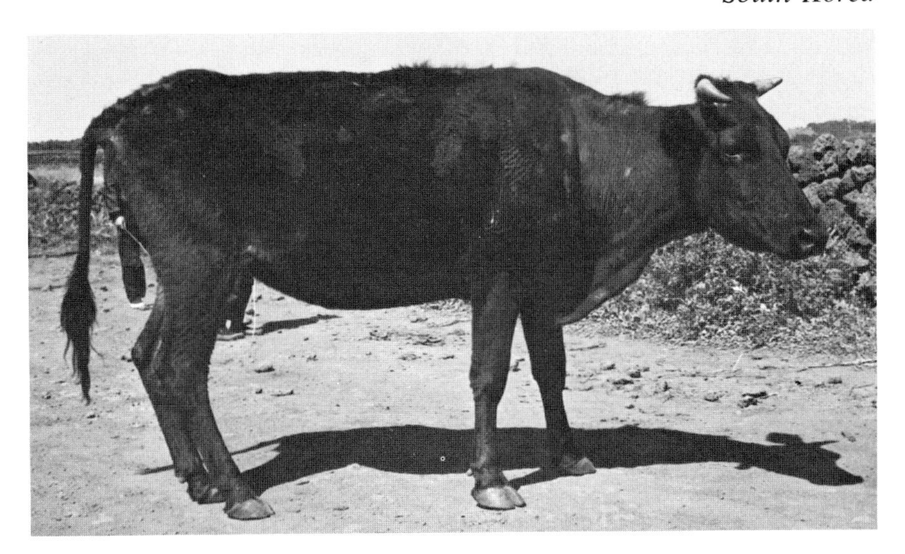

Mature Korean Black cow, 700 pounds. Photographed on Cheju.

ing, mostly for crossing on the Korean Cattle. Nothing of note has developed from this work.

MANAGEMENT PRACTICES

The basic difficulty in raising cattle in Korea is feed. The country is far enough north in the temperate zone that tropical disease and parasites are not the major problem that they are in tropical climates. The Korean is not the meticulous farmer that the Taiwan farmer is; he does, however, plant every level square rod of ground, frequently with two crops. The country is not self-supporting in food staples, and feed for livestock—even coarse roughage—is at a premium. However, the small farmer usually keeps his draft animal, which is practically a member of the household, in fair condition by providing roadside and border grazing and by giving it grasses cut from the hillsides for winter feed. Pasture for even a few head of cattle is unknown except at the government experiment stations and to some extent on Cheju. Even the experiment stations are overstocked to the point that cattle come through the winter in extremely poor condition.

The ambition to produce beef cattle is in evidence in Korean agricultural development plans, as it is in many other countries where meat is scarce because the feed to produce it is lacking. The basic fact that feed, not exotic breeds, is the first requirement for beef production does not seem to be recognized. First-cross yearlings of Santa Gertrudis and

American-Brahman bulls on native cattle, in which some degree of hybrid vigor would normally be expected, are seen without as much growth as the native cattle. This illustrates the excellent adaptability of the Korean Cattle to the coarse feed and environment. The native animal does better on the available feed than those not conditioned to such a meager diet.

The excellence of the Korean Cattle for the purpose they serve does not seem to be appreciated, and the possibility of developing them as a beef animal is receiving no attention. Natural selection over the centuries has developed a hardy, excellent draft animal, tractable and with such a good disposition that bulls are worked throughout their useful years. The development of the forequarters and shoulder muscles is extreme, and the catlike hindquarters are narrow but muscular and heavy-boned. Said to be slow maturing, which they probably are to a degree, they are certainly susceptible to reasonable growth on good nutrition (as shown by the illustration on page 1017 of the bull and the cow raised at the Hwa San Branch Experiment Station).

On Cheju there is a gradual fall in the terrain in all directions from Mount Halla in the center to the sea coast. On the eastern slopes the 2,000-acre Samho Cheju Ranch, near the village of Sangdon, has the largest beef cattle operation in Korea. The base herd consists mostly of Korean Cattle, with a few Korean Blacks. Both are being crossed with Santa Gertrudis and American Brahman bulls imported from the United States. The coastal plain of the island is intensively cultivated; not a level square rod able to hold a bit of soil between the lava outcrops goes unplanted. The mountain slopes at Samho Cheju are being used for pasture, cross-fenced for grazing control. This distinct innovation differs from the practice of the small farmer along the coast, who has his small fields surrounded by lava-rock fences and tethers his cow or ox for feeding. A total of 470 head of all ages is maintained in the herd, and at the beginning of spring the animals are in much better condition than the individual animals of the small farmers in the surrounding area. The ranch, which was formerly a government property, is now operated by the Samho Trading Company, a private concern with large commercial interests. Some of the pastures have been improved. The breeding program of crossing Santa Gertrudis and American Brahman bulls on native cattle does not appear to be accomplishing much in the way of developing an improved beef type.

Slow progress is being made in practicing artificial insemination on the limited number of dairy cattle in the country. Most of these herds are concentrated near the larger cities, and, with few exceptions, are in the hands of farmers owning a maximum of 10 cows. An artificial insemination center was established on the outskirts of Seoul in 1963. Eight hundred cows were bred the first year; this number had increased to 1,336 by 1965. Aided by the Ministry of Agriculture and Forestry, the center is operated as a co-operative and is known as the Artificial Breeding Center of the National Agricultural Co-operative Federation of Seoul. Two very presentable Holstein-Friesian bulls were in the stud; negotiations for a third, a proven $10,000-sire, had been completed recently. Fresh semen was kept for three days, and imported frozen semen from the United States was also used. Korea has exceptionally good transportation facilities to the United States, and frozen semen from good American proven sires for the small number of cows being bred would undoubtedly be less expensive than that processed at the center. The charge to the farmer for a service is $4.00, which, while high for the Korean economy, is much less than would be involved in natural breeding and in maintaining a sufficient number of bulls to serve the widely scattered small dairy herds. Eight inseminators are employed, each of whom travels to the station every three days for a fresh supply of semen. The rectal method is used in breeding. A conception rate of over 90 per cent was claimed on an average of 1.65 services.

MARKETING

There are two types of cattle markets: those in the country, where one farmer sells to another and sales are usually of draft animals or of young stock to grow out; and those where the large slaughterhouses are located and cattle merchants sell to the butchers. These merchants make their purchases in the country from the small farmers, each of whom owns 1 animal, usually a worn-out bull or cow that has served its purpose as a draft animal. The average price in 1966 was equivalent to 27 cents a pound liveweight, actual transactions being by the head.

Cattle merchants sell to the butchers at prices which range from 30 to 35 cents a pound liveweight, these sales also being by the head. Retail butchershops charge a fixed price for dressed meat: 45 cents a pound for Korean-cut beef, 56 cents for better-quality, American-type cuts. These prices are slightly less than those which the butcher paid the merchant for the live animal. The difference, which includes a very small margin of profit, is made up of what the offals, bone, blood, and hide bring. Practically all of the animal except the contents of the digestive tract is used for human consumption. Bones for soup making are particularly prized by the Korean housewife.

At the country market at Ui Jong Bu, a three-year-old Korean bull

Cattle market adjacent to Seoul Municipal Abattoir. A rest barn is in the background.

weighing 900 pounds would be sold by one farmer to another for $220 to be used for draft. This market is operated by a co-operative, the charge for selling being 40 cents for an animal under one year of age, $1.20 for an animal between two and three years, and $2.00 for one over three years.

ABATTOIRS

Most slaughterhouses are operated by the municipality on a custom basis, although there are two privately owned plants in Seoul. Animals are the property of the butchershop owner who purchases them and pays a fee for processing. This amount averages $12.80 a head, including $3.00 to the plant for processing and $9.80 for taxes and for fees to the butcher co-operative which operates the adjoining rest barn and market and to various go-betweens. The large government plant in Seoul kills as many as 150 head of cattle a day, depending on the season. It was built in 1959—a United States AID project. Quite modern facilities are provided but are poorly utilized. Animals are led by halter up a runway to a conventional killing box, the sides of which are not used.

The head is held in a horizontal position by one man on the halter while the slaughterer drives the projection on the killing hammer through the forehead into the brain. Death is practically instantaneous, and the process much more humane than it sounds. The hammer used is an odd instrument consisting of a solid steel cylinder, four inches long and two inches in diameter, with a short piece of one-half-inch pipe welded into the striking end.

From the killing box floor the carcass is moved by hand across the concrete floor and hoisted by the hind legs until the head clears the floor. The flesh of the throat is then slit longitudinally so that the jugular vein can be located and severed. The blood is drained into five-gallon cans and saved to be dried as blood meal for animal food or coagulated for human consumption. After the blood has been collected, the carcass is again lowered to the floor, hoisted by the front legs, and moved on the rail to the skinning cradles. After skinning has been completed except for the back, the carcass is again hoisted to the rail, eviscerated, and the hide removed.

The small chilling room is not used. Most of the dressed meat goes directly into trucks for delivery to the meatshops. Both live animals and carcasses are inspected by veterinarians, not only for disease but to ensure that no pregnant animal is killed. A pregnant cow occasionally reaches the killing box and if diagnosed as with calf, is returned to the owner. This government plant in Seoul processes about one-sixth of the 225,000 head of cattle being slaughtered annually in Korea. The average dressed weight of carcasses is 250 pounds.

A small slaughterhouse adjacent to the Ui Jong Bu cattle market is quite a presentable facility and was provided for the few animals sold for local slaughter. The concrete killing and dressing floor has good drainage and is kept clean. The same kind of killing hammer is used there as in the Seoul slaughterhouse.

CATTLE DISEASES

Disease-control measures are confined mostly to the government stations. Most calves are vaccinated for blackleg and, when prevalent, for anthrax also. Heifer calves are vaccinated against brucellosis when less than one year of age and tested for tuberculosis and brucellosis twice a year. There is said to be some piroplasmosis in the country. The Samho Trading Company Ranch has a dipping vat and the government stations do some spraying. The temperate climate of the country and the isolated conditions under which most cattle are held are the major factors in the control of disease.

GOVERNMENT AND CATTLE

The Office of Rural Development of the National Government has an upgrading program at the experimental station on Cheju. Korean Cattle (called Brown Cattle on the island) and the Korean Black are the base stocks on which American Brahman and Santa Gertrudis bulls are now being used.

The artificial insemination center at the station has 4 imported American Brahman and 2 Santa Gertrudis bulls. The project was started in 1963 and, in addition to breeding the station's herd, 600 cows of farmers in the surrounding area were bred in 1965. Bulls of both breeds are also raised for distribution to villages on the rest of the island.

Deficient nutrition is much in evidence even on the government experiment stations. The Cheju station has ample land to provide adequate feed for its 150 head of cattle; and, although some silage and hay are put up, neither is stored in sufficient quantity to carry the animals decently through the winter. With his wife and children there is always ample labor to scavenge fodder of one kind or another to keep the household beast in fair condition. When a sizable number of cattle are handled as a group, however, this type of individual feeding is not available and the importance of supplying roughage in reasonable quantity does not seem to be recognized.

The Hwa San Branch Experiment Station at Suwan, on the mainland, has only 135 acres of land, 20 acres of which were in yellow corn instead of the usual Korean small white corn and is put up as silage. Here the effort to upgrade the Korean Cattle was made by using Angus and Hereford bulls imported from the United States. The Angus bulls had just been introduced in 1966, but yearlings from the Hereford cross on the Korean Cattle were on the ground. Where some evidence of hybrid vigor would naturally be expected from such a cross, none was apparent; the F_1 crosses actually did not appear to have as much growth as the straight native yearlings of the same age—a fact that has been noted for the American Brahman and the Santa Gertrudis crosses on native cattle. Lack of adequate feed, along with better utilization of the poor-quality roughage by the native cattle, appears to be the explanation for this. The 90 acres of pasture, much of which is of improved type, were sown to a mixture of brome orchard, timothy, fescue, and Ladino clover; but this, even with the 20 acres of corn silage, did not seem to provide sufficient feed for the total herd of 110 animals.

A dairy herd of 80 imported United States Holstein-Friesians at the Song Hwan Experiment Station does not appear to be faring much better. The average milk production was given as 10,000 pounds for a 305-day lactation period. The condition of the cows makes this estimate appear rather optimistic, but the basic quality of the herd was such that a milk yield considerably higher would normally be expected. Corn silage was fed in the winter, and the same type of improved pastures had been established as at Hwa San. The average annual rainfall in this area is 55 inches, largely concentrated in the early summer, and in the dry part of the growing season the pastures are badly overgrazed. Also, the winter feeding season lasts for six months, and sufficient hay and silage are not produced to carry the cattle through properly.

OUTLOOK FOR CATTLE

Although Korea will never be a cattle country, conceivably it could become self-supporting in both beef and milk. Such land as is tillable will always be better utilized for the production of cereals for human consumption than for raising livestock. Only 22 per cent of the land is arable, but much of the remainder could be utilized for grasses if properly handled. Some such areas are being reforested, and it is highly important to the future of the country that this practice should continue; but there are hillsides and rough terrain where, with good control, forage crops could be sustained by seeding to adapted varieties of grasses. These areas are overrun with brush and scrub growth, which now furnish only a meager quantity of fuel for household use. The same type of development which is now being undertaken in Taiwan would be applicable, although the environment is less favorable, with the shorter growing season and the poorer rainfall pattern. Such areas must have had a grass or forest cover in the distant past, but centuries of overgrazing and grubbing out everything that grew for fuel and forage, along with the subsequent erosion, have made such a reseeding program difficult.

The introduction of the American Holstein-Friesian cow for milk production could have a place in Korean agriculture if the necessity of properly feeding her is realized. If sizable dairy herds are eventually established, dual-purpose types might be more useful, augmenting the meat supply as well as furnishing milk. The introduction of the exotic beef breeds for crossing on the native cow is open to question. Good draft animals are the first requirement of the small Korean farmer and he already has these in his Korean Cattle.

Appendix

ORIGIN OF

SOURCES CONSULTED

Bisschop, J. H. R. "Parent Stock and Derived Types of African Cattle," *South African Journal of Science*, Vol. XXXIII (March, 1937).

Epstein, H. "The Origin of the Africander Cattle, with Comments on the Classification and Evolution of Zebu Cattle in General," *Zeitschrift für Tierzuchtung und Zuchtungsbiologie*, Vol. LXVI, No. 2 (1955).

Faulkner, D. E., and H. Epstein. *The Indigenous Cattle of the British Dependent Territories in Africa.*

Food and Agricultural Organization (United Nations). *European Breeds of Cattle.*

Mason, I. L., and J. P. Maule. *The Indigenous Livestock of Eastern and Southern Africa.*

Morse, E. W. *The Ancestry of Domesticated Cattle.*

Zeuner, F. E. *A History of Domesticated Animals.*

This diagram was compiled with the assistance of I. L. Mason.

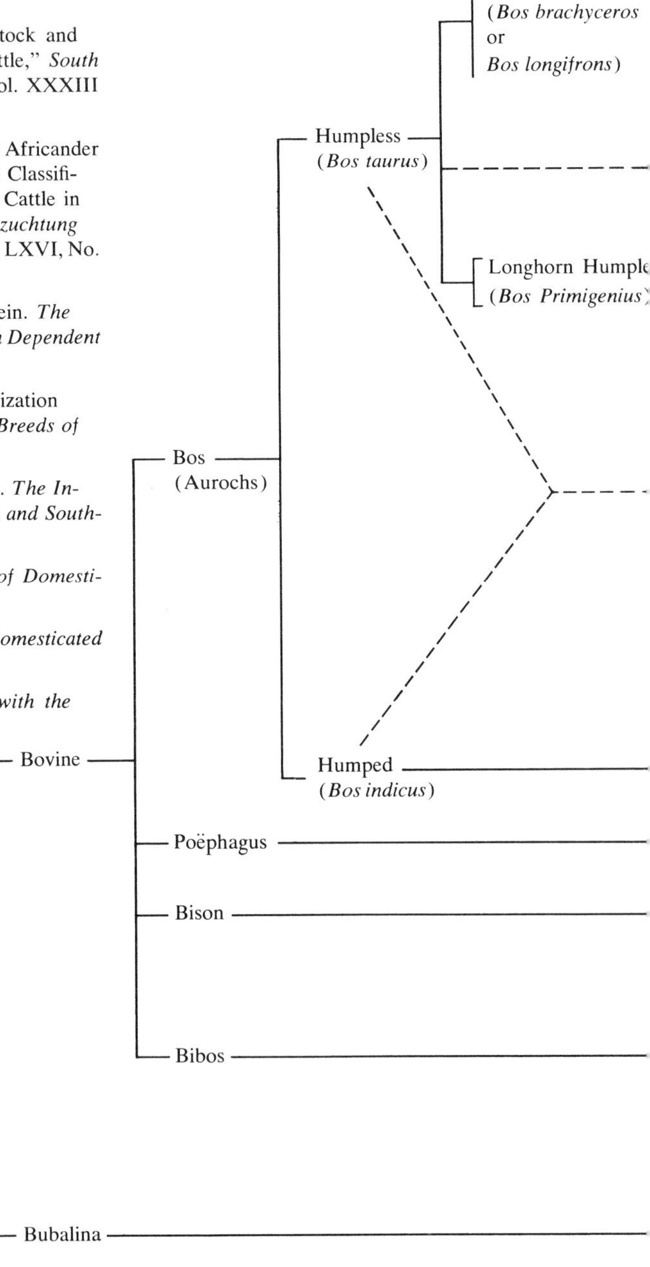

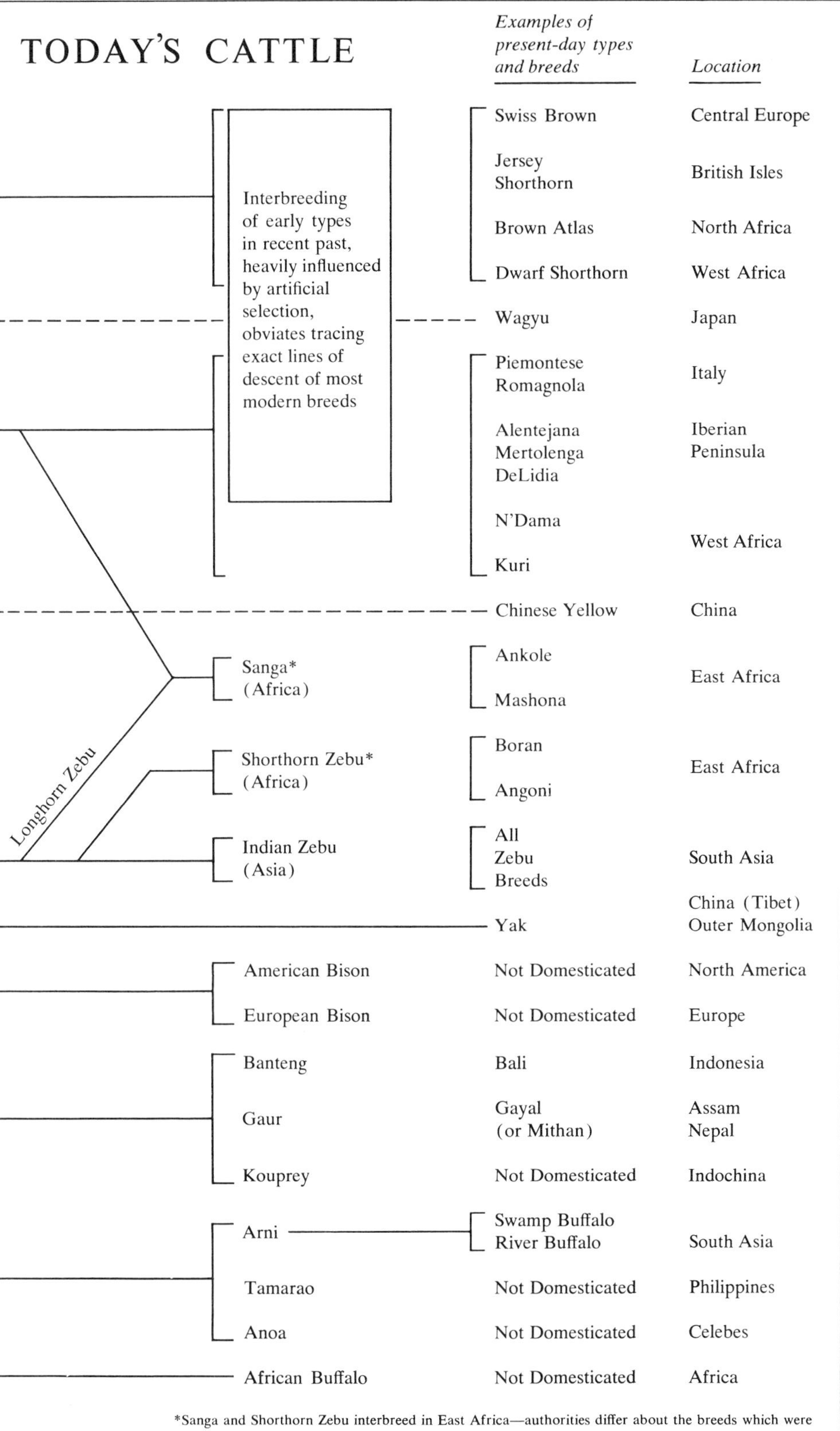

*Sanga and Shorthorn Zebu interbreed in East Africa—authorities differ about the breeds which were derived from this mixture.

Human and Cattle Populations of United States, 1967

		Human Population		Cattle Population (Thousands of Head)			
State	Area, Sq. Mi.	Total	Density per Sq. Mi.*	Dairy	Other (Mostly Beef)	Total	Density per Sq. Mi.†
Alabama	51,609	3,540,000	69	273	1,604	1,877	36
Alaska	586,400	273,000	0.5	3	5	8.3	0.01
Arizona	113,909	1,635,000	14	80	1,054	1,134	10
Arkansas	53,104	1,969,000	37	182	1,471	1,653	31
California	158,693	19,163,000	121	1,324	3,520	4,844	30
Colorado	104,247	1,975,000	19	174	2,872	3,046	29
Connecticut	5,009	2,925,000	585	111	12	123	25
Delaware	2,057	523,000	254	25	10	35	17
District of Columbia	69	809,000	11,700	0	0	0	0
Florida	58,560	5,996,000	102	241	1,636	1,877	32
Georgia	58,876	4,511,000	77	250	1,547	1,797	31
Hawaii	6,424	741,000	115	22	214	236	37
Idaho	83,557	699,000	8	295	1,262	1,557	19
Illinois	56,400	10,894,000	193	647	2,946	3,593	64
Indiana	36,291	4,999,000	138	498	1,521	2,019	56
Iowa	56,290	2,758,000	49	1,076	6,403	7,479	133
Kansas	82,264	2,275,000	28	357	5,149	5,506	67
Kentucky	40,395	3,191,000	79	630	1,913	2,543	63
Louisiana	48,523	3,660,000	75	323	1,451	1,774	36
Maine	33,215	973,000	29	122	23	145	4
Maryland	10,577	3,685,000	349	270	148	418	40
Massachusetts	8,257	5,421,000	655	109	14	123	15
Michigan	58,216	8,584,000	147	883	664	1,547	27
Minnesota	84,068	3,582,000	43	1,932	2,145	4,077	49
Mississippi	47,716	2,348,000	49	410	2,030	2,440	51
Missouri	69,686	4,605,000	66	635	4,020	4,655	67
Montana	147,138	701,000	5	83	2,786	2,869	20
Nebraska	77,227	1,435,000	19	355	6,077	6,432	83

State	Area, Sq. Mi.	Human Population: Total	Human Population: Density per Sq. Mi.*	Cattle Population (Thousands of Head): Dairy	Cattle Population (Thousands of Head): Other (Mostly Beef)	Cattle Population (Thousands of Head): Total	Cattle Population (Thousands of Head): Density per Sq. Mi.†
Nevada	110,540	444,000	4	26	508	534	5
New Hampshire	9,304	685,000	74	70	7	77	8
New Jersey	7,836	7,004,000	895	135	16	151	19
New Mexico	121,666	1,003,000	18	67	1,266	1,333	11
New York	49,576	18,335,000	370	1,737	150	1,887	38
North Carolina	52,712	5,027,000	96	340	631	971	18
North Dakota	70,665	639,000	9	293	1,999	2,292	32
Ohio	41,222	10,462,000	254	822	1,251	2,073	50
Oklahoma	69,919	2,496,000	36	271	4,037	4,308	62
Oregon	96,981	1,999,000	21	222	1,371	1,593	16
Pennsylvania	45,333	11,626,000	257	1,289	474	1,763	39
Rhode Island	1,214	901,000	744	13	2	15	12
South Carolina	31,055	2,603,000	84	119	486	605	20
South Dakota	77,047	674,000	9	337	3,986	4,323	56
Tennessee	42,244	3,888,000	92	561	1,724	2,285	54
Texas	267,338	10,873,000	41	611	10,146	10,757	40
Utah	84,916	1,022,000	12	132	615	747	9
Vermont	9,609	416,000	43	328	16	344	36
Virginia	40,815	4,533,000	111	413	950	1,363	33
Washington	68,192	3,089,000	45	333	1,021	1,354	20
West Virginia	24,181	1,798,000	74	109	318	427	18
Wisconsin	56,154	4,188,000	74	3,309	808	4,117	73
Wyoming	97,914	315,000	3	32	1,333	1,365	14
Total	3,615,210	197,890,000		22,879	85,612	108,491	

* Average: 55.
† Average: 30.

Sources: U.S. Department of Commerce, Bureau of the Census; U.S. Department of Agriculture, Statistical Reporting Service.

ALASKA
HAWAIIAN ISLANDS
WASH.
OREGON
CALIFORNIA
NEVADA
IDAHO
MONTANA
WYOMING
UTAH
ARIZONA
COLORADO
NEW MEXICO
NORTH DAKOTA
SOUTH DAKOTA
NEBRASKA
KANSAS
OKLAHOMA
TEXAS
MINNESOTA
IOWA
MISSOURI
ARKANSAS
LA.
WISCONSIN
ILLINOIS
MICHIGAN
INDIANA
OHIO
KENTUCKY
TENNESSEE
MISS
ALA.
GEORGIA
FLORIDA
N. C.
VIRGINIA
W.V.
PENN.
NEW YORK
VT.
N.H.
MAINE
MASS.
R.I.
CONN.
NEW JERSEY
DELAWARE
MD.

UNITED STATES DATA CHART, 1967

State	Area, Sq. Mi.	Population	Density per Sq. Mi.	Cattle Population
Alabama	51,609	3,540,000	69	1,877,000
Alaska	586,400	273,000	0.5	8,300
Arizona	113,909	1,635,000	14	1,134,000
Arkansas	53,104	1,969,000	37	1,653,000
California	158,693	19,163,000	121	4,844,000
Colorado	104,247	1,975,000	19	3,046,000
Connecticut	5,009	2,925,000	585	123,000
Delaware	2,057	523,000	254	35,000
District of Columbia	69	809,000	11,700	0
Florida	58,560	5,996,000	102	1,877,000
Georgia	58,876	4,511,000	77	1,797,000
Hawaii	6,424	741,000	116	236,000
Idaho	83,557	699,000	8	1,557,000
Illinois	56,400	10,894,000	194	3,593,000
Indiana	36,291	4,999,000	138	2,019,000
Iowa	56,290	2,758,000	49	7,479,000
Kansas	82,264	2,275,000	27	5,506,000
Kentucky	40,395	3,191,000	79	2,543,000
Louisiana	48,523	3,660,000	75	1,774,000
Maine	33,215	973,000	29	145,000
Maryland	10,577	3,685,000	349	418,000
Massachusetts	8,257	5,421,000	655	123,000
Michigan	58,216	8,584,000	148	1,547,000
Minnesota	84,068	3,582,000	43	4,077,000
Mississippi	47,716	2,348,000	49	2,440,000
Missouri	69,686	4,605,000	66	4,655,000
Montana	147,138	701,000	5	2,869,000
Nebraska	77,227	1,435,000	19	6,432,000
Nevada	110,540	444,000	4	534,000
New Hampshire	9,304	685,000	74	77,000
New Jersey	7,836	7,004,000	890	151,000
New Mexico	121,666	1,003,000	8	1,333,000
New York	49,576	18,335,000	370	1,887,000
North Carolina	52,712	5,027,000	96	971,000
North Dakota	70,665	639,000	9	2,292,000
Ohio	41,222	10,462,000	254	2,073,000
Oklahoma	69,919	2,496,000	36	4,308,000
Oregon	96,981	1,999,000	20	1,593,000
Pennsylvania	45,333	11,626,000	256	1,763,000
Rhode Island	1,214	901,000	752	15,000
South Carolina	31,055	2,603,000	84	605,000
South Dakota	77,047	674,000	9	4,323,000
Tennessee	42,244	3,888,000	92	2,285,000
Texas	267,338	10,873,000	41	10,757,000
Utah	84,916	1,022,000	12	747,000
Vermont	9,609	416,000	43	344,000
Virginia	40,815	4,533,000	111	1,363,000
Washington	68,192	3,089,000	45	1,354,000
West Virginia	24,181	1,798,000	74	427,000
Wisconsin	56,154	4,188,000	74	4,117,000
Wyoming	97,914	315,000	3	1,365,000
Total	3,615,210	197,890,000		108,491,300

* Average: 55.

Sources: U.S. Department of Commerce, Bureau of the Census; U.S. Department of Agriculture, Statistical Reporting Service.

Human and Cattle Populations of the World*

Country	Area, Sq. Mi.	Population	Density per Sq. Mi.	Cattle Population	Buffalo Population
Afghanistan	251,000	15,350,000	61	3,051,000	21,000
Albania	11,100	2,000,000	180	402,000	5,000
Algeria	920,000	11,300,000	435†	610,000	
Angola	481,400	5,100,000	11	1,700,000	
Argentina	1,072,000	22,700,000	21	45,000,000	
Australia	2,968,000	11,545,000	4	18,200,000	
Austria	32,374	7,255,000	224	2,500,000	
Belgium	11,781	9,500,000	806	2,730,000	
Bolivia	420,000	4,000,000	10	2,900,000	
Botswana	220,000	593,000	3	1,347,000	
Brazil	3,286,000	84,700,000	26	79,855,000	50,000
Bulgaria	42,830	8,150,000	191	1,430,000	147,000
Burma	262,000	25,200,000	96	6,070,000	1,230,000
Burundi	10,707	3,210,000	300	459,000	
Cambodia	69,900	6,200,000	89	1,402,000	512,000
Cameroon	183,600	5,200,000	28	2,175,000	
Canada	3,852,809	20,334,000	5	11,560,000	
Central African Republic	240,540	1,437,000	6	500,000	
Ceylon	25,300	11,500,000	455	1,851,000	1,002,000
Chad	490,700	3,310,000	7	4,000,000	
Chile	292,000	8,800,000	30	2,900,000	
China (Mainland)	3,691,500	772,000,000	209	65,400,000	20,900,000
China (Taiwan)	13,950	12,765,000	920	105,000	260,000
Colombia	400,000	18,070,000	41	19,500,000	
Congo (Kinshasa)	905,600	16,000,000	18	1,050,000	
Costa Rica	19,652	1,514,000	77	1,480,000	
Cuba	44,218	7,800,000	176	5,772,000	
Czechoslovakia	49,400	14,200,000	288	4,390,000	
Dahomey	43,480	2,462,000	57	370,000	
Denmark	17,144	4,800,000	280	3,375,000	
Ecuador	109,500	5,325,000	49	1,770,000	
El Salvador	8,056	3,037,000	377	1,225,000	
Ethiopia	471,800	22,590,000	48	25,300,000	
Finland	130,100	4,625,000	36	1,895,000	
France	211,200	49,160,000	233	21,400,000	
Germany, East	41,800	17,050,000	408	4,900,000	
Germany, West	96,000	59,295,000	618	13,975,000	
Ghana	92,100	7,740,000	84	538,000	
Greece	50,900	8,600,000	170	1,135,000	68,000
Guatemala	42,042	4,717,000	112	1,340,000	
Guinea	94,925	3,608,000	38	1,500,000	
Honduras	43,277	2,445,000	57	1,450,000	
Hong Kong	400	3,690,000	9,225	12,000	1,000
Hungary	36,000	10,100,000	283	1,975,000	
Iceland	39,700	192,000	5	54,500	
India	1,262,000	499,000,000	395	176,000,000	51,000,000
Indonesia	575,900	108,000,000	147	6,348,000	2,893,000
Iran	636,000	22,900,000	36	5,640,000	335,000
Iraq	167,600	8,260,000	49	1,500,000	225,000
Ireland	27,100	2,880,000	106	5,500,000	
Israel	8,000	2,635,000	330	210,000	
Italy	116,000	52,930,000	455	9,960,000	40,000

Country	*Area, Sq. Mi.*	*Population*	*Density per Sq. Mi.*	*Cattle Population*	*Buffalo Population*
Japan	142,700	98,900,000	688	2,810,000	
Jordan	37,300	2,000,000	54	66,000	
Kenya	225,000	9,645,000	43	7,235,000	
Korea, North	46,540	12,800,000	275	679,000	
Korea, South	38,000	28,815,000	758	1,321,000	
Laos	91,430	2,440,000	27	1,363,000	
Lebanon	4,000	2,490,000	620	110,000	
Liberia	43,000	1,065,000	25	28,000	
Libya	679,000	1,675,000	142‡	147,000	
Malagasy Republic	227,000	6,335,000	28	7,500,000	
Malaysia	128,000	9,395,000	73	325,000	350,000
Mali	478,800	4,745,000	10	3,923,000	
Mauritania	397,700	1,100,000	3	2,000,000	
Mauritius	720	722,000	1,000	44,000	
Mexico	761,600	45,671,000	60	24,000,000	
Mongolia	604,200	1,156,000	2	1,050,000	
Morocco	174,500	13,300,000	76	2,900,000	
Mozambique	303,000	6,900,000	23	1,130,000	
Nepal	54,000	10,300,000	189	5,600,000	1,600,000
Netherlands	13,967	12,375,000	886	3,920,000	
New Zealand	103,700	2,670,000	26	7,600,000	
Nicaragua	50,193	1,685,000	34	1,600,000	
Niger	459,075	3,546,000	8	3,600,000	
Nigeria	356,700	57,500,000	161	9,500,000	
Norway	125,000	3,740,000	30	1,040,000	
Pakistan	365,000	100,000,000	274	33,500,000	8,400,000
Panama	29,208	1,329,000	46	1,011,000	
Paraguay	157,000	2,100,000	14	4,500,000	
Peru	496,000	12,000,000	24	3,500,000	
Philippine Islands	116,000	33,500,000	289	1,600,000	3,200,000
Poland	120,400	31,550,000	261	10,400,000	
Portugal	35,500	9,156,000	258	1,100,000	
Rhodesia	150,000	4,330,000	29	3,855,000	
Romania	91,700	19,100,000	208	5,030,000	171,000
Rwanda	10,169	3,239,000	319	509,000	
Saudi Arabia	872,700	6,700,000	8,375§	56,000	
Senegal	76,100	3,500,000	46	2,200,000	
Sierra Leone	27,925	2,403,000	86	200,000	
Somali Republic	246,155	2,750,000	11	842,000	
South Africa	471,820	18,300,000	39	12,000,000	
South-West Africa	318,000	610,000	2	2,550,000	
Soviet Union	8,650,000	231,870,000	27	97,100,000	
Spain	194,000	31,900,000	164	3,700,000	
Sudan	967,500	13,900,000	14	7,000,000	
Sweden	173,665	7,790,000	45	2,160,000	
Switzerland	15,940	5,880,000	369	1,835,000	
Syria	71,000	5,100,000	72	387,000	2,000
Tanzania	363,000	10,500,000	29	10,000,000	
Thailand	198,500	30,600,000	154	6,100,000	7,100,000
Tunisia	63,400	4,565,000	72	565,000	370,000
Turkey	301,000	32,900,000	109	13,175,000	1,200,000
Uganda	91,000	7,550,000	83	3,627,000	
United Arab Republic	386,900	30,055,000	2,246‖	1,590,000	1,590,000

Country	*Area, Sq. Mi.*	*Population*	*Density per Sq. Mi.*	*Cattle Population*	*Buffalo Population*
United Kingdom	94,800	55,145,000	584	12,030,000	
United States of America	3,615,210	198,065,000	65#	108,491,000	
Upper Volta	105,870	5,054,000	48	2,000,000	
Uruguay	69,000	2,750,000	40	8,700,000	
Venezuela	352,000	9,030,000	26	6,840,000	
Vietnam, North	61,300	18,300,000	299	796,000	1,508,000
Vietnam, South	67,000	16,000,000	239	1,185,000	850,000
Yugoslavia	98,800	19,800,000	200	5,750,000	
Zambia	291,000	3,780,000	13	1,300,000	

Total cattle population of countries listed above: 1,022,556,000
Total buffalo population of countries listed above: 105,030,000

* Some countries with nominal numbers of cattle are omitted.
† Arable area (26,000 sq. mi.).
‡ Arable area (11,800 sq. mi.).
§ Arable area (800 sq. mi.).
|| Inhabited area (13,600 sq. mi.).
Excluding Alaska.

Sources: *Encyclopaedia Britannica, Production Year Book* of the FAO (Foreign Agricultural Services of USDA publications), and data obtained locally by the author.

Bibliography

Bisschop, J. H. R. "Parent Stock and Derived Types of African Cattle," *South African Journal of Science*, Vol. XXXIII (March, 1937).

Breeding Beef Cattle for Unfavorable Environments: A Symposium Presented at the King Ranch Centennial Conference. Austin, University of Texas Press, 1955.

Briggs, Hilton M. *Modern Breeds of Livestock*. New York, The Macmillan Company, 1958.

Cattle Country: An Illustrated Survey of the Australian Beef Cattle Industry. Sydney, F. H. Johnston Publishing Company Proprietary, Ltd., 1960.

Cattle of Britain. Ministry of Agriculture, Fisheries and Food. London, Her Majesty's Stationery Office, 1963.

Cole, H. H., ed. *Introduction to Livestock Production: Including Dairy and Poultry*. Agricultural Science Series. San Francisco, W. H. Freeman and Company, 1962.

Compendio de Prototipos Raciales Españoles. Ministerio de Agricultura, Dirrección General de Ganadería. Madrid, 1953.

Curson, H. H., and J. H. R. Bisschop. "Some Comments on the Hump of African Cattle," *Onderstepoort Journal of Veterinary Science and Animal Industry*, Vol. V, No. 2 (October, 1935).

Curson, H. H., and R. W. Thornton. "A Contribution to the Study of African Native Cattle," *Onderstepoort Journal of Veterinary Science and Animal Industry,* Vol. V, No. 2 (October, 1935).

Epstein, H. "The Origin of the Africander Cattle, with Comments on the Classification and Evolution of Zebu Cattle in General," *Zeitschrift für Tierzuchtung und Zuchtungsbiologie*, Vol. LXVI, No. 2 (1955).

Faulkner, D. E., and H. Epstein. *The Indigenous Cattle of the British Dependent Territories in Africa, with Material on Certain Other African Countries.* Colonial Advisory Council of Agriculture, Animal Health and Forestry Series. London, Her Majesty's Stationery Office, 1957.

Food and Agricultural Organization (United Nations). *European Breeds of Cattle.* 2 vols. Rome, 1966.

———. *Production Yearbook 1965.* XIX. Rome, 1965.

———. *Types and Breeds of African Cattle.* Rome, 1957.

———. *Zebu Cattle of India and Pakistan.* Rome, 1953.

Kelley, R. B. *Native and Adapted Cattle.* Sydney, Angus and Robertson, Ltd., 1959.

Lerner, I. Michael, and H. P. Donald. *Modern Developments in Animal Breeding.* London, Academic Press, 1966.

Mason, I. L. *A World Dictionary of Breeds, Types and Varieties of Livestock.* Slough, England, Commonwealth Agricultural Bureaux, 1960.

Mason, I. L., and J. P. Maule. *The Indigenous Livestock of Eastern and Southern Africa.* Farnham Royal, England, Commonwealth Agricultural Bureaux, 1960.

Ministry of Agriculture, Fisheries and Food. *British Livestock Breeding.* London, Her Majesty's Stationery Office, 1962.

Morse, E. W. *The Ancestry of Domesticated Cattle*, Twenty-seventh Report, Bureau of Animal Industry, U.S. Department of Agriculture. Washington, Government Printing Office, 1912.

Phillips, Ralph W., Ray G. Johnson, and Raymond T. Moyer. *The Livestock of China.* Report of Department of State. Washington, Government Printing Office, 1945.

Phillips, Ralph W., and Leslie T. C. Kuo. *Sciences in Communist China.* Washington, American Association for the Advancement of Science, 1961.

Prentice, E. Parmalee. *Hunger and History.* Caldwell, Idaho, The Caxton Printers, Ltd., 1951.

Quittet, E. *Races Bovines Francaises.* Paris, La Maison Rustique, 1963.

Ruane, J. B. "Notes on Animal Breeding." Kildare, Ireland. Unpublished presentation.

Sanders, A. H. "The Taurine World—Cattle and Their Place in the Human Scheme—Wild Types and Modern Breeds in Many Lands," *National Geographic Magazine*, Vol. XLVIII, No. 6 (December, 1925).

Schmid, A. *Rassenkunde des Rindes.* 2 vols. Bern, Benteli A. G., 1942.

Sinclair, James, ed. *History of Shorthorn Cattle.* London, Vinton and Company, Ltd., 1908.

Stanford, J. K. *British Friesian, a History of the Breed.* London, Max Parrish, 1956.

Statesman's Year-Book 1964–1965. London, Macmillan and Company, Ltd., 1964.

Stonaker, H. H. "Origin of Animal Germ Plasm Presently Used in North America," *Germ Plasm Resources*. Washington, American Association for the Advancement of Science, 1961.

Towne, C. W., and E. N. Wentworth. *Cattle and Men*. Norman, University of Oklahoma Press, 1955.

Tankerville, Earl of, F. Z. S. *The Wild White Cattle of Chillingham*. Pamphlet.

Uesaka, Shoji. *Wagyu Cattle Breeding in Japan*. Kyoto, Kyoto University, 1958.

Whitehead, G. Kenneth. *The Ancient White Cattle of Britain and their Descendants*. London, Faber and Faber Ltd., 1953.

Williamson, G., and W. J. A. Payne. *An Introduction to Animal Husbandry in the Tropics*. London, Longmans, 1959.

Zeuner, F. E. *A History of Domesticated Animals*. London, Hutchinson, 1963.

Index of Cattle